PROBABILISTIC APPROACHES TO DESIGN

Probabilistic Approaches to Design

EDWARD B. HAUGEN

Aerospace & Mechanical Engineering Department
University of Arizona
Formerly with North American Aviation
Space Division

John Wiley & Sons, Inc., London · New York · Sydney

R
620
H 371 P

Copyright © 1968 by John Wiley & Sons, Inc.

All rights reserved.
No part of this book may be reproduced by any means,
nor transmitted, nor translated into a machine language
without the written permission of the publisher.

Library of Congress Catalog Card Number: 67-31377
GB 471 36025X
Printed in the United States of America

This Book is Dedicated

to

My Wife Mercedes

and to

Johnnie and Cecilia

Preface

The concern of engineers with design adequacy and optimization in the various disciplines has grown rapidly during recent years. This concern suggests the realization that a critical re-evaluation of the foundations of design, based on empirical multipliers such as safety factors, is needed. Present analyses result in designs in which adequacy or safety is neither balanced nor clearly specified.

The recognition that design parameters generally are characterized by spectra of values (i.e., are statistical in nature), rather than by unique values points directly to probabilistic methods and statistics as a logical design approach. Conventional methods appear as special cases of the more general probabilistic methods; for example, those design situations in which all variability is zero are rare cases indeed.

I wish to express my indebtedness to the publisher's reviewers whose comments were very helpful; to the class in Design by Reliability at the University of Arizona "Summer Institute For College Teachers" (1966) for uncovering errors in the original manuscript and for many valuable comments (particularly those of Dr. Charles O. Smith of the University of Detroit); to North American Aviation Company, Space Division, in the persons of Mr. W. K. Warner (Director of Systems Effectiveness Engineering) and Mr. H. L. Steverson (Manager of Reliability Engineering), for making available the results of several probabilistic studies (and for friendly encouragement); and to Dr. Leo Kovalevsky for his many helpful suggestions. I am grateful to Professor Benjamin Solomon of California State College at Los Angeles for reviewing the final manuscript and for his detailed critique of Chapters 2, 3, and 4. Special thanks are due to my friend Dr. Dimitri Kececioglu of the University of Arizona who influenced the development of this book, made possible the initial classroom exposure of the text, and generously exerted continual encouragement. Finally, a word of thanks to Miss Agnes L. Quinn for many hours of devotion and labor in typing and correcting the text.

Edward B. Haugen

Tucson, Arizona
January 1968

Contents

1. **Introduction** 1

PART I FOUNDATION 11

2. **Mathematical Considerations** 13
   - 2.1 Basic Probability Concepts 13
     - 2.1.1 Classical Theory 14
     - 2.1.2 Relative Frequency Theory 15
     - 2.1.3 Properties and Theorems 19
   - 2.2 Random Variables: Discrete 23
     Definition of Random Variable, Distributions, Probability Function, Distribution Function
     - 2.2.1 The Binomial Distribution 25
     - 2.2.2 The Poisson Distribution 29
   - 2.3 Random Variables: Continuous 32
     - 2.3.1 Density Functions 36
     - 2.3.2 Gamma Distribution 39
     - 2.3.3 Evaluation of Poisson from Incomplete Gamma Function 40
     - 2.3.4 Sums of Gamma Variates 41
     - 2.3.5 Beta Distribution 42
     - 2.3.6 Normal Distribution 44
       (Stirling's Approximation, Derivation-Normal Function)
     - 2.3.7 Log Normal Distribution 47
     - 2.3.8 Central Limit Theorem 53
     - 2.3.9 Normal Approximation of Binomial . . . 54
     - 2.3.10 Gamma Approximation of the Normal Distribution . 55
   - 2.4 Joint Distributions 56
   - 2.5 Point Estimation 58

ix

x Contents

- 2.6 Expected Values and Moments 61
  - 2.6.1 Averages 63
  - 2.6.2 Properties of the Mean Values 64
  - 2.6.3 Measure of Dispersion 65
  - 2.6.4 Extreme Value Probability 68
- 2.7 Moments of Random Variables 69
  - 2.7.1 Moment Generating Function 70
  - 2.7.2 Multivariate Moments 72
- 2.8 Distributions of Functions 75
  - 2.8.1 Simple Monotonic Types 75
  - 2.8.2 The Basic Distribution Transformation . . . 79
  - 2.8.3 Density Function of Powers 81
  - 2.8.4 Method of Convolutions 81
- 2.9 Moments of Functions of Random Variables . . . 88
  - 2.9.1 Maximum Likelihood Estimators 89
  - 2.9.2 Partial Derivative Methods 90
  - 2.9.3 Moment Generating Function Methods . . . 93
  - Problems 102

3. Algebra of Normal Functions 105

- 3.1 Independent Binary Operations 106
  - 3.1.1 Sums (Maximum Likelihood Estimators) $Z = X + Y$ 107
  - 3.1.2 Differences (Maximum Likelihood Estimators) $Z = X - Y$ 113
  - 3.1.3 Products (Maximum Likelihood Estimators) $Z = XY$ 116
  - 3.1.4 Quotients (Maximum Likelihood Estimators) $Z = X/Y$ 120
  - 3.1.5 Summary: Binary Operations 123
- 3.2 Moment Generating Functions 123
  - 3.2.1 Moments of the Sum $Z = X + Y$ 124
  - 3.2.2 Moments of the Difference $Z = X - Y$ . . . 124
  - 3.2.3 Moments of the Product $Z = XY$ . . . 125
  - 3.2.4 Moments of a Square, $Z = X^2$ 126
  - 3.2.5 Moments of a Root, $Z = X^{1/2}$ 127
  - 3.2.6 Moments of the Quadratic Form 128
- 3.3 Method of Partial Derivatives 129
  - 3.3.1 Sums and Differences 129
  - 3.3.2 Products 129
  - 3.3.3 Quotients 129

Contents

- 3.4 Special Correlated Combinations 131
  - 3.4.1 Binary Operations—Correlated Variates . . . 131
  - 3.4.2 Sum, $Z = S + S$ 131
  - 3.4.3 Difference, $Z = S - S$ 131
  - 3.4.4 Quotient, $Z = S/S$ 132
  - 3.4.5 Product, $Z = SS$ 132
  - 3.4.6 Product of Variate (S) and Its Inverse $(1/S)$, $Z = S(1/S)$ 133
  - 3.4.7 Summary (Nonindependent Combinations) Binary Operations 133
  - 3.4.8 Correlated Functions: General 134
  - 3.4.9 Constraints 134
  - 3.4.10 Coefficient of Variation 135
- 3.5 Laws of Combination 136
  - 3.5.1 Characteristics 136
  - 3.5.2 Cummutative Law Addition and Multiplication . . 137
  - 3.5.3 Associative Law 137
  - 3.5.4 Existence of the Zero [(0, 0) is in the set S] . . 138
  - 3.5.5 Existence of the Unity [(1, 0) is in the set S] . . 138
  - 3.5.6 Closure 138
  - 3.5.7 Distributive Law 139
- 3.6 Mathematical Structure 139
  - 3.6.1 Field 139
  - 3.6.2 Vector Space 140
  - 3.6.3 Abelian Group 140
  - 3.6.4 Limit of Convergence 140
  - Problems 142

4. Determination of Reliability 145

- 4.1 Generally Distributed Allowable and Applied Stresses . . 145
- 4.2 Determination of Reliability when Strength and Stress Distributions are Normal 148
  - 4.2.1 Derivation by Difference Function 148
  - 4.2.2 Derivation by Convolutions 151
- 4.3 Nonnormal Distributions: Transform Method for Determining Reliability 154
- 4.4 Normal Correlated Allowable and Applied Stress Random Variables 156

5. Numerical Methods 158

- 5.1 Numerical Integration 161
  - 5.1.1 Simple Integration Formula (for Equidistant Ordinates) 162
  - 5.1.2 Simpson's Rule 163
  - 5.1.3 Errors in Simpson's Rule. 167
- 5.2 Double Numerical Integration 170
  - 5.2.1 Two Way Differences 171

6. Monte Carlo Methods 178

- 6.1 Random Numbers 179
- 6.2 Nonrectangular Distribution Sampling 182
  - 6.2.1 Direct Simulation 183
- 6.3 Monte Carlo General Principles 194
  - 6.3.1 Crude Monte Carlo 194
  - 6.3.2 Hit or Miss Monte Carlo. 195

PART II—APPLICATIONS 197

7. Mechanical Elements 199

- 7.1 Elements of Force Systems 199
- 7.2 Centroids 200
- 7.3 Moment of Inertia 201
- 7.4 Radius of Gyration 204
- 7.5 Methods of Analysis 204
  - 7.5.1 Estimating Variance 205
  - Problems 205

8. Elements in Tension 208

- 8.1 Design of a Tension Element 208
- 8.2 Analysis of Tension Elements 210
- 8.3 Consideration of Length 213
- 8.4 Generalization of Section 214
  - Problems 215

9. Simple Beams: Concentrated Loading 217

- 9.1 Simple Beam: Single Load. 217

9.2	Stresses in Beams	219
9.3	Multiforce Beam	225

10. Simple Beams: Distributed Loads 226

10.1	Reactions	228
10.2	Shear	229
10.3	Moment	229
	10.3.1 Bending Moment Signs	229
	10.3.2 Shearing Force Signs	230
10.4	Uniform Load on Part of Span	234
10.5	Design Computation	236
10.6	Deflection	238
10.7	Beam: Both Ends Built-In	241

11. Cantilever Beams 244

11.1	Uniformly Loaded Cantilever	246
11.2	Propped Cantilever	248

12. Column Design 254

12.1	The Compression Block	255
12.2	The Short Column	256
12.3	The Short Column, Eccentric Loading	260
12.4	Intermediate Columns—Eccentric Compression	262
12.5	Critical Load or Euler Load	264
12.6	The Beam Column	270

13. Torsion and Combined Torsion and Bending . . . 274

13.1	Torsion	274
13.2	Combined Stresses (Bending and Torsion in a Circular Member)	279

14. Statistical Study of Distortion 283

14.1	Model and Initial Assumptions	283
14.2	Statistical Nature of the Problem	284
14.3	Mechanics of the Distortion	286
14.4	Likelihood of Critical Distortion	288
14.5	Most Severe Configuration	291

xiv   Contents

15. Electromechanical Devices 293
    15.1 Statistical Approach to Analysis 293
    15.2 Electromechanical Switches 296
    15.3 Statistical Considerations 297
    15.4 Mechanical Circuit 298
    15.5 Electrical Circuit 305
    15.6 Magnetic Circuit 307
    15.7 Reliability Estimation 311
    15.8 Fatigue Considerations 311
    15.9 Design to Specified Reliability 311

References 315

Index 317

PROBABILISTIC APPROACHES TO DESIGN

CHAPTER 1

Introduction

During the last few years engineers and designers have shown increasing concern with problems of design adequacy (reliability) in the various disciplines. Such concern indicates that a critical reevaluation of the foundation of design† is needed. According to Dr. A. Freudenthal, "Careful and rigorous analyses may be largely deprived of their merits if the accuracy of results is diluted by the employment of empirical multipliers—selected rather arbitrarily on the basis of considerations not always rational or even relevant."

An example of this thought is expressed in the following quotation [2]:

"At the George C. Marshall Space Flight Center, requirements are usually stated as a single value—maximum load or minimum strength—with no indication of the variation to be expected. It appears that each group inflates its computed requirement before release as a design parameter. In a sequence of such steps, the amount of inflation of requirements can become quite large, and in any event, the consequence is certainly an unknown amount of inflation. The net result is (often) overdesign.

"The reason for emphasis is not that results have been bad; quite the contrary, the failure probability of structural subsystems (for example) is quite low, if the analysis of tension failure mode is indicative of design practice as it affects other modes. The real reason for emphasis is that the analyst cannot know what stresses are actually involved. An analysis may or may not be realistic; there is no way of knowing which it is."

This quotation, concerned with elements of structures, may be restated for other disciplines with relevance.

It is well known that conventional practice often results in overly conservative designs. It is also known that the degree of conservatism (reflected, sometimes, in excess weight) cannot readily be estimated by current methods.

† Dependent on empirical multipliers such as safety factors, placard values, and margin of safety.

2 Introduction

The growing need for answers to such questions—How safe is design? How safe should a design be? How can designs be made to specified levels of adequacy (reliability)?—leads directly to probabilistic methods and statistics.

Operating in a domain fraught with uncertainty, that of statistics, and with tools not ideally suited to the task, the set of real numbers, conventional methods have performed surprisingly well in the past. However, with a rational approach (probabilistic considerations) and better tools (statistical algebra—see Chapters 3, 5, and 6) considerable improvement in designs can be expected. It is suggested that one factor motivating conventional design approach may be found in the nature of the real (decimal) number system. The algebra of real numbers requires unique single-value representation of each variable in order to compute results. We shall see that a more versatile (and more applicable) mathematical system now exists and that this system takes into account both mean value and variance (through the solution of problems) for the parameters under consideration.

The statistical nature of design variables is usually ignored in conventional practice, as is demonstrated by the efforts made to find representative unique values such as minimum guaranteed values, limit loads, or ultimate loads. The conventional approach in design practice may be compared to a kind of worst-case analysis. The maxima of loading and the minima of strength are treated not only as representative of design situations, but also of simultaneous occurrence. This is the basis on which unknown parameters are computed. Actually, magnitude and frequency relationships for both load and strength must be considered to avoid unrealistic results. If an extremely large load (of rare occurrence) must act on an extremely low value of strength (of rare incidence) to induce a failure, then the probability of such simultaneous occurrences is very important. Figure 1.1 pictures the type of comparison discussed above [10].

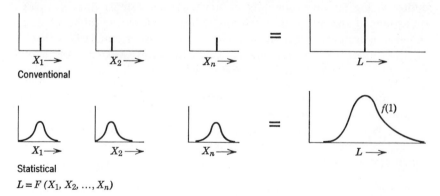

Figure 1.1 Parameter representation.

Introduction 3

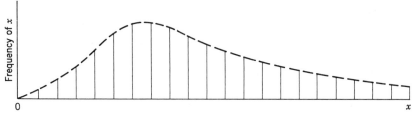

Figure 1.2 Empirical distribution.

Consideration of physical systems, in general, leads to the observation that the magnitude of almost every measurable parameter tends to vary in a random fashion. Variables are characterized by spectra of values rather than by unique values. This observation appears to be true for a large class of parameters, including such diverse entities as height of individuals in a crowd, a strength property of a material, a load magnitude, the area of a structural section, thermal conductivity, or resistance. The result of a series of measurements is a population of values:

$$x_1, x_2, \ldots, x_{n-1}, x_n.$$

When a typical population of values is plotted on a graph as magnitude versus frequency, such a plot generally tends toward a stable, predictable distribution as the size of the sample is increased. A variable displaying such a characteristic is called a random variable or variate (Fig. 1.2).

A typical plot of values usually approximates some known distribution type such as the normal, gamma, log-normal, or beta. Thus, most engineering variables may be realistically described by defining statistics, such as mean values and standard deviations; for example, the cross-sectional area of a tension member may be described by a mean value ($\bar{A}$ in.²) and a measure of variation about the mean, called the standard deviation (s_A in.²), or the tensile yield strength of the aluminum alloy in a tension element may be described by a sample mean value,

$$\bar{f}_{ty} \text{ psi,}$$

and a measure of variation about the mean value, the sample standard deviation

$$s_{f_{ty}} \text{ psi.}$$

Consequently, the variable is described by the couple,†

$$(\bar{f}_{ty}, s_{f_{ty}}) \text{ psi}^2\ddagger$$

---

† Couple notation is used to describe normally distributed random variables (see Section 2.3.6 and Chapter 3).
‡ *Note:* $\bar{x}$ and s_x are estimates of the true mean μ_x and standard deviation σ_x, since they are computed from samples of finite size.

4 Introduction

Important design variables are usually functionally related—they may be related as sums or products, for example. Since each variable is a random variable, the design process involves functional relationships among random variables. Thus, the deflection of a beam is a function of length, L, moment of inertia, I, and modulus of elasticity, E, of the material. Each of L, I, and E is a random variable, and the deflection (of a beam) $= f(L, I, E)$.

It follows that $f(L, I, E)$ is a random variable determined by the statistical characteristics of L, I, and E. The random variable $f(L, I, E)$ is called a multivariate. The applied stress resulting from the action of loading is also a random variable, often a multivariate (Section 2.4).

Since the safety factor (SF) is of central importance in conventional design and is also the focus of much criticism, we list several of the more frequently encountered definitions of safety factor [11].

1. Ratio of ultimate or yield strength in a component to the actual working stress.

2. Ratio of maximum safe load to normal service load.

3. Safety factor is the ratio of computed strength, S, to the corresponding computed load, R, that is, $n = S/R$. Inherent variability accounts for variation of strength, ΔS, and of load, ΔR, from the computed values. The lowest probable strength $(S = S - \Delta S)$ and the largest probable load $(R = R + \Delta R)$ must satisfy the inequality $S - R > 0$ to avoid failure.

$$S - \Delta S > R + \Delta R \quad \text{or} \quad S\left(1 - \frac{\Delta S}{S}\right) \geq R\left(1 + \frac{\Delta R}{R}\right)$$

Minimum safety factor is

$$n = \frac{S}{R} = \frac{1 + \Delta R/R}{1 - \Delta S/S}.$$

If the maximum variations are 25 percent of the computed values, the minimum safety factor is

$$n = \frac{1 + 0.25}{1 - 0.25} = 1.67.$$

4. Ratio of damaging stress (e.g., fatigue limit) to maximum (known) working stress.

5. Ratio of mean strength to mean load.

6. Ratio of significant strength to significant stress.

These definitions vary widely and several are probabilistically inaccurate.

Theoretically, safety factor is the ratio of mean failure governing strength to mean failure governing stress:

$$\text{safety factor} = SF = \frac{\bar{S}}{\bar{s}}.$$

The concept a safety factor conveys is that of separation of mean strength from mean stress, with no indication of the failure probability of the component. Some designers believe that designing to a safety factor above a preconceived magnitude, usually 2.5, eliminates component failure. In reality with such safety factors failure probability may vary from a low value to an intolerably high one.

Fallacies in Designing by Safety Factors. The nature of the variability of nominal stress and of the stress factors that affect component stress, and of nominal strength and the strength factors that affect component strength, explain the existence of stress and strength distributions. In each design problem it is these distributions with which the designer should be concerned. The safety factor concept completely ignores the facts of variability that result in different reliabilities for the same safety factor.

To understand this the concept of failure probabilities must be explored. When the stress and strength distributions for a component are known, adequacy may be determined from (a) the probability that stress exceeds strength,† or (b) the probability that strength is less than stress; (a) and (b) exist since an overlap of the stress and strength distributions is unavoidable (Fig. 1.3).

The shaded area in Fig. 1.3 indicates a finite probability of failure (see Chapter 4) whose magnitude is a function of the degree of overlap of the two distributions. As the overlap increases, the shaded area and, consequently, failure probability increase proportionately. Thus there is a rational explanation for changes in failure probability which accompany changes in the stress–strength distribution overlap. In Chapter 4, mathematical relationships quantifying failure probabilities are presented.

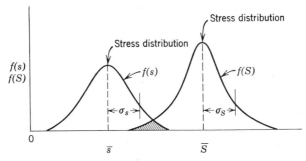

Figure 1.3 Stress (s) and strength (S) distributions, including probability of failure (shaded area) [11] (commonly called "the Warner diagram").

† By definition, reliability = 1 − failure probability. Roughly, reliability is a measure of adequacy of a design in its intended environment.

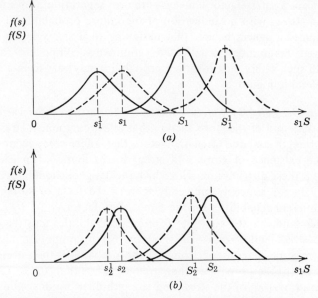

Figure 1.4 Effects on failure probability due to proportional changes in stress and strength mean values, resulting in a constant safety factor [11]: (a) case of $S_1/s_1 = k_1S_1/k_1s_1 = S_1^1/s_1^1 = k$ and k_1^1; (b) case of $S_2/s_2 = k_1S_2/k_1s_2 = S_2^1/s_2^1 = k$ and k_1^1.

Three possibilities exist in which a safety factor may be maintained, as failure probability varies:

1. Mean stress and mean strength may be changed in the same proportion with no change in the standard deviations. Thus

$$SF = \frac{S}{s} = \frac{K_1 S}{k_1 s} = K. \tag{a}$$

In Eq. a, K_1S reflects a shift in mean strength, either right or left depending on k_1 (greater or less than one); similarly for k_1s. Figure 1.4 a illustrates the shift for $k_1 > 1$; Fig. 1.4 b illustrates the shift for $k_1 < 1$. This demonstrates the limitation of safety factor as an indicator of failure incidence and of design safety [11].

2. Failure probability varies if the mean values of stress and strength distributions are held constant and the standard deviations are varied. As shown in Fig. 1.5, (Curves 3) decreasing the standard deviation of both distributions reduces overlap of the curves (3). Furthermore, shrink in overlap is observed when the standard deviation of but one distribution is reduced (Curves 2). Curves 1 or 2 show the effect of increasing one or both standard deviations, that is, overlap and probability of failure increase.

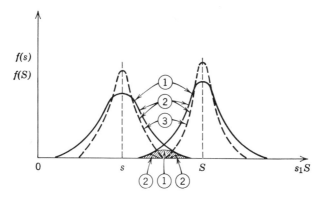

Figure 1.5 Effects on failure probability due to changing stress and strength standard deviations while maintaining a constant safety factor [11]. Curves 1. Original distributions. Curves 2. One of the standard deviations decreased. Curves 3. Both standard deviations decreased.

Since the mean values are not changed, the safety factor remains the same, but failure probability does change.

3. It is possible to change both mean values and standard deviations without affecting the safety factor.

It is now clear that the probability of failure may vary from a relatively low value to near unity because of such variations even though the safety factor remains constant (see Table 1.1).

Philosophic Reasons for the Probabilistic Approach

1. The variable nature of parameters—encountered in the physical sciences.
2. Variability—the particular domain of probability theory and statistics.
3. The need for a rational approach in design and analysis.
4. The economy suggested by uniform reliability.

The ever increasing demands for performance—resulting in operation near "limit" conditions—places increasing emphasis on accuracy and realism in design.

The Probabilistic Approach to Analysis

1. Loads estimated (including variability). Load function = (mean load, standard deviation).
2. Load distribution = f(mean load, geom$_m$, orientation$_m$) + f(variations).
3. Applied stress = f(mean parameters) + (parameter variations).
4. Allowable stress computed = f(mean strength parameters) + (parameter variations).
5. Reliability. $R = P$ (strength > stress).

Table 1.1 Safety Factors, Safety Margins, and Associated Reliabilities for Specific Stress and Strength Distributions [11]

Case Number	Mean Strength $\bar{S}$ (psi)	Mean Stress $\bar{s}$ (psi)	Strength Standard Deviation σ_S (psi)	Stress Standard Deviation σ_s (psi)	Safety Factor SF	Safety Margin SM	Reliability R
1	A 25,000	B 10,000	C 1,000	D 1,500	2.5	8.25	$0.9_{(16)}6^a$
2	A 25,000	B 10,000	k_2C 5,000	k_3D 3,000	2.5	0.3	0.9949
3	A 25,000	B 10,000	k_4C 8,000	k_3D 3,000	2.5	0.188	0.95994
4	A 25,000	B 10,000	k_2C 5,000	k_2D 7,500	2.5	−3.76	0.95254
5	A 25,000	B 10,000	k_4C 8,000	k_2D 7,500	2.5	−2.35	0.91466
6	A 25,000	B 10,000	k_5C 10,000	k_6D 6,000	2.5	−1.2	0.8997
7	A 25,000	B 10,000	k_7C 25,000	k_8D 25,000	2.5	−4.0	0.6628
8	k_1A 12,500	k_1B 5,000	C 1,000	D 1,500	2.5	0.75	$0.9_{(4)}85^a$
9	k_3A 50,000	k_3B 20,000	C 1,000	D 1,500	2.5	23.25	1.0
10	k_3A 50,000	k_3B 20,000	k_7C 25,000	k_8D 25,500	2.5	−3.4	0.7995
11	k_1A 12,500	k_1B 5,000	k_2C 5,000	k_3D 3,000	2.5	−1.2	0.90147
12	k_3A 50,000	B 10,000	C 1,000	D 1,500	5.0	33.25	1.0
13	A 25,000	k_1C 5,000	k_7C 25,000	k_8D 25,500	5.0	−3.8	0.7132
14	A 25,000	k_3B 20,000	C 1,000	D 1,500	1.25	−1.75	0.9973
15	A 10,000	B 10,000	C 1,000	D 1,500	1.0	−6.75	0.50

$k_1 = 0.5$
$k_2 = 5$
$k_3 = 2$
$k_4 = 8$
$k_5 = 10$
$k_6 = 4$
$k_7 = 25$
$k_8 = 17$

Means
A = 25,000 psi
B = 10,000 psi

Standard deviations
C = 1,000 psi
D = 1,500 psi

$SF = \bar{S}/s$

$SM = \dfrac{\bar{S} - s_{max}}{\sigma_S}$

$s_{max} = s + 4.5\,\sigma_s$
4.5 chosen arbitrarily

[a] $0.9_{(16)}6$ means 0 followed by 16 nines, then 6 (in Case No. 1), and 0 followed by 4 nines, then 85 in Case No. 8.

Introduction 9

Practical Results of Statistical Methods

1. Quantify adequacy.
2. Optimized design.
3. Uniform reliability.

As in conventional design practice, the failure governing stresses (maximum stress determining factors)† must be determined by some kind of "mode of failure" analysis. Among such maximum stress determining factors may be listed the following (in structures and mechanics) [11]:

1. Stress concentration factors can increase the nominal stress to a higher level at a point in the component until failure becomes likely to occur. In the case of a shaft, such a point is at the fillet joining the nominal diameter to a shoulder (where the bending moment is maximum).

2. Load factors, such as static, quasi-static, dynamic, impact, shock, and energy. These are factors which are applied to the axial, bending, and torsional loads to estimate the actual loads acting on the shaft under operational conditions.

3. Temperature stress factors. The temperature distribution near critical points in a component significantly changes the failure governing stress. In a shaft the effect of bearing temperature on magnitude of fillet stress may be significant. The resulting increase in the failure governing stress, at the critical point, should be accounted for either directly by thermal stress additions (reductions) or by thermal stress factors.

4. Forming or manufacturing stress factors—the residual stresses imparted into the component during forming operations such as machining, grinding, extruding, and drawing—should be added to previously calculated stresses.

5. Surface treatment stress factors resulting from shot peening, cold working, plating, and the like.

6. Heat-treatment stress factors—the stress induced in a component by distortion, the result of temperature gradients introduced during the heat-treatment process—should be added to the previously calculated stresses.

7. Assembly stress factors—shrink and press fits and other assembly practices—introduce stresses which, with existing stresses, increase the failure governing stress.

Additional stress factors may be applicable in special cases and should be recognized and properly accounted for in the stress-determination methodology. The effort to determine stresses has the objective of arriving at the maximum stresses which accumulate at a point in the component and result in the failure governing stress. Since it is difficult, at the outset, to identify

† The stresses relevant to the discipline being considered.

the point in the component at which the maximum failure governing stress will occur, stress probing is necessary. This involves the selection (by inspection and intuition) of several points at which stresses might be relatively higher than at adjacent points, based on consideration of section sizes, discontinuities in the component, total normal force, shear force, bending moment, and torque. If study of this information does not lead to identification of the critical stress point, the stresses in two or three of the most likely points are calculated. This is to ensure identification of the point where failure will most likely occur.

Efforts in the past were primarily limited to reliability predictions and reviews. These reviews resulted in changes in system configuration rather than change in elements or components. Such changes presumably upgraded reliability, but more effective reliability improvement results from changes in component design. Improvement cannot be effective unless the methodology of designing a specified reliability into a component is known.

This is the general problem to be studied, that is, design by the probabilistic approach with reliability criteria.

PART ONE

FOUNDATION

CHAPTER 2

Mathematical Considerations

In this section concepts necessary to the designing processes by the probabilistic approach are discussed. This discussion includes (a) elementary probability spaces, (b) discrete random variables, (c) continuous random variables, and (d) combinations of random variables. The distributions of functional combinations are derived, and the methods of determining the defining moments (such as mean values) of functional combinations of random variables are developed. The material presented is not intended to be a course in probability theory, for only those concepts that have a bearing on the subsequent work are covered. Students not acquainted with probability theory and statistics may profit from supplemental reading of standard texts.

2.1 BASIC PROBABILITY CONCEPTS † [12]

A familiar phenomenon to engineers is that of a variable (quantitative) of a well-defined range of possible values, but random (or haphazard) in the order in which the values appear, that is, possessed of no obvious law of ordering; for instance, in the physical sciences, despite the absence of a law describing order, we may frequently notice the similarity in proportionate makeup of pairs of records of a specific phenomenon provided the records are of reasonable length. Such proportionwise regularity of random series provides a basis for the necessary laws of the subject of probability theory.

The axioms of probability amount to idealizations of observation and experience. Two of the original approaches to probability theory (called the classical and relative frequency theories) are discussed briefly before distributions of random variables are introduced. Classical theory presumes a

† For further reading see [12] to [15].

priori knowledge. In contrast, relative frequency theory is developed in terms of observed data. In certain applications of Monte-Carlo methods† applications of relative frequency theory may be observed. The following definitions are needed:

Experiment—denotes a defined process (tossing a coin, rolling a die, selecting a resistor, measuring a column, compressing a spring, or measuring a weight).
Trial—denotes a single carrying out of the specified experiment (or measurement).
Event—denotes the outcome of a single trial (such as the value of a measurement).

Probability may be defined in terms of the likelihood of a specific event (E): $\bar{E}$ will indicate failure of E to occur. Thus E and $\bar{E}$ are complementary and mutually exclusive events. As is conventional, $p(E)$ signifies the probability of occurrence of E, and $p(\bar{E})$ signifies the probability that E will not occur.

2.1.1 Classical Theory

The undefined notion of equally likely events provides the foundation for the classical theory. The possible outcomes of an experiment are

$$a_1, a_2, \ldots, a_{n-1}, a_n.$$

Example 1. An octagonal, numbered vernier is given a random spin, and the number at the stopping point is noted. The elementary events are $E = i$ ($i = 1, 2, 3, 4, 5, 6, 7,$ and 8). $E_1, E_2, \ldots,$ and E_8 is an exhaustive set.
It follows that $p(a_1) = p(a_2) = \cdots = p(a_n) = 1/n$.

Example 2. Assuming that the elementary events (E_i) in Example 1 are equally likely,

$$p(E_1) = p(E_2) = \cdots = p(E_8) = \frac{1}{n} = \frac{1}{8}.$$

If E occurs "iff"‡ one element of a subset of elementary events occurs, then such elementary events are said to be favorable to E. In this way the definition of probability is extensible beyond equally likely events. Thus, if θ of the equally likely events

$$a_1, a_2, \ldots, a_{n-1}, a_n$$

are favorable to E,

$$p(E) = \frac{\theta}{n}. \tag{2.1}$$

† See Chapter 6 for Monte-Carlo methods.
‡ If and only if.

Example 3. Let E be the event that the number showing is 2 or 6 after casting a true die

$$p(E) = \tfrac{2}{6} = \tfrac{1}{3} \quad \text{and} \quad p(\bar{E}) = \tfrac{2}{3}.$$

In general, θ may vary in magnitude from 0 to n, so that $p(E)$ may take on values between 0 and 1. With θ elementary events favorable to E, it follows that $n - \theta$ are favorable to $\bar{E}$, from which

$$p(\bar{E}) = \frac{n - \theta}{n} \tag{2.2}$$

and

$$p(E) + p(\bar{E}) = \frac{\theta}{n} + \frac{n - \theta}{n} = \frac{\theta + n - \theta}{n} = 1. \tag{2.3}$$

Equation 2.1 is not a rigorous definition of probability since $p(E)$ depends on the number of observations. It is, however, a simple and useful definition. One theoretically possible definition of probability is

$$p(E) = \lim_{n \to \infty} \left(\frac{\theta}{n}\right).$$

A precise definition is the following:
The probability of an event is a unique real number p:

$$0 \le p \le 1.$$

A practical objection to the classical theory is its dependence on a priori analysis. The theory is applicable only when mathematical models that incorporate all possibilities can be constructed. Furthermore, the analysis must lead to the identification of equally likely events. In simple situations (not often in the physical sciences) this is feasible, but a priori knowledge is usually not sufficient to determine the probabilities of various possible events—these must be empirically estimated.

2.1.2 Relative Frequency Theory† [12]

The relative frequency theory is based on observational concepts. With θ_1 occurrence of the event E in n trials of a specified experiment, θ_1/n is the relative frequency of E, and $(n - \theta_1)/n$ is the relative frequency of $\bar{E}$. If the two relative frequencies are denoted $R(E)$ and $R(\bar{E})$, then

$$R(E) + R(\bar{E}) \equiv 1. \tag{2.4}$$

Relative frequencies tend to stabilize with increasing n.

The existence of a limiting value is postulated, and $p(E)$ of the event E is defined as the limit of the relative frequency as the number of trials is increased indefinitely.

$$p(E) \equiv \lim_{n \to \infty} \left(\frac{\theta_1}{n}\right) \equiv \lim_{n \to \infty} R(E)$$

† See Section 6.2.1, Example 1.

and

$$p(\bar{E}) \equiv \lim_{n \to \infty} \left(\frac{n - \theta_1}{n}\right) \equiv \lim_{n \to \infty} R(\bar{E})$$

The classical and relative frequency theories, although different in concept, both define probability as a ratio.

Consider the rectangles, each of known area xy, in Figs. 2.1a and 2.1b. If, in some manner, random values x_i and y_j are selected such that

$$0 \leq x_i \leq x$$

and

$$0 \leq y_j \leq y,$$

random pairs (x_i, y_j) are formed. Each (x_i, y_j) represents the position of a point p_k. If n random pairs are formed, these correspond to n random points. Of the n points, some number k_1, $0 \leq k_1 \leq n$ are inside the inscribed circle (Fig. 2.1).

The area of the circle A may be estimated

$$A = xy\left(\frac{k_1}{n}\right).$$

The accuracy of estimating the area A is proportional to the number of sample points n.

The area of any plane figure, such as that shown in Fig. 2.1b, may be estimated in a similar way. By a simple extension relative frequencies may be utilized to estimate such constants as π.

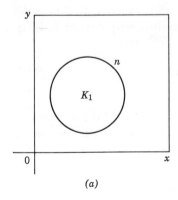
(a)

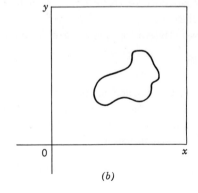
(b)

Figure 2.1

Properties of Relative Frequency and Probability

Example 1. For a single event E, $R(E)$ is a real number, $0 \leq R(E) \leq 1$,† for which Eq. 2.4 is valid.

Next consider two events E_1 and E_2, assumed to exist (or occur) simultaneously. The possible results of a specified experiment may be classified under one of four mutually exclusive categories: (E_1, E_2), $(E_1, \bar{E}_2)$, $(\bar{E}_1, E_2)$, and $(\bar{E}_1, \bar{E}_2)$. If the number of occurrences in each category is denoted by θ_1, θ_2, θ_3, and θ_4 with the total,

$$\theta_1 + \theta_2 + \theta_3 + \theta_4 = n,$$

the results may be summarized as in Table 2.1. Thus the number of occurrences of E_1 is $\theta_1 + \theta_2$ and the number of occurrences of E_2 is $\theta_1 + \theta_3$. The corresponding frequencies are

$$R(E_1) = \frac{\theta_1 + \theta_2}{n}$$

$$R(E_2) = \frac{\theta_1 + \theta_3}{n}$$

It may happen that a sufficient condition for the occurrence of some contingent result is the single or simultaneous occurrence of *two events* E_1 and E_2. The composite event (union)

Table 2.1

(E_1, E_2)	$(E_1, \bar{E}_2)$	$(\bar{E}_1, E_2)$	$(\bar{E}_1, \bar{E}_2)$	Total
θ_1	θ_2	θ_3	θ_4	n

defined by the occurrence of either E_1 or E_2 or both is signified by the symbol $E_1 + E_2$ (e.g., a joint picnic of two professions is open to all persons who belong to either and to any who belong to both). Because the union $E_1 + E_2$ is satisfied by each of (E_1, E_2), $(E_1, \bar{E}_2)$, and $(\bar{E}_1, E_2)$, the relative frequency of $E_1 + E_2$ is

$$R(E_1 + E_2) = \frac{\theta_1 + \theta_2 + \theta_3}{n} = \frac{\theta_1 + \theta_2}{n} + \frac{\theta_1 + \theta_3}{n} - \frac{\theta_1}{n}$$

Thus we have the frequency of the union

$$R(E_1 + E_2) = R(E_1) + R(E_2) - R(E_1, E_2)\ddagger \tag{2.5}$$

or

$$p(E_1 U E_2) = p(E_1) + p(E_2) - p(E_1, E_2)$$

Example 2. If the probability of live loading E_1 on a structure is $p(E_1) = 0.2$, the probability of a high wind loading E_2 is $P(E_2) = 0.001$, and $p(E_1, E_2) = 0.002$; the probability of human loading and/or of a high wind is

$$p(E_1 U E_2) = p(E_1) + p(E_2) - p(E_1, E_2)$$
$$= 0.2 + 0.001 - (0.2)(0.001) = 0.2008$$

Note. If E_1 and E_2 are mutually exclusive, then the intersection§ is empty,

$$p(E_1, E_2) = 0 \tag{2.6}$$

and

$$p(E_1 U E_2) = p(E_1) + p(E_2). \tag{2.7}$$

† See [12] and [15].
‡ *Notation.* $E_1 + E_2 = E_1 U E_2$
§ See [15] and [16].

Example 3. Consider a specific square foot of floor area in an office. If occupancy at a given time by a desk is defined as E_1 and occupancy by a filing cabinet as E_2, these events may be considered mutually exclusive. Let $p(E_1) = 0.2$ and $p(E_2) = 0.1$. The probability of desk or filing cabinet occupancy is

$$p(E_1 U E_2) = 0.2 + 0.1 = 0.3.$$

Conditional relative frequency (conditional probability). Conditional implies restricton to a special class or subset as compared with unconditional which includes all possibilities (the whole set). An event E occurring under a condition K may be indicated $E \mid K$, which reads the event E given K. Conditional relative frequency (probability) is computed in the same way as unconditional relative frequency, except that computation is confined to those events that satisfy the restriction. Conditional relative frequency E_1 given E_2, indicated by

$$R(E_1 \mid E_2),$$

is the proportionate number of occurrences of event E_1 among all occurrences of event E_2. This is the number of simultaneous occurrences of events E_1 and E_2 to the total number of occurrences of event E_2; thus

$$R(E_1 \mid E_2) = \frac{\theta_1}{\theta_1 + \theta_3} = \frac{(\theta_1/n)}{(\theta_1 + \theta_3)/n}$$

$$R(E_1 \mid E_2) = \frac{R(E_1, E_2)}{R(E_2)}$$

and similarly

$$R(E_2 \mid E_1) = \frac{\theta_1}{\theta_1 + \theta_2} = \frac{(\theta_1/n)}{(\theta_1 + \theta_2)/n}$$

$$R(E_2 \mid E_1) = \frac{R(E_1, E_2)}{R(E_1)}.$$

from which follows

(c) $\quad R(E_1, E_2) = R(E_1)R(E_2 \mid E_1) \equiv R(E_2)R(E_1 \mid E_2).$ \hfill (2.8)

Example 4. Consider a mare, the mother of two colts. Let b and g stand for male and female, and the first letter in each pair below for the older colt. There are four possibilities: b-b, b-g, g-b, and g-g. These pairs represent four sample points, and since $p(b) = p(g) = \frac{1}{2}$ associated with each is a probability $P = \frac{1}{4}$. Given that a mare has a male colt (event S), what is the probability that both colts are males (event T)?

The event ST is interpreted b-b and the events S is b-b or b-g or g-b. Thus $p(b$-$b) = p(ST) = \frac{1}{4}$, and $p(b$-b or b-g or g-$b) = p(S) = \frac{3}{4}$. Therefore

$$p(T \mid S) = \frac{p(ST)}{p(S)} = \frac{1/4}{3/4} = \frac{1}{3}.$$

Example 5. Audubon and Winkler City each possess weather stations. Let A and W denote the occurrence of snow at Audubon and Winkler City, respectively, during a one-day period in January. Observations determined that $p(A) = p(W) = 0.45$ and $p(A, W) = 0.30$. What are the conditional probabilities $p(A \mid W)$ and $p(W \mid A)$?

$$p(A \mid W) = \frac{p(A, W)}{(W)} = \frac{0.30}{0.45} = 0.666,$$

$$p(W \mid A) = \frac{p(A, W)}{p(A)} = \frac{0.30}{0.45} = 0.666.$$

2.1.3 Properties and Theorems [12]

A set of axioms for the subject of probability may be formulated in several ways. The following three properties, suggested by the behavior of relative frequencies, are sufficient for most physical sciences applications (as axioms for probability).

Property A. The probability $p(E)$ of an event E is a real number, $0 \leq p(E) \leq 1$. The probability of a certain event, is 1; that of an impossible event is 0.
$$p(E) + p(\bar{E}) = 1.$$

Example 1. Let E be the event "6 appears" after the cast of a die, then $\bar{E}$ represents the appearance of any other number. Thus $p(E) = \frac{1}{6}$; $p(\bar{E}) = \frac{5}{6}$
$$\tfrac{1}{6} + \tfrac{5}{6} = 1.$$

Property B (Total Probability).
$$p(E_1 + E_2) = p(E_1) + p(E_2) - p(E_1, E_2).$$

Example 2. From the data given in Example 5, Section 2.1.2, determine the total probability, $p(A + W)$.
$$p(A + W) = p(A) + p(W) - p(A, W)$$
$$= 0.45 + 0.45 - 0.30 = 0.60.$$

Property C (Joint Probability).

1. If $p(E_1) > 0$ and $p(E_2) > 0$,
$$p(E_1, E_2) = p(E_1) \cdot p(E_2 \mid E_1) = p(E_2) \cdot p(E_1 \mid E_2).$$
2. If $p(E_1) = 0$ or $p(E_2) = 0$,
$$p(E_1, E_2) = 0.$$

A, B, and C are called properties rather than axioms, because a set of axioms is usually reduced to a minimal number of statements. Here, an extended form of property C is used as a matter of convenience.

Concept of Independent Events. If $p(E_1 \mid E_2) = p(E_1)$, (2.9)
then E_1 does not depend on E_2.

If $p(E_2 \mid E_1) = p(E_2)$, then E_2 does not depend on E_1. For independent events,
$$p(E_1, E_2) = p(E_1 \mid E_2) \cdot p(E_2) = p(E_1) \cdot p(E_2)$$
$$p(E_2, E_1) = p(E_2 \mid E_1) \cdot p(E_1) = p(E_1) \cdot p(E_2).$$

Example 3. If (as in Example 2, Section 2.1.2) we have live loading E_1 on a structure with probability $p(E_1) = 0.2$ and high wind loading E_2 with probability $p(E_2) = 0.001$, the probability of live loading at the time of a high wind is
$$p(E_1, E_2) = p(E_1 \cap E_2) = p(E_1) \cdot p(E_2)$$
$$= (0.2)(0.001) = 0.0002 \quad \text{(since } E_1, E_2 \text{ are independent).}$$

Notation. $p(E_1, E_2) \equiv p(E_1 \cap E_2)$.

20 Foundation

Example 4. The two events in Example 2 above are not independent,
$$p(A \mid W) = 0.66 \text{ and } p(A) = 0.45,$$
thus
$$p(A \mid W) \neq p(A).$$

Example 5. The modulus of elasticity of the material of construction of a beam is statistically independent from the moment of inertia.

Example 6. The resistance value in a resistor is statistically independent from the voltage value from a power source.

Example 7. Consider the test results of 15 nominally identical beams (reinforced concrete) loaded progressively (increasing) until the first crack appeared.

Beam Number	First Crack (lb)
1	10,335
2	8,450
3	7,200
4	5,100
5	6,500
6	10,600
7	6,000
8	6,000
9	9,500
10	6,500
11	9,300
12	6,000
13	6,000
14	5,800
15	6,500

E_1 consists of tests 1–5; $E_1 \cap E_2$ consists of tests 4, 5; E_2 consists of tests 4, 5, 6, 7; $E_1 \cup E_2$ covers tests 1–7.

Consider two subsets E_1 and E_2. E_1 consists of tests 1, 2, 3, 4, and 5, and E_2 consists of tests 4, 5, 6, and 7.

The data belonging to subsets E_1 and E_2 is called the intersection of E_1 and E_2: $(E_1 \cap E_2) = (E_1, E_2)$. The data belonging to at least one of the subsets are called the union: $(E_1 \cup E_2) = (E_1 + E_2)$, thus $E_1 \cup E_2$ contains all data from tests 1, 2, 3, 4, 5, 6, and 7. The intersection $E_1 \cap E_2$ contains the data from tests 4 and 5.

Basic Theorems.† The quantity $p(E_1 + E_2)$ is called the total probability of E_1 and E_2. If events E_1 and E_2 are mutually exclusive then their simultaneous occurrence is impossible, and $p(E_1, E_2) = 0$. In such a case property B yields

$$p(E_1 + E_2) = p(E_1) + p(E_2).$$

Example 8. In a coin-tossing experiment, let E_1 denote the appearance of heads and E_2 the appearance of tails after one toss. Assume $p(E_1) = p(E_2) = \frac{1}{2}$, then

$$p(E_1 + E_2) = \tfrac{1}{2} + \tfrac{1}{2} = 1.$$

Total probability may be generalized to any finite number of mutually exclusive events. Let E_1, E_2, and E_3 be mutually exclusive events, where

$$W = E_1 + E_2,$$

$$E_1 + E_2 + E_3 = W + E_3,$$

and

$$\begin{aligned} p(W + E_3) &= p(W) + p(E_3) \\ &= p(E_1) + p(E_2) + p(E_3). \end{aligned}$$

Theorem A (Total Probability—Mutually Exclusive Events). If E_1, E_2, ..., E_n are mutually exclusive events, then

$$p(E_1 + E_2 + \cdots + E_n) = p(E_1) + p(E_2) + \cdots + p(E_n). \quad (2.10)$$

From Property B a theorem may be deduced for any finite number of events, mutually exclusive or not. Let E_1, E_2, and E_3 be any three events, not necessarily mutually exclusive, and let $W = E_1 + E_2$. Then

$$\begin{aligned} p(E_1 + E_2 + E_3) &= p(W + E_3) = p(W) + p(E_3) - p(W, E_3) \\ &= p(E_1) + p(E_2) - p(E_1, E_2) + p(E_3) - p(W, E_3) \end{aligned}$$

$$\begin{aligned} p(E_1 + E_2 + E_3) = p(E_1) + p(E_2) + p(E_3) - p(E_1, E_2) \\ - p(E_1, E_3) - p(E_2, E_3) + p(E_1, E_2, E_3). \end{aligned} \quad (2.11)$$

Exercise. Write the derivation above including all steps in the proof.

Theorem B (Law of Total Probability). The probability $p(E_1 + E_2 + \cdots + E_n)$ is the sum of the probabilities of the events in all possible combinations; singles, pairs, triples, ..., n-tuples. The sign is plus for the odd combinations (singles, triples, etc.) and minus for even combinations (pairs, quadruples, etc.).

The probability of simultaneous occurrences of two or more events is called the joint probability. The relations stated in Property C may be generalized to any finite number of events.

† See [12] and [15].

Consider the simultaneous occurrence of three events E_1, E_2, and E_3 and let W denote the simultaneous occurrence of E_1 and E_2. Then,

$$(E_1, E_2, E_3) \equiv (W, E_3)$$
$$p(E_1, E_2, E_3) = p(W)p(E_3 \mid W)$$
$$= p(E_1)p(E_2 \mid E_1)p(E_3 \mid E_1, E_2). \quad (2.12)$$

Exercise. Develop the derivation of Eq. 2.12 in detail.

Similarly, the joint probability of any number of events (finite) E_1, $E_2, E_3, \ldots, E_n$ is the product of n factors, the first of which is the unconditional probability of any particular one of the events, selected arbitrarily, and the second the conditional probability of any particular one of the remaining events, given the occurrence of the first selected. The general term is the conditional probability of any particular one of the remaining events given the occurrence of those already selected. There are $n!$ equivalent expressions for this joint probability.

Theorem C (General Law of Compound Probability)

$$p(E_1, E_2, E_3, \ldots, E_n) = p(E_1) \cdot p(E_2 \mid E_1) \cdots$$
$$\cdot p(E_3 \mid E_1, E_2) \cdots p(E_n \mid E_1, E_2, \ldots, E_{n-1}). \quad (2.13)$$

A special case arises when the events are independent. In the probability sense, two events E_1, E_2 are said to be independent if and only if neither affects the probability of occurrence of the other. A definition of independence is
$$p(E_2 \mid E_1) = p(E_2)$$
and
$$p(E_1 \mid E_2) = p(E_1).$$

Each of these expressions implies the other. If either is valid, the joint probability is
$$p(E_1, E_2) = p(E_1) \cdot p(E_2).$$

The most suitable definition of the independence of two events is that their joint probability equals the product of their respective unconditional probabilities. This definition does not fail if one of the events has probability zero.

Example 9. Let E_1 be the event "heads on first toss of a coin," E_2 the event "heads on second toss." Assume that $p(E_1) = p(E_2) = \frac{1}{2}$; then
$$p(E_1, E_2) = (\tfrac{1}{2})(\tfrac{1}{2}) = \tfrac{1}{4}.$$

Note. Mutually exclusive events are not independent unless the probability of one is zero. The probability of simultaneous occurrence of mutually exclusive events must be zero.

n events are independent if the probability of each event is unaffected by the occurrence or nonoccurrence of any of the others (either alone or in combination).

Theorem D (Multiplicative Law for Independent Events)

$$p(E_1, E_2, E_3, \ldots, E_n) = p(E_1)p(E_2)p(E_3) \cdots p(E_n). \qquad (2.14)$$

Example 10. Consider a series system comprised of three independent components A, B, and C. The probability that each (A, B, and C) will perform according to design intent is given as

$$p(A) = 0.99,$$
$$p(B) = 0.99,$$
$$p(C) = 0.97.$$

Compute the probability that the system will perform as intended.

$$p(A, B, C) = p(A)p(B)P(C)$$
$$= (0.99)(0.99)(0.97) = 0.95.$$

Example 11.† Since the 2^n possible n-fold combinations of $E_1, E_2, \ldots, E_n$ and $\bar{E}_1, \bar{E}_2, \ldots, \bar{E}_n$ represent a complete system of mutually exclusive categories, their total probability is equal to one.

2.2 RANDOM VARIABLES: DISCRETE

In Section 2.1, basic probability concepts were discussed in terms of events. Events were defined as the possible outcomes of an experiment (many times a measurement) and often interpreted as numerical quantities. The emphasis was of a qualitative nature. In this section the emphasis is on quantitative data. Analysis and design applications of probability theory involve numerical values of appropriate random variables. Most random variables of interest (in engineering) are capable of taking on any of an infinite number of possible values. If probability can be defined at all (for the results of a given type of experiment), then the total probability of all possible results will necessarily be equal to one, since some result is certain to occur. Our next concern is with sets of numerical values, such as the entire range of values a random variable may display.

Generally, a variable that eludes predictability in assuming its different values is called a random variable or a variate. To be precise a random variable must have a specific range of values and a definite probability associated with each value.

Example.

Event	Random Variable
Yield	X
	$X = f_t$ psi

f_t—tensile stress

† See step 5 of Example 1, Section 6.2.1.

Thus every experiment is characterized by a set of numbers (values):

$$x_1, x_2, x_3, \ldots, x_{n-1}, x_n.$$

Examples of random variables:

1. The number of people who report for work at North American Aviation on a given day.
2. The weekly expenditure of a family.
3. The weight of a machined metal part.
4. The cross-sectional area or moment of inertia of a standard steel beam.

A random variable is called discrete if it may take a finite or denumerably infinite (countable) number of unique values. A variate is called continuous (in a specified range) if it can take any value (real value) in that range. The first two random variables above are discrete, whereas 3 and 4 are continuous.

Distribution. The distribution of a random variable is the mathematical definition of its behavior. It is the specification of its values together with their respective probabilities. Continuous random variables involve subtlety since the range of values (the real numbers) cannot be placed in one-to-one correspondence† with any countable series of events. Examples of real valued random variables:

1. The value of a resistor, within its range.
2. The value of tensile yield strength, within its range.
3. The value of thermal conductivity of a collection of nominally identical metal alloy plates.

Probability Density Function. Relationships such as $x \geq s$ or $t < x < v$ are often of interest, and probability notation takes cognizance of this. Thus such symbols as $p(x \geq s)$ and $p(t < x < v)$ read "the probability that x equals or exceeds s" and "the probability that x is greater than t and less than v."

The function $f(x)$, called the "probability density function" of X, expresses the probability of the variate associated with any one of its admissible values. The distribution of X is completely defined by $f(x)$ and a statement of the range of values which X may assume.

Cumulative Distribution Function. In addition to the probability of a specific (possible) value of x, it is often necessary to know the probability that x exceeds a stated value or alternatively the probability that x is equal to or less than a stated value. The probability $p(x \leq x_i)$ that x is less than or equal to x_i is given by $F(x)$,

$$F(x_i) = p(x \leq x_i).$$

† Not isomorphic, see [16].

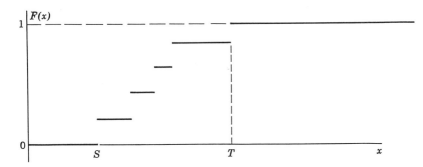

Figure 2.2 Discrete random variable cumulative distribution function.

More generally, the function $F(x)$, called the cumulative *distribution function* of X, expresses the probability that the variate displays some value less than or equal to a stated value of x. If the least possible value of a variate X is S and the greatest possible value is T, then $F(x) = 0$ for all values $x < S$ and $F = 1$ for all values $x > T$ (see Fig. 2.2).

Typical Distribution of a Discrete Variate. Discrete variates possess cumulative distribution functions of the discontinuous type (called step function), shown in Fig. 2.2. Values of x that do not actually occur add nothing to the probability that a random value of the variate falls within a stated range; hence $F(x)$ displays horizontal intervals between admissible values, with jumps at each point that represents an admissible value, since $F(x)$ is discontinuous, its value at some allowable values of x is not defined explicitly.

The remainder of this section is devoted to the consideration of two discrete distributions.

2.2.1 The Binomial Distribution

The binomial distribution is probably the most frequently used discrete distribution in applications of probability theory. It is the distribution associated with repeated trials of the same event (see Distortion in Flat Plates, Chapter 14).

Assume that the occurrence of an event E is random and that the number x of occurrences of E in n (independent) trials is a discrete random variable, with possible values $x = 0, 1, 2, 3, \ldots, n-1, n$. If the probability ($p$) of E at an individual trial is constant, then X is called a binomial random variable and its distribution is the binomial distribution. The failure of E to occur (at a given trial) implies the occurrence of $\bar{E}$, with probability $q = (1 - p)$.

26 Foundation

Any sequence of n trials is comprised of x instances of E and $(n - x)$ instances of $\bar{E}$, $(x = 0, 1, 2, 3, \ldots, n - 1, n)$. Thus

$$(p + q)^n = \sum_{x=0}^{n} C(n, x) p^x q^{n-x} = 1\dagger \tag{2.15}$$

where

$$C(n, x) = \frac{Pm(n, x)}{x!} = \frac{n!}{x!\,(n - x)!} = \binom{n}{x}. \tag{2.16}$$

$$Pm(n, x) = n(n - 1)(n - 2) \cdots (n - x + 1) = \binom{n}{x} x!$$

$$= \frac{n!}{(n - x)!}.$$

Thus the density function for the binomial distribution is

$$f(x) = \binom{n}{x} p^x (1 - p)^{n-x}, \quad (x = 0, 1, 2, 3, \ldots, n). \tag{2.17}$$

The parameters n and p account for the multiplicity of geometrical forms. In words, $\binom{n}{x}$ is an expression of the number of ways in which there may be x successes and $n - x$ failures (in n trials), and the probability of any particular x is $p^x(1 - p)^{n-x}$.

Example 1. Consider the toss of a true die, and let the occurrence of 1 or 6 be considered a success. If the die is tossed three times, the number of successes can assume any one of four values: 0, 1, 2, or 3. The probability of a success is $p(S) = \frac{2}{6} = \frac{1}{3}$ at any single trial. The probability of failure $p(F) = \frac{2}{3}$.

If x denotes the number of successes, the probabilities (assuming independence) are

$$f(0) = p(x = 0) = 1(\tfrac{2}{3})^3 = \tfrac{8}{27},$$
$$f(1) = p(x = 1) = 3(\tfrac{2}{3})^2(\tfrac{1}{3}) = \tfrac{12}{27},$$
$$f(2) = p(x = 2) = 3(\tfrac{2}{3})(\tfrac{1}{3})^2 = \tfrac{6}{27},$$
$$f(3) = p(x = 3) = 1(\tfrac{1}{3})^3 = \tfrac{1}{27},$$

and

$$f(0) + f(1) + f(2) + f(3) = \tfrac{27}{27} = 1.$$

Figure 2.3 shows that the binomial distribution is symmetrical when $p = 0.5$ and otherwise is nonsymmetrical (skewed). If the long tail is to the right, the distribution is said to be skewed to the right. The interchange of p and $q = (1 - p)$ in any binomial distribution results in its mirror image. As is apparent in Fig. 2.4, skewness in the binomial distribution is less pronounced as n increases.

† See density functions, Section 2.3.1.

Mathematical Considerations 27

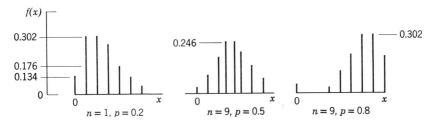

Figure 2.3 Binomial distribution, $n = 9$ (values from [43], p. 364) [12].

The successive values of the F's are given by the corresponding progressive sums of the f's except where rounding errors in the latter figures violate the logical requirements that $F(x) < 1$ when $x < n$, and $F(n) = 1$.

A probability distribution must equal unit total probability† over the set of admissible values. This fact serves as a check on the validity of alleged distributions. For the binomial distribution:

$$\sum_{x=0}^{n} f(x) = q^n + nq^{n-1}p + \cdots + nqp^{n-1} + p^n = (p+q)^n = 1^n = 1$$

Example 2. In Fig. 2.5 the slotted disk (A) makes four revolutions in five minutes. Small parts on a conveyor (B) may pass the barrier when the slot is over the conveyor belt, otherwise their progress is arrested. Consider the geometry of each part negligible, and time of arrival at the barrier random. Thus passing the barrier is a chance event. Find the distribution for five independent parts (trials).

This becomes a problem in geometrical probability in the time domain. A complete revolution of the disk requires 1.25 minutes of which the barrier is open $(1.25)72°/360° = 0.25$ minute. Thus

$$p(\text{barrier open}) = \frac{0.25(60)}{1.25(60)} = 0.20$$

and

$$1 - p = q = 0.80.$$

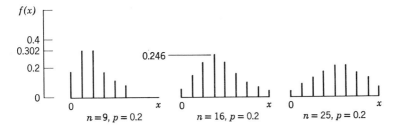

Figure 2.4 Binomial distribution, $p = 0.2$ (values from [43], pp. 364–365) [12].

† See density function, Section 2.3.1.

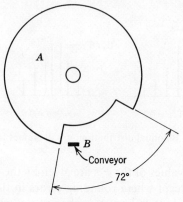

Figure 2.5

For any number n of trials the number x of successes has the binomial distribution with $p = 0.20$. The distribution of successes in five independent trials is

x	0	1	2	3	4	5
$f(x)$	0.3277	0.4096	0.2048	0.0512	0.0064	0.0003
$F(x)$	0.3277	0.7373	0.9421	0.9933	0.9997	1.0000

(For values see [43] p. 363.)

Example 3. A common application of the binomial distribution in manufacturing is in lot acceptance inspection. A lot is usually large compared with sample size; p is the fraction of defectives, n is the size of sample, and x is the observed number of defectives.

Assume that $n = 18$ and $p = 0.10$. The probability of obtaining x defectives ($x = 0, 1, 2, \ldots, 17, 18$) is

x	Probability of x Defectives
0	0.1501
1	0.3002
2	0.2835
3	0.1680
4	0.0700
5	0.0218
6	0.0052
7	0.0010
.	.
.	.
.	.
18	0.0000

(Values from tables in [19])

2.2.2 The Poisson Distribution†

Many random phenomena of interest in the physical sciences and engineering yield a discrete random variable X having an infinite number of integral possible values $(0, 1, 2, 3, \ldots)$ and satisfying conditions which lead to the Poisson Distribution [20].

The Poisson Distribution may be derived as an approximation of the binomial distribution. When p is small and n is large, the Poisson and Binomial distributions closely approximate one another. It is often convenient to make use of this fact. The Poisson approximation holds when the following three conditions in the binomial distribution are satisfied:

1. n very large.
2. p very small.
3. The product np moderate in magnitude.

$$np \ll \sqrt{n}$$

Derivation of the Poisson distribution may start with Eq. 2.17:

$$f(x) = p(x; n, p) = \binom{n}{x}(p)^x(q)^{n-x},$$

where

$$p + q = 1,$$
$$np = \mu \quad \text{(Section 2.5, Example 1)},$$
$$p = \frac{\mu}{n}.$$

Then,

$$p(x; n, p) = \binom{n}{x}p^x q^{n-x} = \frac{n(n-1)(n-2)\cdots(n-x+1)}{x!}$$

$$\times \left(\frac{\mu}{n}\right)^x \left(1 - \frac{\mu}{n}\right)^{n-x}$$

$$= n^x \frac{1(1-1/n)(1-2/n)\cdots[1-(x-1)/n]}{x!}\left(\frac{\mu}{n}\right)^x \frac{(1-\mu/n)^n}{(1-\mu/n)^x}$$

$$p(x; n, p) = \frac{1(1-1/n)(1-2/n)\cdots[1-(x-1)/n]}{(1-\mu/n)^x} \cdot \frac{\mu^x}{x!}\left(1 - \frac{\mu}{n}\right)^n.$$

Under the Poisson conditions (n very large, p very small, and the product np of moderate magnitude) the first term in the equation above approximates

† See [15].

one:
$$\frac{1(1-1/n)(1-2/n)\cdots[1-(x-1)/n]}{(1-\mu/n)^x} \approx 1.$$

The last factor may be approximated by an exponential. Use

$$e^x = \frac{1}{0!} + \frac{x}{1!} + \frac{x^2}{2!} + \frac{x^3}{3!} + \cdots$$

and the binomial expansion of $(1-\mu/n)^n$

$$\left(1-\frac{\mu}{n}\right)^n = 1 - \frac{n(\mu/n)}{1!} + \frac{n^2(1-1/n)(\mu/n)^2}{2!} - \frac{n^3(1-1/n)(\mu/n)^3(1-2/n)}{3!}$$
$$+ \cdots \pm \left(\frac{\mu}{n}\right)^n,$$

where the sign of the last term depends on whether n is even or odd. Now by comparison it is apparent that

$$\lim_{n\to\infty}\left(1-\frac{\mu}{n}\right)^n = 1 - \frac{\mu}{1!} + \frac{\mu^2}{2!} - \frac{\mu^3}{3!} + \cdots = e^{-\mu},$$

with the first factor of $p(x; n, p)$ set equal to one and the last factor an exponential $e^{-\mu}$, the binomial probability becomes the Poisson probability; thus

$$p(x;\mu) = \frac{\mu^x e^{-\mu}}{x!}. \tag{2.18}$$

This is called the Poisson density function.

The cumulative distribution function is given by ending the following summation at a desired value of x:

$$\sum_{x=0}^{n} p(x;\mu) = \sum_{x=0}^{n} \frac{\mu^x e^{-\mu}}{x!} = 1.$$

The equation above is a density function,† since

$$\sum_{x=0}^{\infty} \frac{\mu^x}{x!} = \frac{\mu^0}{0!} + \frac{\mu^1}{1!} + \frac{\mu^2}{2!} + \cdots = e^{\mu};$$

and

$$e^{-\mu} e^{\mu} = 1.$$

Stirling's formula may be employed in evaluating $x!$.

For a Poisson random variable the probability of no successes is $e^{-\mu}$, of one success, $\mu e^{-\mu}$, of two successes, $\mu^2 e^{-\mu}/2!$, and so on (using Eq. 2.18).

† See Section 2.3.1.

Example 1. **The Poisson as an approximation of the binomial distribution.** What is the probability of accepting a lot of resistors that includes 4 percent out of tolerance (defectives)? The criterion is that in a sample of size 30 no more than one defective is found.

Binomial computation:

$$p(0.04) = \sum_{x=0}^{1} \binom{30}{x} (0.04)^x (0.96)^{30-x}$$

$$= (0.04)^{30} + \frac{30!}{(29)!(1)!} (0.04)(0.96)^{29}$$

$$= 0.661.$$

Poisson approximation:

$$np = (30)(0.04) = 1.20$$

and

$$p(0.04) = \sum_{x=0}^{1} \frac{1.20^x e^{-1.20}}{x!} = 0.663.$$

Example 2. A certain electrical servomotor design is known to have a probability of failure of $p_f = 0.005$, for a routine operation. Employing a Poisson model, compute the approximate probability of at least one failure in 1000 operations of the system and at least two failures in 1000 operations of the system.

Solution. Probability (servomotor failure) $= p_f = 0.005$, and $\lambda_{1000} = np = 1000 \cdot (0.005) = 5$.

1. Set up a model for a Poisson approximation. Since the probability of at least one failure equals one minus the probability of no failures,

$$p(x \geq 1) = 1 - p(x = 0).$$

Thus

$$p(x \geq 1) = 1 - \frac{e^{-\lambda_{1000}} \lambda_{1000}^x}{x!} = 1 - \frac{e^{-5} 5^0}{0!} = 1 - e^{-5}$$

$$= 1 - 0.0067 = 0.993.$$

2. Using the model of Part 1 observe that

$$p(x \geq 2) = 1 - p(x = 0) - p(x = 1)$$
$$= 1 - e^{-5} - 5e^{-5}$$
$$= 1 - 0.0067 - 0.0337 = 0.9596.$$

Example 3. In an engineering drawing the frequency of error in dimensioning (such as misplaced decimal point, inversion of numerals in a number, a wrong number, etc.) is observed to have a frequency of 0.10%. How many dimensions may a drawing contain if the probability of finding one or more dimensional errors is to be less than 0.05? (Use a Poisson approximation.)

Solution. Probability of a dimensional error $= p(de) = 1/1000$. We are required to find the number n of dimensions such that

$$p(x_{\text{d.e.}} \geq 1) = 0.05.$$

The probability of finding no errors $x = 0$,

(1) $$p(x = 0) = 0.95 = \frac{e^{-\lambda} \lambda^x}{x!}.$$

Setting $x = 0$,

(2) $$0.95 = \frac{e^{-n/1000}\lambda^0}{0!} = e^{-n/1000}$$

where
$$\lambda = np = n(1/1000) = n/1000,$$

and taking the natural logs of (2),
$$\ln_e 0.95 = \frac{n}{1000} \ln_e e,$$

$$-10 + 9.994 = -\frac{n}{1000},$$

we obtain
$$n = 0.051(1000) = 51 \text{ dimensions}.$$

Shapes of Poisson distributions are similar to those of the binomial distribution for n large and p small. The skewness is positive, but approaches

Table 2.2 Poisson Approximation of the Binomial [12]

x	Binomial	Poisson
0	0.3675	0.3679
1	0.3682	0.3679
2	0.1841	0.1839
3	0.0612	0.0613
5	0.00303	0.00307
6	0.00050	0.00051
7	0.000071	0.000073

symmetry as μ increases. When μ is rather large, the distribution is in the transition region between the Poisson and the normal distributions. In this region the binomial may usually be approximated by either.

An example of the Poisson approximation to the binomial appears in Table 2.2 in which $n = 500$, $p = 1/500$, and $np = 500(1/500) = 1.0$.

2.3 RANDOM VARIABLES: CONTINUOUS

Frequently, the statement of the range of a continuous random variable may be accomplished by defining the range (or ranges) of continuous variation, such as $\alpha \leq x \leq \beta$.

A one-to-one correspondence between permissible values of the continuous random variable and any infinite ordered series of events, $E_1, E_2, E_3, \ldots$ cannot be established, for the values are nondenumberable [17 and 18]. The problem of defining the distribution of a continuous variate requires use of a special device. This may be accomplished in the following way:

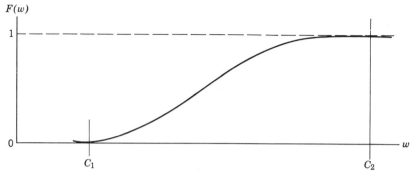

Figure 2.6 Use of cumulative distribution function in defining distributions.

First, an arbitrary number "α" is selected. Next, two mutually exclusive sets of values are defined:

1. $x \leq \alpha$,
2. $x > \alpha$.

With set 1 is associated event E_α in the sense that any value of the variate which belongs to set 1 corresponds to an occurrence of E_α; similarly, the event $\bar{E}_\alpha$ is associated with set 2. In this way, $p(E_\alpha)$ and $p(\bar{E}_\alpha)$ may be rigorously defined. Such construction may be applied to any value of α, the result being a function of α. Thus $p(E_\alpha)$ may be replaced by

$$F(\alpha) \equiv p(E_\alpha) \equiv p(x \leq \alpha). \tag{2.19}$$

Since α is allowed to take on all possible values, it is convenient to use the notation $F(x)$ in place of $F(\alpha)$. $F(x)$ denotes the probability that the variate in question assumes a value equal to or less than x. A mathematical formula giving $F(x)$ for all values of x uniquely defines the distribution of x as a variate. The variable x in the mathematical expression for $F(x)$ or $f(x)$ is a real number. If the variate assumes a value of 30 ohms, then x is the real number 30.

The cumulative distribution function of a discrete random variable is a step function, whereas that of a continuous random variable is generally a continuous function over its range (of definition). Discontinuities may occur, but these are usually confined to isolated points. Typically, the cumulative distribution function of a continuous random variable W possesses a range that is written as $C_1 \leq w \leq C_2$ (see Fig. 2.6).

The distribution of a discrete random variable was defined by stating the equation of its point probability density function $f(w)$ together with the range of admissible values. A point probability density function† is without meaning when the cumulative distribution function‡ is continuous. Instead

† See density functions, Section 2.3.1.
‡ See Section 2.2.

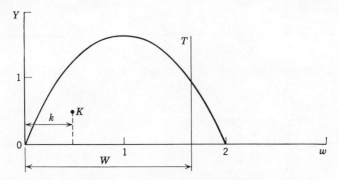

Figure 2.7 Parabola $y = 2w - w^2$ [12].

of a probability density function, the distribution of a continuous random variable may be defined by stating the equation of its cumulative distribution function $F(w)$ and a specified range of admissible values (see Fig. 2.7). The permissible region is indicated by the parabola $y = 2w - w^2$ and the w-axis. If all points within the region are equally likely, let us find the distribution of the random points between the Y-axis, and $w = 2$. At any distance $w \leq 2$ to the right of the Y-axis passes the straight line T parallel to OY. The distance of a random point K from OY is equal to k and the probability that $k \leq w$ is the same as the probability that K lies to the left of T. Hence

$$F(w) = p(k \leq w) = \frac{\text{area to left of } T}{\text{total area}} = \frac{\int_0^w y\, dx}{\int_0^2 y\, dx}$$

$$= \frac{3w^2 - w^3}{4}.$$

Thus the distribution of the random variable W is (see Fig. 2.8):

$$F(w) = \frac{3w^2 - w^3}{4} \qquad (0 \leq w \leq 2).$$

Figure 2.8 Cumulative distribution function.

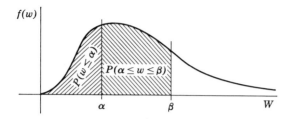

Computation of Probabilities. The probability that w lies in any particular range $\alpha < w < \beta$, is computed from the cumulative distribution function of W, since the event $w \leq \beta$ may be partitioned into two mutually exclusive events $w \leq \alpha$ and $\alpha < w \leq \beta$. From Section 2.1.3 Theorem A it follows that $p(w \leq \beta) = p(w \leq \alpha) + p(\alpha < w \leq \beta)$.

$$F(w) = P(K \leq w) = \frac{\text{area to left of } T}{\text{total area}} = \frac{\int_0^W y\, dx}{\int_0^2 y\, dx} = \frac{3w^2 - w^3}{4}$$

Thus

$$p(\alpha < w < \beta) = p(w \leq \beta) - p(w \leq \alpha)$$
$$p(\alpha < w < \beta) = F(\beta) - F(\alpha). \qquad (2.20)$$

Implied is that $F(\beta) \geq F(\alpha)$, when $\beta > \alpha$. The left side of Eq. 2.20 cannot be negative (negative probability is undefined).

Next, consider the probability of x in an open interval† (end points excluded), usually denoted

$$p(\alpha < x < \beta),$$

and in a closed interval† (end points included), usually denoted

$$p(\alpha \leq x \leq \beta).$$

It can be shown that

$$p(\alpha < x \leq \beta) = p(\alpha \leq x \leq \beta) = p(\alpha \leq x < \beta)$$
$$= p(\alpha < x < \beta) = F(\beta) - F(\alpha) \qquad (2.21)$$

Thus in the parabolic distribution (Fig. 2.9) the probability that a random value of x in the open interval $1 < x < 2$ and the probability that it is in

```
         α                                           β
    ─────┼───────────────────────┬───────────────────┼─────
        K-ε                      K                  K+ε
```
Figure 2.9 Probability interval.

† Also see discussion in [12].

the closed interval $1 \leq x \leq 2$ are both given by

$$F(2) - F(1) = 1 - \tfrac{1}{2} = \tfrac{1}{2}.$$

Note. The zero probability for $x = K$ does not mean that the value $x = K$ is impossible, but that it is insignificant as compared with the total inexhaustible domain of possibilities.

2.3.1 Density Functions† [12]

The concept of density has proved useful in the physical sciences and also in the study of random variables. In a nominally homogeneous rod, for example, the linear mass density may be considered constant, and is mass per unit length. For a nonhomogeneous rod density varies as a function of the distance x measured from an origin (taken as one end of the rod). If $R(x)$ represents the mass of a rod segment from an origin to any point x, the linear mass density $\rho(x)$ is defined as the limit of the ratio of the mass increment $\Delta R(x)$ to the length increment Δx as $\Delta x \to 0$. It follows that $\rho(x)$ is the derivative of the mass function and

$$R(x) = \int_0^x \rho(K)\, dK.$$

Provided that the distribution of a random variable is differentiable, probability density may be defined similarly to that above. The probability that a random value k of the random variable lies between x and $x + \Delta x$ is interpreted as the probability increment at the point x—represented by ΔP. Thus

$$\Delta P = P(x \leq k \leq x + \Delta x) = F(x + \Delta x) - F(x) = \Delta F(x)$$

The average concentration of probability in the neighborhood of x is given by

$$\frac{\Delta P}{\Delta x} = \frac{\Delta F(x)}{\Delta x}$$

If $F(x)$ is assumed to be differentiable, then this ratio approaches a limit as $\Delta x \to 0$. Let

$$f(x) = \frac{d\,F(x)}{dx}, \qquad (2.22)$$

then

$$d\,F(x) = f(x)\,dx$$

In the discrete case $f(x)$ is used to compute the probability that the random variable assumes a particular value of x. In the continuous case it is the probability density function of the random variable at the point x.

† See Section 2.2.

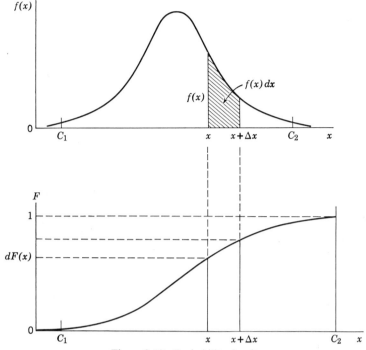

Figure 2.10 Probability density.

Density Function: Properties. From the fact that (Fig. 2.10)

$$F(x + \Delta x) \geq F(x) \quad \text{and} \quad \Delta F(x) \geq 0 \quad \text{for all} \quad \Delta x \geq 0,$$

the density function $f(x)$ is nonnegative. If C_1 and C_2 represent the minimum and maximum values respectively of a random variable X,

$$F(C_1) \equiv p(x \leq C_1) = p(x = C_1),$$

and

$$F(C_2) \equiv p(x \leq C_2) = 1.$$

(a) $\qquad\qquad f(x) \geq 0 \quad \text{for all} \quad x.$

(b) $\displaystyle\int_{C_1}^{x'} f(x)\,dx = F(x') - F(C_1) = F(x') = p(C_1 < x < x').$

(c) $\displaystyle\int_{C_1}^{C_2} f(x)\,dx = F(C_2) = 1 = p(C_1 < x < C_2).$

(d) $\displaystyle p(a \leq x \leq b) = \int_a^b f(x)\,dx = p(a < x < b)$

(e) $\displaystyle p(x = R) = \int_R^R f(x)\,dx = 0 \quad$ (in particular $F(C) = 0$.)

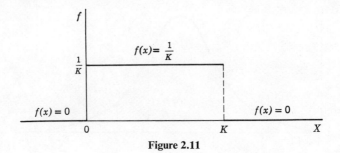

Figure 2.11

Density Functions as Defining Functions. If the density function exists, its equation, together with a specification of range, may be employed to define the distribution.

Example 1. Consider the rectangular distribution shown in Fig. 2.11. Given

$$F(x) = \frac{x}{K} \quad (0 \leq x \leq K),$$

$$f(x) = 0 \quad \text{for} \quad x < 0 \quad \text{and for} \quad x > K,$$

$$f(x) = \frac{dF(x)}{dx} = \frac{1}{K}.$$

Then the definition of the distribution is

$$f(x) = \frac{1}{K} \quad (0 \leq x \leq K).$$

Example 2. The following distribution could conceivably describe a useful property of some parameter:

$$F(w) = \frac{3w^2 - w^3}{4} \quad (0 \leq w \leq 2).$$

$$f(w) = \frac{dF(w)}{dw} = \frac{6w - 3w^2}{4} = \tfrac{3}{4}(w)(2 - w)$$

The definition of the distribution is (see Fig. 2.12):

$$f(w) = (\tfrac{3}{4})w(2 - w), \quad (0 \leq w \leq 2).$$

Definition of a distribution by the (cumulative) distribution function or the probability density function always requires a statement of the range. Without the range, the definition is incomplete, and the possibility of values leading to contradictions, such as $f(x) < 0$ or $f(x) > 1$, occurs.

The remainder of this section is devoted to a discussion of specific continuous distributions.

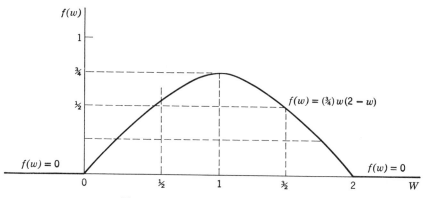

Figure 2.12 Graph of Example 2 [12].

2.3.2 Gamma Distribution

The gamma (as also the normal distribution) is a two parameter family (in α and β). The ordinary form of the gamma distribution describes a continuous random variable x with range of values $0 \leq x < \infty$. The gamma density function is

$$f(x)\dagger = \frac{1}{\beta^{\alpha+1}\Gamma(\alpha + 1)} \cdot x^\alpha e^{-x/\beta}, \qquad (\alpha > -1, \beta > 0; 0 \leq x < \infty). \quad (2.23)$$

The gamma distribution is skewed to the right for all values of parameters α, β. As the value of α increases, however, skewness becomes less pronounced.

The special case, when $\alpha = 0$, is known as the exponential distribution (Fig. 2.13), which may be defined in terms of its density function as

$$f(x) = \frac{1}{\beta} e^{-(x/\beta)}, \qquad (\beta > 0; 0 \leq x < \infty). \quad (2.24)$$

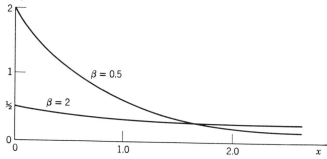

Figure 2.13 Exponential distribution [12].

† For discussion of Γ functions, see [12] and footnote, Section 2.6.3.

The first two derivatives in the exponential (Eq. 2.24) are

$$f'(x) = \frac{1}{\beta^2} e^{-(x/\beta)}$$

$$f''(x) = \frac{1}{\beta^3} e^{-(x/\beta)}$$

Another special case of the gamma distribution is called the chi-square distribution. (For derivation see [18], p. 200.)

Gamma Distribution Properties. With the restriction that negative values of x are excluded, the following conclusions may be drawn.

1. The curve approaches the x-axis asymptotically for large values of x.
2. If α has a negative value $-1 < \alpha < 0$, the ordinate increases without limit for smaller and smaller values of x.
3. The ordinate remains finite for $\alpha \geq 0$. If $\alpha = 0$, the curve assumes its highest value, $1/\beta$ at $x = 0$.
4. The curve is tangent to the vertical axis at the origin if α lies in the range $0 < \alpha < 1$, it has a slope $1/\beta^2$ for $\alpha = 1$ and is tangent to the x-axis at the origin for $\alpha > 1$.

The density function of the gamma distribution is integrable exactly for α an integer. The process may be lengthy if α is very large. For an arbitrary value of α, not an integer, numerical integration is required.

Example 1.† Consider a device with useful service life of 1000 hours. Its probability of performing satisfactorily for any operating time within 1000 hours may be estimated by the equation

$$R(t) = e^{-\lambda t}.$$

Given that the failure rate is $\lambda = 0.0001/\text{hr}$, with λ constant, what is its reliability for any 10-hour period of operation (within the 1000-hour useful life)? $R(10) = p(t \geq 10)$.

$$R(10) = e^{-0.0001(10)} = e^{-0.001} = 0.9990.$$

What is the probability $R(1000)$ that the device will perform satisfactorily over its specified useful life of 1000 hours?

$$R(1000) = e^{-0.0001(1000)} = e^{-0.1} = 0.9048.$$

2.3.3 Evaluation of Poisson from Incomplete Gamma Function [12]

When α is an integer, a convenient relationship exists between the upper tail area of the gamma distribution and the Poisson distribution function. Consider the gamma distribution in which α is an integer and $\beta = 1$. Now

† Adapted from *Reliability Theory and Practice*, Igor Bozavsky, Prentice-Hall, 1961.

let $H(\phi)$ denote the probability $P(x \geq \mu)$ that x will assume a value equal to or greater than an arbitrary quantity $(+)$ μ:

$$H(\phi)\dagger = \frac{1}{\Gamma(\alpha+1)} \int_\phi^\infty x^\alpha e^{-x}\,dx.$$

Integrating by parts yields

$$H(\phi) = \frac{1}{\Gamma(\alpha+1)} \phi^\alpha e^{-\phi} + \frac{1}{\Gamma(\alpha)} \int_\phi^\infty x^{\alpha-1} e^{-x}\,dx.$$

With α an integer, the exponent of x can be reduced by successive integrations to zero. The gamma functions appear as factorials of integers.

$$H(\phi) = e^{-\mu}\left[\frac{\phi^\alpha}{\alpha!} + \frac{\phi^{\alpha-1}}{(\alpha-1)!} + \frac{\phi^{\alpha-2}}{(\alpha-2)!} + \cdots + \frac{\phi^2}{2!} + \phi + 1\right]$$

$H(\phi)$ is the distribution function of a discrete random variable V, of Poisson distribution (with parameter ϕ). And,

$$p(V \leq \alpha) = \sum_{V=0}^{\alpha} e^{-\phi} \frac{\phi^V}{V!} = \frac{1}{\Gamma(\alpha+1)} \int_\phi^\infty x^\alpha e^{-x}\,dx = H(\phi) \qquad (2.25)$$

2.3.4 Sums of Gamma Variates [12]

If $z_1, z_2, \ldots, z_n$ are independent random variables of the same general distribution type

$$K(z_i) = \frac{1}{\beta} e^{-z_i/\beta}, \qquad (0, \infty).$$

The sum random variable

$$z = z_1 + z_2 + \cdots + z_n$$

is gamma distributed with identical β and with $\alpha = n - 1$. Thus

$$f(z) = \frac{1}{\beta^n \Gamma(n)} z^{n-1} e^{-z/\beta}, \qquad (0, \infty). \qquad (2.26)$$

Summing theorem:

If $X_1, X_2, X_3, \ldots, X_n$ are independent gamma random variables with a common parameter β and integral values of α, $(\alpha_1, \alpha_2, \ldots, \alpha_n)$, then their sum $x = x_1 + x_2 + \cdots + x_n$ is a gamma variate with β and $\alpha = \alpha_1 + \alpha_2 + \cdots + \alpha_n + n - 1$.

Example 1 (see Example 1, Section 2.3.10). Precise values of the binomial distribution are obtainable from the incomplete beta function, and approximate values (for large n) from either the Poisson or normal distributions. The gamma distribution is a useful approximation for intermediate values of n (beyond existing tables of the binomial but smaller than required for reasonable approximations of the normal). In the tabulated range of n, the gamma distribution is useful in circumstances where a continuous variate is appropriate.

† For discussion of the Γ function see footnote, Section 2.6.3, and [12].

Example 2. Gamma approximation of the binomial, where

$$\alpha = \frac{4(1+n)(1-p)p}{(1-2p)^2} \quad \text{(see [12] p. 100)}$$

$$\beta = \frac{1-2p}{2} \quad \text{and} \quad A = \frac{2np+1}{2(1-2p)}$$

$$G(x) = \frac{1}{\beta^{\alpha+1}\Gamma(\alpha+1)}(x+A)^{\alpha}e^{-(x+A)/\beta} \quad (-A \le x < \infty)$$

with $0 < p < 0.5$. In applying the approximation, the gamma distribution is evaluated at $x = r + \frac{1}{2}$ in order to estimate the binomial distribution at r.

Comparison. Exact binomial distribution (F) and approximation (Γ) using the gamma distribution (Table 2.3).

Table 2.3

	$n = 16; p = 0.2$	
r	F	Γ
0	0.0281	0.04
1	0.1407	0.16
2	0.3518	0.357
3	0.5981	0.595
4	0.7982	0.791
5	0.9183	0.909
6	0.9733	0.967
7	0.9930	0.990
8	0.9985	0.997
9	0.9997	0.9993
10	0.9999	0.99985

[a] For values, see [43], p. 365, and [12], p. 94.

2.3.5 Beta Distribution [12]

The beta distribution describes a continuous random variable, with range of values between zero and one, and probability density function:

$$f(x) = \frac{(\alpha+\beta+1)}{\alpha!\,\beta!}x^{\alpha}(1-x)^{\beta} \quad (0 < x < 1), \tag{2.27}$$

$f(x) = 0$ elsewhere.

From Fig. 2.14 it is seen that when parameters α and β are equal, the beta distribution is symmetrical. When α and β are equal, the curve rises higher and is sharper with increasing values (see [12], p. 102, Fig. 4.18).

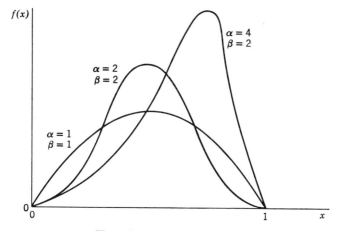

Figure 2.14 Beta distribution [12].

A *special case* of the beta distribution occurs when parameters α and β are zero. The resulting function is called the *rectangular distribution*. The density function becomes $f = 1$ and the curve is a rectangle, (the two ends are not actually part of f). A second special case arises when one parameter is zero and the other equals one. The curve is a straight line with slope $\tan \theta = +2$ if $\alpha = 1$ or $\tan \theta = -2$ if $\beta = 1$. If one parameter equals unity and the other exceeds unity, there is a single point of inflection. If $\alpha = 1$, $\beta > 1$, then the inflection point is located at

$$x = \frac{2}{1+\beta}.$$

If $\alpha > 1$ and $\beta = 1$, the inflection point occurs at

$$x = \frac{\alpha - 1}{\alpha + 1}.$$

The beta density function cannot be integrated from zero to x (except numerically), unless α and β are each integral multiples of $\frac{1}{2}$.

Utility of the beta distribution can be increased by permitting any range of finite limits S, T. The density becomes

$$f(w) = \frac{1}{w}(w - S)^{\alpha}(T - w)^{\beta}(S, T),$$

with

$$w = (T - S)^{(\alpha+\beta+1)} \cdot \beta(\alpha + 1, \beta + 1)$$

2.3.6 Normal Distribution

This distribution was discovered by the French mathematician, De Moivre, about 1730. It was derived as a limiting form of the binomial distribution. The normal distribution is prominent in the field of probability (see Chapter 4). In addition to having very convenient properties, it serves as a useful approximation to many other distributions that are less conveniently manipulated (see Chapter 3). Because the normal distribution is of great importance in this work, the derivation that follows is presented in detail.

In order to derive the normal probability density,† the first term in Stirling's approximation of a factorial,

$$n! = \sqrt{2\pi n} \cdot e^{-n} \cdot n^n \left(1 + \frac{1}{12n} + \frac{1}{288n^2} + \frac{139}{51{,}840 \cdot n^3} + \cdots \right)$$

$$n! \approx \sqrt{2\pi n}\, e^{-n} \cdot n^n$$

is employed to represent each factorial coefficient in the binomial expression. Equation 2.17,

$$f(x) = \frac{n!}{x!\,(n-x)!}\, p^x \cdot q^{n-x}$$

$$\approx \frac{\sqrt{2\pi n} \cdot e^{-n} \cdot n^n}{\sqrt{2\pi x}\, e^{-x} x^x \sqrt{2\pi(n-x)}\, e^{-(n-x)} (n-x)^{n-x}} \cdot p^x q^{n-x}$$

$$f(x) \approx \left(\frac{n}{2\pi x(n-x)}\right)^{1/2} \frac{e^{-n} n^n}{e^{-x} \cdot x^x \cdot e^{-(n-x)} \cdot (n-x)^{(n-x)}} \cdot p^x q^{n-x}$$

$$\approx \left(\frac{n}{2\pi x(n-x)}\right)^{1/2} \frac{n^n}{x^x (n-x)^{(n-x)}} \cdot p^x \cdot q^{(n-x)}.$$

Since

$$n^n = n^x \cdot n^{(n-x)}$$

$$f(x) \approx \left(\frac{n}{2\pi x(n-x)}\right)^{1/2} \cdot \left(\frac{np}{x}\right)^x \cdot \left(\frac{nq}{n-x}\right)^{(n-x)}$$

The variable change x to z is made as follows:

$$z \equiv x - \mu = x - np \approx x - x_0.$$

(Recall that, in the binomial distribution $\mu = np$, x_0 is the expected value of x.)

$$n - x = nq - z,$$

since $p + q = 1$.

$$f(x)_{[n\text{ large}]} \approx \left[2\pi npq\left(1 + \frac{z}{np}\right)\left(1 - \frac{z}{nq}\right)\right]^{-1} \left(\frac{1}{1 + z/np}\right)^{np-z} \cdot \left(\frac{1}{1 - z/nq}\right)^{nq-z}$$

† See [15] and [20].

For large values of n

$$\frac{z}{np} \text{ and } \frac{z}{nq} \ll 1,$$

and may be neglected in the first factor, but not in those factors raised to higher powers. The expression above is rewritten in logarithmic form to take advantage of the power expansion of a logarithm. An expansion for x, such that $2 \geq x > 0$, is

$$\log e^x = (x - 1) - \tfrac{1}{2}(x - 1)^2 + \tfrac{1}{3}(x - 1)^3 - \cdots$$

$$\log_e f(x) \approx -\tfrac{1}{2} \log_e (2\pi npq) - (np + z)\left(\frac{z}{np} - \frac{z^2}{2n^2p^2} + \cdots + \frac{z^3}{3n^3p^3} - \cdots\right)$$

$$+ (nq - z)\left(\frac{z}{nq} + \frac{z^2}{2n^2q^2} + \frac{z^3}{3n^3q^3} + \cdots\right)$$

$$\approx -\tfrac{1}{2} \log_e (2\pi npq) - \frac{z^2}{2n}\left(\frac{1}{p} + \frac{1}{q}\right) + \frac{z^3}{6n^2}\left(\frac{1}{p^2} + \frac{1}{q^2}\right)$$

$$- \frac{z^4}{12n^3}\left(\frac{1}{p^3} + \frac{1}{q^3}\right) + \cdots.$$

Several approximations have already been made and (for large n) another is made now:

$$\frac{z^3(q^2 - p^2)}{6n^2p^2q^2} - \frac{z^4(p^3 + q^3)}{12n^3p^3q^3} \approx 0$$

or, for the binomial distribution,

$$\sigma \equiv \sqrt{npq} \text{ (Section 2.6.3)},$$

$$\frac{z^3(q^2 - p^2)}{6\sigma^4} - \frac{z^4(p^3 + q^3)}{12\sigma^6} \approx 0.$$

Which of these terms is more important depends on the values of p and q. With p of moderate value, the net effect of all z terms of power higher than 2 is small. As an approximation, $f(x)$ is changed to $f(z)$. Since $p + q = 1$,

$$\log_e f(z) = -(\tfrac{1}{2}) \log_e (2\pi npq) - \frac{z^2}{2npq}$$

and

$$f(z) = \frac{1}{\sqrt{2\pi npq}} \exp\left[-\left(\frac{z^2}{2npq}\right)\right]$$

46 Foundation

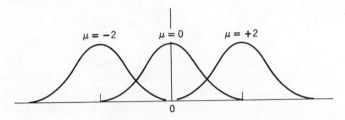

Figure 2.15 Normal distribution with $\sigma = 1$ and selected values of μ.

or, in more familiar form, with $\sigma^2 = npq$ and $z = x - \mu_x$,

$$f(x) = \frac{1}{\sqrt{2\pi}\,\sigma_x} \exp\left[-\frac{1}{2}\left(\frac{x - \mu_x}{\sigma_x}\right)^2\right] \quad (-\infty, \infty) \quad (2.28)$$

$$F(x) = \frac{1}{\sqrt{2\pi}\,\sigma_x} \int_{-\infty}^{\infty} \exp\left[-\left(\frac{x - \mu_x}{\sigma_x}\right)^2\right] dx \quad (2.28a)$$

The normal distribution is unimodal, the mode being located at $x = \mu$. There are two points of inflection, located at a distance of σ on either side of the modal† value. The curve is symmetrical about μ and falls off rapidly as the magnitude of $x - \mu$ increases. Although the range is infinite $(-\infty, \infty)$, the probability of a large deviation from μ becomes very very small and may be neglected (many times) for practical purposes (as in some physical sciences applications). The normal distribution may be used to approximate other distributions whose ranges are finite.

Example 1. The theoretical range of admissible values of a normally distributed random variable is $(-\infty < x < \infty)$. Of what import is the negative density?

Consider the tensile yield strength of a common structural material such as 2024-T36 aluminum alloy sheet. The mean tensile yield strength estimate is

$$\bar{x}_{ty} = 68{,}000 \text{ psi,}$$

and the standard deviation estimate is

$$s_{ty} = 3400 \text{ psi.}$$

The number of standard deviations z from $\bar{x}_{ty}$ to the ordinate (0) is

$$z = \frac{\bar{x}_{ty}}{s_{ty}} = \frac{68{,}000}{3400} \approx 20.$$

† Mean value is located at the mode in the normal distribution.

With values from Table 2.4, it is seen that:

z	Negative Density	$P(x \leq 0)$
3	$\dfrac{0.002700}{2} = 0.00135$	0.00135
4	$\dfrac{0.000063}{2} = 0.000031$	0.000031
5	$\dfrac{0.0000006}{6} = 0.0000003$	0.0000003
.		
.		
.		
20		0.000000...

Since the normal density function cannot be integrated in closed form, the evaluation of probabilities would involve numerical integration except that extensive tables are available.† These tables have been constructed from the unit normal distribution but can be adapted for any other normal distribution by the change of variable

$$z = \left(\frac{x-\mu}{\sigma}\right).$$

For instance, if x is a normal variate with arbitrary parameters,

$$P(a \leq x \leq b) = \int_a^b \frac{1}{\sigma\sqrt{2\pi}} \cdot \exp\left[-\frac{1}{2}\left(\frac{x-\mu}{\sigma}\right)^2\right] dx$$

$$= \int_{a'}^{b'} \frac{1}{\sqrt{2\pi}} e^{-(u^2/2)} du, \quad u = \frac{x-\mu}{\sigma}$$

and

$$a' = \frac{a-\mu}{\sigma}, \quad b' = \frac{b-\mu}{\sigma}.$$

2.3.7 Log Normal Distribution‡

The lognormal distribution can be developed in the same way as the normal distribution. The logarithm of x is the random variable instead of x, and the density function is derived from the cumulative distribution function of the normal distribution. The normal cumulative distribution function, Eq. 2.28a, is

$$F(x) = \int_{-\infty}^{x} f(x)\, dx$$

† (See Table 2.4 and [21]).
‡ *Statistics for Civil Engineers*, Dr. Jack R. Benjamin, to be published by McGraw-Hill Book Co., New York, adapted by permission of the author.

Table 2.4 Standard Normal Distribution

z	$f(z)$	$F(z)$	$R(z)$	$2R(z)$
0.0	0.398942	0.500000	0.500000	1.000000
0.1	0.396952	0.539827	0.460173	0.920344
0.2	0.391043	0.579260	0.420740	0.841481
0.3	0.381388	0.617911	0.382089	0.764177
0.4	0.368270	0.655422	0.344578	0.689156
0.5	0.352065	0.691465	0.308535	0.617075
0.6	0.333225	0.725747	0.274253	0.548506
0.7	0.312254	0.758036	0.241964	0.483927
0.8	0.289692	0.788145	0.211855	0.423711
0.9	0.266085	0.815940	0.184060	0.368110
1.0	0.241971	0.841345	0.158655	0.317311
1.1	0.217852	0.864334	0.135666	0.271332
1.2	0.194186	0.884930	0.115070	0.230139
1.3	0.171369	0.903195	0.096805	0.193601
1.4	0.149727	0.919243	0.080757	0.161513
1.5	0.129518	0.933193	0.066807	0.133614
1.6	0.110921	0.945201	0.054799	0.109599
1.7	0.094049	0.955435	0.044565	0.089131
1.8	0.078950	0.964069	0.035931	0.071861
1.9	0.065616	0.971284	0.028716	0.057433
2.0	0.053991	0.977250	0.022750	0.045500
2.1	0.043984	0.982136	0.017864	0.035729
2.2	0.035475	0.986097	0.013903	0.027807
2.3	0.028327	0.989276	0.010724	0.021548
2.4	0.022395	0.991803	0.008197	0.016395
2.5	0.017528	0.993791	0.006209	0.012419
2.6	0.013583	0.995339	0.004661	0.009322
2.7	0.010421	0.996533	0.003467	0.006934
2.8	0.007915	0.997495	0.002505	0.005110
2.9	0.005952	0.998134	0.001866	0.003732
3.0	0.004432	0.998650	0.001350	0.002700
3.5	0.000873	0.999768	0.000232	0.000465
4.0	0.000134	0.999968	0.000032	0.000063
4.5	0.000016	0.999996	0.000004	0.000008
5.0	0.0000015	0.9999997	0.0000003	0.0000006

(Table 2.4 developed from data appearing in [21]), [12].

Let

$$x = \log_e x_1,$$

then

$$dx = \frac{1}{x_1} dx_1$$

and

$$F(\log_e x_1) = \int_0^{x_1=e^x} \frac{1}{x_1} f(\log_e x_1)\, dx_1.$$

The probability of $\log x$ is identical to that of x_1.

$$f(\log x_1) = f(x_1)$$

$$f(x_1) = \frac{1}{x_1} \cdot f(\log_e x_1)$$

$$f(x_1) = \frac{1}{x_1 \sigma \sqrt{2\pi}} \exp\left\{ -\frac{1}{2} \cdot \left[\frac{\log_e x_1 - \overline{\log_e x_1}}{\sigma} \right]^2 \right\}$$

The mean value, $\overline{\log_e x}$, is a logarithm, and the standard deviation, σ, is also a logarithm. The standard deviation of $\log_e x$ can be computed, and is sometimes denoted as $\log_e \sigma$ (this is not to be confused with the logarithm of the standard deviation of the normal distribution). Note that the mean, $\overline{\log_e x}$, is *not* the logarithm of the mean value of x.

If logarithms to the base 10 are employed, the expression for $f(x)$ is modified by $\log_{10} e = 0.4343$. Thus

$$f(x_1) = \frac{0.4343}{x_1 \sigma \sqrt{2\pi}} \exp\left\{ -\frac{1}{2}\left[\frac{\log_{10} x_1 - \overline{\log_{10} x_1}}{\sigma} \right]^2 \right\}$$

$\overline{\log_e x_1}$ or $\overline{\log_{10} x_1}$ corresponds to a cumulative distribution function value of 0.50. The median value of x_1 has a logarithm identical to the mean logarithm of x_1. If $\check{x}_1$ denotes the median of the x_1 values, then

$$f(x_1) = \frac{0.4343}{x_1 \sigma \sqrt{2\pi}} \exp\left[-\frac{1}{2\sigma^2} \log^2_{10}\left(\frac{x_1}{\check{x}_1} \right) \right]$$

The logarithm of 1 is zero, the logarithm of a value less than 1 is negative, and the logarithm of 0 is $-\infty$. Thus, the probability density function has zero value at $x = 0$ (see Fig. 2.16).

If a random variable, x, is lognormally distributed, then $\log_{10} x$ is normally distributed. Table 2.4, for the normal distribution, yields values of the density

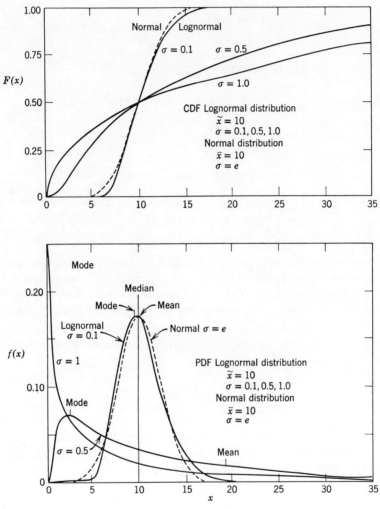

Figure 2.16 Three lognormal distribution curves and one normal distribution curve.

function for the standardized variable z;

z	Normal Density	$\log x$	x	Lognormal Density
-3	0.0044	0.7	5.012	0.0038
-2	0.0540	0.8	6.310	0.0372
-1	0.2420	0.9	7.943	0.1325
0	0.3989	1	10.000	0.1733
$+1$	0.2420	1.1	12.59	0.0836
$+2$	0.0540	1.2	15.85	0.0159
$+3$	0.0044	1.3	19.95	0.0010

for example,

$$y = \frac{\log x - \log \breve{x}}{\sigma}$$

$$\log x = \log \breve{x} + y\sigma.$$

Assume $\sigma = 0.1$ and $\log \breve{x} = 1$, then using the base, 10,

$$\log x = 1 + 0.1y.$$

The lognormal density at $x = 10$ is then calculated from

$$f(x) = \frac{0.4343}{x\sigma} \left(\frac{1}{\sqrt{2\pi}} e^{-\frac{1}{2}y} \right)$$

$$f(x = 10) = \frac{(0.4343)(0.3989)}{(10)(0.1)} = 0.1733.$$

These parameters are plotted in Fig. 2.17, with the same distributions for $\sigma = 0.5$ and 1.0 and $\log \breve{x} = 1$.

The curves of Fig. 2.16 show that the probability density function changes as the standard deviation varies.

The logarithm of a random variable is characterized by a much smaller variation than the random variable itself. With very large variations, the curve of x will be likely to lie between the curves for $\sigma = 0.1$ and 0.5, at least for values of x between $x = 1$ and $x = 25$. The logarithm of x varies between 0 and 1.398 with a standard deviation of $\log x$ from 0.1 to 0.3.

If x is large and the variation in x is small, $\log x$ is nearly proportional to x, in the range of observed values. Under these circumstances, $\log x$ is essentially normally distributed if x is normally distributed (see Fig. 2.17).

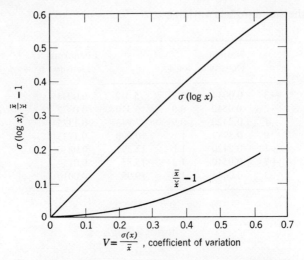

Figure 2.17 Lognormal distribution characteristics.

The mode, median, and mean of the lognormal distribution occur at values of x such that

(a)   mode: $\log x = \log \breve{x} - 2.3026\sigma^2$,

(b)   median: $\log x = \log \breve{x}$,

(c)   mean: $\log x = \log \breve{x} + 1.1513\sigma^2$;

(a), (b), and (c) show that, as the standard deviation decreases, the lognormal distribution approaches the normal distribution; for example, if $\log x = 1$ and $\sigma = 0.1$,

         mode: $\log x = 1 - (2.3026)(0.1)(0.1)$

                      $\log x = 0.97697$

                      $x = 9.484$

         mean: $\log x = 1 + (1.1513)(0.1)(0.1)$

                      $\log x = 1.011513$

                      $x = 10.27$

Products or quotients of random variables are sometimes needed (see Chapter 3). Consider the density function of deflection, in which load, span, moment of inertia, and modulus of elasticity are random variables.

If the coefficient of variation (see Section 3.4.10), of the random variable is small, then the difference between the mean and the median is small, and also the difference between $\sigma_{ln\,x}$ and the coefficient of variation is small.

With X normally distributed,

$$x = \xi + y$$

$$x = \xi\left(1 + y\frac{\sigma}{\xi}\right)$$

$$x = \xi(1 + yv).$$

Then

$$\log x = \log \xi + \log(1 + yv)\dagger$$

$$\xi \approx \breve{x}$$

$$\log(1 + yv) \approx yv$$

$$\log x \approx \log \breve{x} + yv.$$

The lognormal expression is

$$\log x = \log \breve{x} + y\sigma(\log x).$$

With v small

$$\sigma(\log x) \approx v$$

and

$$\sigma^2(\log x) \approx v^2 \quad \text{(see Fig. 2.17)}.$$

These approximate relationships are valid to $v = 0.50$.

Example 1. To illustrate the use of these expressions, consider $F = fA$ with f and A normally distributed. Then

$$\log F = \log f + \log A.$$

2.3.8 Central Limit Theorem

Let n independent observations (measurements)

$$x_1, x_2, \ldots, x_{n-1}, x_n$$

of the same random variable X be taken (at random), and the sample mean $\bar{x}$ computed by adding the numbers together and dividing by n. From each such sample, a particular value of $\bar{x}$ is computed. Then, if the same sample size is n each time, the computed averages will be characterized by a spectrum of values.

A theorem of great importance, called *the central limit theorem* which deals with the above situation, may be stated as follows [18].

If a population has a finite variance σ^2 and mean value μ, then the distribution of the sample mean approaches the normal distribution with variance σ^2/n and mean μ as the sample size n increases.‡

† v equals the coefficient of variation.
‡ See Section 2.6.3.

The surprising thing about this theorem is that nothing is said about the form of the population distribution function. For a proof of the central limit theorem, see [19], p. 223.

2.3.9 Normal Approximation of Binomial

The normal is the limiting form for a great many distributions. It is the limiting form approached by the binomial for large n (for application of this property see Chapter 14). Thus, if X is binomially distributed, then

$$f(x) = \binom{n}{x} p^x q^{n-x} \quad (x = 0, 1, 2, \ldots).$$

As n increases,

$$f(x) \to \frac{1}{\sqrt{npq}\sqrt{2\pi}} \cdot \exp\left[-\frac{(x-np)^2}{2npq}\right] (-\infty, \infty),$$

clearly the normal density when $\mu = np$ and $\sigma^2 = npq$. For moderately large n, the binomial distribution is adequately approximated, for many purposes, by the normal (with $\mu = np$ and $\sigma^2 = npq$) provided $np \geq 5$. In applying the normal approximation, a compensation is made for the fact that the variate is actually discrete and n is finite. The conventional correction consists of changing the numerical deviation from μ by $\frac{1}{2}$ unit before dividing by σ, on the argument that the discrete value $x = a$ should correspond on a continuous scale to the interval.

$$a - \tfrac{1}{2} \leq x \leq a + \tfrac{1}{2}.$$

Hence to estimate the probability

$$p(x \geq a) = \sum_{x=a}^{n} \binom{n}{x} p^x q^{n-x} \quad \text{(Eq. 2.17)},$$

where $a = \mu + \frac{1}{2}$, the value needed is

$$u = \frac{a - \tfrac{1}{2} - \mu}{\sigma}$$

and the probability value $R(u)$ is found in Table 2.4. To estimate the probability

$$p(x \leq a) = \sum_{x=0}^{a} \binom{n}{x} p^x q^{n-x}$$

where $a \leq \mu - \frac{1}{2}$, the value used is

$$\mu' = \frac{\mu - (a + \tfrac{1}{2})}{\sigma}.$$

Finally, to estimate $p(a \leq x \leq b)$, the interval is extended $\frac{1}{2}$ unit to either side before dividing by σ.

Table 2.5 Comparison of Normal and Poisson Approximations of the Binomial Probability Function $f(x)$ with $n = 2500$ and $p = 0.02$

x†	Binomial	Normal	Poisson
25	0.0000	0.0001	0.0000
35	0.0052	0.0057	0.0054
45	0.0460	0.0442	0.0458
55	0.0424	0.0442	0.0422
65	0.0061	0.0057	0.0063
75	0.0002	0.0001	0.0002

† For binomial values see [12], Table 3-19; for normal values, see [42] and, for Poisson values, see [12], Table 3-19.

As suggested in Table 2.5, the Poisson distribution (with $\mu = np$) approximates the binomial for sufficiently small p, while the normal approximates the binomial for sufficiently large n, provided np is not too small. This implies that the normal approximates the Poisson when μ is sufficiently large. The normal approximation employs the same μ as the Poisson distribution and σ is taken as $\sqrt{\mu}$.

2.3.10 Gamma Approximation of the Normal Distribution

The gamma distribution approaches normality as α increases. The relationships employed are

$$\mu = \beta(1 + \alpha)$$

and

$$\sigma = \beta\sqrt{1 + \alpha}.$$

A comparison between exact values of $F(x)$ for the gamma distribution with $\alpha = 15, \beta = 1$, first treating the gamma itself as normal, is given in Table 2.6.

Table 2.6 Normalizing Transformations

x	$F(x)$	Direct Normal	Fishers Transform
8.180	0.01	0.025	0.013
10.035	0.05	0.068	0.055
11.135	0.10	0.112	0.103
13.152	0.25	0.238	0.247
15.668	0.50	0.467	0.492
18.487	0.75	0.733	0.746
21.293	0.90	0.907	0.902
23.098	0.95	0.962	0.953
26.744	0.99	0.996	0.992

56 Foundation

Other normalizing transformations have been devised for special cases. In addition to the advantages that existing probability tables may be applied, normalizing transformations serve the purpose of reducing many difficult analytical problems to solvable forms.

Example 1. Loads Model. The gamma distribution with limits $0 \leq x < \infty$, tending to skew toward the right (as do certain loads distribution) and being expressed in two parameters, α and β, is well suited as a model for certain loads distributions. The gamma distribution mates the normal distribution when $\alpha \geq 14$. Thus several load components, each highly skewed and gamma distributed, may sum algebraically to yield a resultant loading. Such a resultant loading may be very nearly normally distributed and consequently suited to the statistical algebra of Chapter 3. For example, given

$$\text{load component 1}: (\alpha_1, \beta) = (5, \beta)$$
$$\text{load component 2}: (\alpha_2, \beta) = (6, \beta)$$
$$\text{load component 3}: (\alpha_3, \beta) = (8, \beta)$$

where

$$\beta = 0.40, \, k = 3$$
$$\alpha = \alpha_1 + \alpha_2 + \alpha_3 + k - 1$$
$$\alpha = 5 + 6 + 8 + 3 - 1 = 21$$

Then, by the equations:

$$\mu = \beta(\alpha + 1) = 0.4(22.0) = 8.800,$$
$$\sigma = \beta\sqrt{\alpha + 1} = 0.40(4.68) = 1.872.$$

Thus the normal distribution approximation is

$$(\mu, \sigma) = (8.800, 1.872),$$

where the gamma distribution of the sum is

$$(\alpha, \beta) = (21.0, 0.4).$$

2.4 JOINT DISTRIBUTIONS

In the work for which a foundation is being prepared the distributions of importance are almost exclusively of the continuous type. In the design process it is rarely the case that only one random variable is of interest. It is with combinations of random variables that we must deal.

Up to this point, the discussion has been confined to distributions of single random variables, called univariates. Distinguished from univariates are bivariates, which involve two random variables, and multivariates, which involve two or more random variables. After the bivariate concept is developed, extension to finite higher dimensions does not involve any new concepts.

A joint distribution can be derived (from marginal distributions) only if the variates are independent. The joint distribution of independent random variables is definable if enough information exists to determine the compound probability (either directly or by analysis).

Consider the joint distribution of random variables $x_1, x_2, \ldots$. $F(\alpha, \beta, \gamma, \ldots)$ gives the probability that the unequalities

$$x_1 \leq \alpha, \quad x_2 \leq \beta, \quad x_3 \leq \gamma, \ldots$$

are simultaneously satisfied. The joint cumulative distribution function is $F(x_1, x_2, x_3, \ldots)$, since $\alpha, \beta, \gamma, \ldots$ denote specific values of the variates.

$$F(x_1, x_2, x_3, \ldots, x_n) = F(x_1) F(x_2) F(x_3) \cdots F(x_n). \qquad (2.29)$$

Equation 2.29 states the general definition of independence for variates (note the analogy to Eq. 2.14).

The idea of density (for continuous random variables) is extendable to n dimensions. If the joint distribution function is continuous and differentiable, the joint density function is defined as its nth partial derivative. (Each random variable is differentiated once in sequential order.)

$$f(x_1, x_2, \ldots, x_n) = \frac{\partial^n}{\partial x_1 \, \partial x_2 \cdots \partial x_n} [F(x_1, x_2, \ldots x_n)]. \qquad (2.30)$$

Interpretation of the joint density function is as an extension of the univariate density function. Thus

$$f(x_1, x_2, \ldots, x_n) \, dx_1 \, dx_2 \cdots dx_n$$

represents the probability that a random point

$$x_1', x_2', \ldots, x_n'$$

will fall in an infinitesimal region such that

$$x_1 < x_1' < x_1 + dx_1; \; x_2 < x_2' < x_2 + dx_2; \ldots; x_n < x_n' < x_n + dx_n.$$

Thus, the integral over the "space" of definition equals one. For independent random variables, the joint density function factors into a product of marginal density functions:

$$f(x_1, x_2, \ldots, x_n) = f_1(x_1) \cdot f_2(x_2) \cdots f_n(x_n). \qquad (2.31)$$

The joint density function of independent random variables is the product of the respective density functions.

Marginal density functions of each random variable are obtained (from the joint density function) by integration with respect to the other random variables. For example, let the joint density function of x and y be denoted by $f(x, y)$ and the marginal density function of x by $f_1(x)$. Take the infinitesimal rectangle with corners located at the points (x, y), $(x + dx, y)$, $(x, y + dy)$, $(x + dx, y + dy)$. The probability of a random point (x', y') falling in this rectangle is $f(x, y)\, dx\, dy$. The integral of such probability elements with respect to y (for a fixed value of x) is the sum of probabilities of all the mutually exclusive ways of obtaining the points lying between x and $x + dx$. Unless $f(x, y)$ is defined for the entire plane, the admissible region is bounded by lines or curves, and the integral with respect to y is taken from the bottom to the top of the vertical strip (representative of the fixed value of x). The lower limit of y (ordinarily a point on a curve) is expressed in terms of x as $a_1(x)$ and the upper limit as $b_1(x)$. Thus

$$p(x \leq x' \leq x + dx) = \left[\int_{a_1(x)}^{b_1(x)} f(x, y)\, dy\right] dx = f_1(x)\, dx$$

and

$$f_1(x) = \int_{a_1(x)}^{b_1(x)} f(x, y)\, dy. \tag{2.32}$$

By similar reasoning [12, 15, and 18]:

$$f_2(y) = \int_{a_1(y)}^{b_1(y)} f(x, y)\, dx,$$

where $a_2(y)$ and $b_2(y)$ are the functional values of x in terms of y at the two ends of the horizontal strip, corresponding to a fixed value of y.

2.5 POINT ESTIMATION

A point estimate is a single value that is used to estimate the parameter† (such as mean value) in question. Such a point estimate rarely exactly agrees with the unknown parameter to be estimated. It must usually suffice that the estimate be close to the unknown parameter value. Any random variable may be considered an estimate of a parameter; for instance, if the expected value, μ, is to be estimated from a sample of size n, each of the statistics—the sample mean, sample median, largest observation, etc.—may be considered as estimates of μ. The question is—How is one selected? The desirable estimate is that which is usually closest to the true value of μ.

From n independent trial measurements of the random variable X, the task is to find the most likely estimate (or estimator) g of a true parameter θ in a

† Note the connotation on the word parameter as used in this chapter versus that given in Chapter 1, footnote.

known mathematical functional form $h(x, \theta)$. For the moment assume that there is only one parameter to be estimated. First establish a function

$$g = g(x_1, x_2, \ldots, x_n)$$

of the trial values of X from which the estimate g is to be deduced. The function g is also a random variable. The function desired is one that possesses the narrowest possible scatter about the true value θ.

There are several methods of writing such g functions, and each method gives a different degree of accuracy of fit in the estimate of g. If N sets of samples, each of size n, are taken from a parent population, N different values of g are obtained. These N values of g form a distribution from which the standard deviation is computed. This process is repeated for each method. Among them the method with the smallest standard deviation has its g values clustered most compactly, and therefore is called the most efficient. With any method, if the mean of the g distribution for a sample tends to a value different from θ, the estimate is called biased. If the estimate converges to θ as $N \to \infty$, the estimate is called consistent.

For most parametric estimation problems, the method of estimation called the "method of maximum likelihood" is most efficient, and with large n, is usually satisfactorily consistent. The likelihood function, the product of all n values of $\phi(x_1; \theta)$, is written

$$L(x_1, x_2, \ldots, x_n; \theta) = \phi(x_1; \theta) \cdot \phi(x_2; \theta) \cdots \phi(x_n; \theta). \quad (2.33)$$

Now consider the general case in which there is a continuum of possible values of g (i.e., a parameter that is a continuous random variable). The relative probability of any two different values of g is given by the likelihood ratio, in which the likelihood functions are of the form given in Eq. 2.33 with one value of g in place of θ for the numerator of the likelihood ratio and with the other value in place of θ for the denominator. Imagine each of N possible values of g, that is, $g_1, g_2, g_3, \ldots, g_n$, inserted in the L function and each of the N values of L_j computed. These N values of L_j form a distribution which, as $N \to \infty$, can be shown to approach a normal distribution, whose mean value at the maximum of the distribution corresponds to the desired estimate of g.

To find the value of L_j that makes L a maximum, L is differentiated with respect to θ and the derivative set equal to zero. Since L is a maximum when $\log L$ is a maximum, the logarithmic form is used when sums are more convenient to manipulate than products. Thus

$$\left(\frac{\partial \log L}{\partial \theta}\right)_{\theta=g} = 0 = \sum_{i=1}^{n} f_i \frac{\partial}{\partial g} \log \phi(x_i; g) \quad (2.34)$$

and a solution is sought for g. This value for g is the most likely estimate of θ (not always unbiased). Solutions of Eq. 2.34 are quite often explicit and free of multiple roots. When multiple roots do occur, the most significant root is selected. The procedure may be generalized to treat more parameters than one.

The maximum likelihood method is generally considered to be unsurpassed as a statistical approach to the majority of measurement problems encountered in the physical sciences. This method uses all of the experimental information in the most direct and efficient fashion possible to give an unambiguous estimate. Its main disadvantage is that the functional relationship must be known or assumed.

Example 1. To illustrate the method, consider a set of r measurements that are known to be fitted by a binomial distribution. It is desired to find the maximum likelihood estimate of the success probability p. Call p^* this estimate of p. Suppose success is observed w times and failure $r - w$ times. Thus the likelihood function is (by Eq. 2.33)

$$L\dagger = \binom{r}{w} p^w (1-p)^{r-w}.$$

Taking the natural log of both sides,

$$\log_e L = \log_e \binom{r}{w} + w \log_e p + (r - w) \log_e (1 - p).$$

Applying Eq. 2.34,

$$\frac{\partial}{\partial p} \log L = 0 = \frac{w}{p^*} - \frac{r-w}{1-p^*}.$$

The most likely estimate of the true value of p is

$$p^* = \frac{w}{r}.$$

Example 2. Let the mean value μ and the standard deviation σ of a normal distribution be estimated from a sample of size r. Call the estimates $\bar{x}$ and s_x respectively. With density,

$$f(x) = \frac{1}{\sqrt{2\pi}\sigma_x} \exp\left[-\frac{1}{2}\left(\frac{x_i - \mu_x}{\sigma_x}\right)^2\right],$$

the likelihood function is (by Eq. 2.33):

$$L = \prod_{i=1}^{r} \frac{1}{\sigma\sqrt{2\pi}} \exp\left[-\frac{1}{2}\left(\frac{x_i - \mu}{\sigma}\right)^2\right]$$

Applying Eq. 2.34,

$$\left(\frac{\partial}{\partial \mu} \log L\right)_{\mu=\bar{x}} = 0 = \sum_{i=1}^{r} \frac{x_i - \bar{x}}{\sigma^2}$$

† Note that, in this case, n in Eq. 2.33 is 1, the observation of w successes in r trials.

from which

$$0 = r\bar{x} - \sum_{i=1}^{r} x_i.$$

$$\bar{x} = \frac{1}{r}\sum_{i=1}^{r} x_i = \frac{x_1 + x_2 + \cdots + x_r}{r}.$$

To estimate the standard deviation σ

$$\left(\frac{\partial}{\partial \sigma} \log L\right)_{\sigma = s_x} = 0 = \sum_{i=1}^{r}\left(-\frac{1}{s} + \frac{(x_i - x)^2}{s}\right)$$

and the variance is

$$s_x^2 = \frac{1}{r}\sum_{i=1}^{r}(x_i - \bar{x})^2.$$

From which, by Eq. 2.43, the standard deviation is

$$s_x = \sqrt{\frac{1}{r}\sum_{i=1}^{r}(x_i - \bar{x})^2}.$$

The principle of maximum likelihood essentially assumes that the sample is representative of the population.

2.6 EXPECTED VALUES AND MOMENTS

The concept of mathematical expectation is of importance both in the mathematical development and the practical use of probability theory.

Consider the cumulative distribution function $F(x)$ of any variate X, either discrete or continuous. Also, consider a function $\phi(x)$ which is continuous over the entire region of definition of $F(x)$. Recalling that $F(x)$ is always monotonic, the ordinate range from 0 to 1 is divided into n divisions, with each part $\Delta F(x_i)$. The divisions need not be equal, but the sum $\sum \Delta F(x_i) = 1$. Lines parallel to the abscissa from two consecutive points on the vertical axis will ordinarily intersect the curve $F(x)$ in two points (see Fig. 2.18). The abscissas of these two points may be the same or different. If a point x_i is selected anywhere between the two abscissas, it will determine a value of

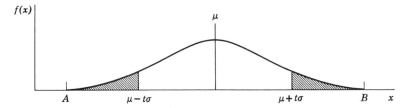

Figure 2.18 Camp-Meidall inequality.

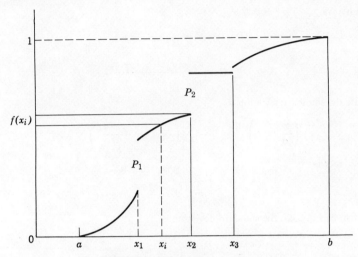

Figure 2.19 Distribution function [12].

$\phi(x)$, for instance $\phi(x_i)$. The sum $\sum \Delta(x_i) \Delta F(x_i)$ with the summation from $i = 1$ to $i = r$ can be shown to approach a limit as the number of divisions $\Delta F(x_i)$ become infinite, regardless of the choice of size of each $\Delta F(x_i)$, provided $\Delta F(x_i) \to 0$ as $n \to \infty$. This (unique) limit is defined as the expected value of $\phi(x)$ and is expressed by

$$E[\phi(x)] = \int_a^b \phi(x) \, dF(x). \tag{2.35}$$

The contribution to the integral at points x_1 and x_2 (Fig. 2.19) would be $\phi(x_1)p_1 + \phi(x_2)p_2$. The expected value may be a summation such as

$$E[\phi(x)] = \sum_{i=1}^{r} \phi(x_i) f(x_i), \tag{2.36}$$

where the x_i's are the values of x for which the probability is defined. If $F(x)$ is differentiable, then x is a continuous variate with density function $f(x)$, and, by Eq. 2.22,

$$E[\phi(x)] = \int_a^b \phi(x) f(x) \, dx. \tag{2.37}$$

The case shown in Fig. 2.20 is a combination of the two concepts, so (except at two points) the function is differentiable, and the integral may be computed by Eq. 2.37, while at two points (x_1 and x_2) the integral must be evaluated as

$$\sum_{i=1}^{r} \phi(x_i) f(x_i).$$

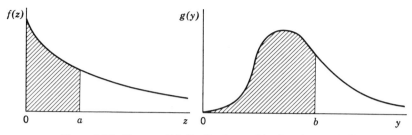

Figure 2.20 Exponential distribution and its fourth root [12].

The sum of the two integrals gives the expected values (see Properties of Density Functions, Section 2.3.1):

$$\int_a^b \phi(x)f(x)\,dx$$

with $a = p_1$ and $p_2 = b$, this gives a value equal to unity.

Example 1. The income of a certain irrigating company is a function of the seasonal rainfall (in acre feet) which falls on the watershed under its control. The income value has been found to be closely estimated as $3.65(1.15 - e^{-x})$ hundreds of thousands (10^5) dollars. The variable X is a real number which expresses the rainfall in 10^4 of acre feet. It is assumed that the distribution of X is

$$f(x) = e^{-x}\ (0 < x < \infty).$$

Determine the expected value of rainfall to the irrigating company.

If the value of rainfall is represented by $\phi(x)$,

$$E[\phi(x)] = \int_0^\infty \phi(x)f(x)\,dx = \int_0^\infty 3.65(1.15 - e^{-x})e^{-x}\,dx$$

$$= 4.20\int_0^\infty e^{-x} - 3.65\int_0^\infty e^{-2x}\,dx$$

$$= 4.20\left[e^{-x}\right]_0^\infty - 3.65\left[\frac{e^{2-x}}{2}\right]_0^\infty$$

$$= 4.20 - \frac{3.65}{2} - 2.375(10^5\ \$)$$

$$= \$237{,}500.$$

2.6.1 Averages

The average of a set of N numbers is obtained by summing the set and then dividing by N. Since all of the numbers need not be different, assume that there are r distinct values, $x_1, x_2, \ldots, x_r$, each occurring $N_1, N_2, \ldots, +N_n$ times, respectively. Then

$$N_1 + N_2 + \cdots + N_r = N.$$

Thus the average is given as follows:

$$\text{average} = \frac{\sum_{i=1}^{r} N_i x_i}{N}. \tag{2.38}$$

Since each N_i is divided by N, the relative frequency N_i/N (see Eq. 2.1) is multiplied by each x_i, but N_i/N may be replaced by a probability. Thus

1. For a discrete variate X with density function $f(x)$ and corresponding probabilities $f(x_i)$

$$\text{average of } x = \sum_{i=1}^{r} x_i f(x_i). \tag{2.39}$$

2. For a continuous variate X with density function $f(x)$ defined in the range (α, β)

$$\text{average of } x = \int_{\alpha}^{\beta} x f(x) \, dx. \tag{2.40}$$

3. For a random variate X with a distribution function $F(x)$ defined over the region (α, β)

$$\text{average of } x = \int_{\alpha}^{\beta} x \, dF(x). \tag{2.41}$$

The averages defined in 1, 2, and 3 above are usually denoted by the letter μ. By comparing Eq. 2.41 with the definition of expected value, it is seen that, for the special case $\phi(x) = x$, the two are equivalent, and

$$E(x) = \int_{\alpha}^{\beta} x \, dF(x) = \text{average of } x.$$

Expected values may always be thought of as species of averages.

Example 1. Compute the average value $E(w)$ of the continuous random variable W, given that

$$f(w) = (3w - 3w^2)2 \quad (0 < w < 1).$$

From Eq. 2.40,

$$E(w) = \int_0^1 w f(w) \, dw$$

$$E(w) = 2 \int_0^1 w(3w - 3w^2) \, dw = 2 \int_1^0 (3w^2 - 3w^3) \, dw = \tfrac{1}{2}.$$

2.6.2 Properties of the Mean Values

Two intuitively obvious properties are the following:

1. If $S \leq z \leq T$, then $S \leq E(z) \leq T$.
2. If $\alpha(z) = c_0 + c_1 \cdot \alpha_1(z) + \cdots + c_n \cdot \alpha_n(z)$, then

$$E[\alpha(z)] = c_0 + c_1 E[\alpha_1(z)] + \cdots + c_n E[\alpha_n(z)].$$

Property 1 states that the mean value does not fall outside the range of definition of the variate. Property 2 states that if $\alpha(x)$ is a linear function of any number of components, then its expected value is given by the same linear function of the expected values of these components.

Example 1. By Eq. 2.36, compute the mean of the binomial distribution.

$$\mu = E[\phi(x)] = \sum_{i=0}^{r} \phi(x_i) f(x_i)$$

Here $\phi(x) = x$.

$$\mu = E[\phi(x)] = E(x) = \sum_{x=0}^{r} \phi(x_i) f(x_i)$$

Thus with $f(x) = \binom{r}{x} p^x q^{r-x}$,

$$\mu = E(x) = \sum_{x=0}^{r} x \frac{r!}{x!\,(r-x)!} p^x q^{r-x} = \sum_{x=0}^{r} \frac{r(r-1)!\,pp^{x-1}q^{r-x}}{(x-1)!\,(r-x)!},$$

put $m = r - 1$ and let $u = x - 1$, and $r - x = r - (u + 1)$

$$= (r - 1 - u) = m - u$$

Then

$$\mu = E(x) = rp \sum_{u=0}^{m} \frac{m!}{u!\,(m-u)!} p^u q^{(m-u)} = rp$$

since

$$\sum_{u=0}^{m} \frac{m!}{u!\,(m-u)!} p^u q^{m-u} = 1.$$

2.6.3 Measure of Dispersion

Expected value (mean) is a measure of central tendency, indicating the location of a distribution on some coordinate axis. Other possible measures are the median and the mode. There remains the question "How completely is the distribution described?" A concept of concentration applies (equally) to each side of the mean; the function to be averaged should depend upon the magnitude of difference between the mean and a value x_i. A simple measure satisfying intuition and analytic requirements is the mean square deviation from the mean. When the mean (μ) is used, the measure of dispersion is called the variance, σ^2. The (positive) square root of the variance is called the standard deviation, σ. Thus

$$\text{variance} \equiv \sigma^2 \equiv E[(x - \mu)^2]. \tag{2.42}$$

$$\text{standard deviation} \equiv \sigma \equiv \sqrt{\sigma^2} \tag{2.43}$$

Expanding Eq. 2.42,

$$\sigma^2 \equiv E(x^2 - 2x\mu + \mu^2).$$

66 Foundation

Applying property 2 of E† from Section 2.6.2,

$$\sigma^2 \equiv E(x^2) - 2\mu E(x) + \mu^2 \equiv E(x^2) - \mu^2 \tag{2.44}$$

Equation 2.44 sometimes simplifies the determination of σ^2.

Example 1. Find the mean, variance, and standard deviation of the variate Z, where the density function of z is

$$f(z) = \frac{z}{\beta^2} e^{-(z/\beta)} \qquad (0, \infty; \beta > 0).$$

The mean (μ_z) is given by (see Eq. 2.40):

$$\mu_z = \int_0^\infty z f(z)\, dz = \beta \Gamma(3)\ddagger = 2\beta$$

The variance (σ_z^2) is given by (see Eq. 2.44):

$$\sigma_z^2 = E(z^2) - \mu_z^2 = \int_0^\infty z^2 f(z)\, dz - 4\beta^2$$
$$= \beta^2\, \Gamma(4) - 4\beta^2 = 2\beta^2.$$

Standard deviation (σ_z) is (see Eq. 2.43):

$$\sigma_z = \sqrt{\sigma_z^2} = \sqrt{2}\beta.$$

Example 2. Find the variance of the Poisson distribution, given that the density of Z is (Eq. 2.18):

$$f(z) = \frac{e^{-\mu}\mu^z}{z!} \qquad (0 \leq z < \infty).$$

Initially, let

$$z^2 \equiv z(z-1) + z.$$

Then

$$E(z^2) = e^{-\mu} \sum_{z=0}^{\infty} z(z-1) \frac{\mu^z}{z!} + e^{-\mu} \sum_{z=0}^{\infty} \frac{\mu^z}{z!}.$$

---

† Also see properties of E given in Section 2.9.3.
‡ The gamma function $\Gamma(r)$ is given by

$$\Gamma(r) = \int_0^\infty x^{r-1} e^{-x}\, dx.$$

Thus

$$\Gamma(1) = \int_0^\infty e^{-x}\, dx = 1; \qquad \Gamma(2) = 1$$

and the recurrence formula is

$$\Gamma(r) = (r-1)\Gamma(r-1).$$

Let $u = z - 2$, $v = z - 1$

$$e^{-\mu}\left(2\sum_{u=0}^{\infty}\frac{\mu^u}{u!} + \mu\sum_{v=0}^{\infty}\frac{\mu^v}{v!}\right)$$
$$= e^{-\mu}(\mu^2 e^\mu + \mu e^\mu) = \mu^2 + \mu$$
$$\sigma_z^2 = E(z^2) - [E(z)]^2 = (\mu^2 + \mu) - \mu^2 = \mu.$$

Thus the mean and variance of the Poisson distribution are

$$\mu_z = \sigma_z^2 = np.$$

Example 3. Find the mean and variance of the normal distribution, given the normal density function:

$$f(z) = \frac{1}{\beta\sqrt{2\pi}} \exp -\frac{1}{2}\left(\frac{z-\alpha}{\beta}\right)^2 \qquad (-\infty, \infty).$$

Let $u = z - \alpha$. Applying Eq. 2.40,

$$\mu = E(z) = \int_{-\infty}^{\infty} z\, f(z)\, dz = \int_{-\infty}^{\infty} (u + \alpha)\, f(z)\, dz$$
$$= \frac{1}{\beta\sqrt{2\pi}}\int_{-\infty}^{\infty} u e^{-u^2/2\beta^2}\, du + \alpha \int_{-\infty}^{\infty} f(z)\, dz = 0 + \alpha = \alpha.$$

Put $\mu = \alpha$; then the variance is (see Eq. 2.48)

$$\sigma^2 = \int_{-\infty}^{\infty} (z - \mu)^2 f(z)\, dz = \frac{2}{\beta\sqrt{2\pi}} \cdot \int_{-\infty}^{\infty} u^2 e^{-u^2/2\beta^2}\, du$$
$$= \frac{2\beta^2}{\sqrt{\pi}}\left(\frac{3}{2}\right) = \beta^2.$$

The parameters α and β of the normal distributions are, respectively, equal to the mean μ and standard deviation σ.

Example 4. Determine the variance of the binomial distribution. Recall that $\sigma^2 = E(x^2) - \mu^2$ (Eq. 2.44). Let $x^2 \equiv x(x-1) + x$; $s = r - 2$. Employing Eq. 2.42,

$$E(x^2) = \sum_{x=0}^{p} \frac{x(x-1)!\, r!}{x!\,(r-x)!} p^x q^{r-x} + \sum_{x=0}^{p} \frac{xr!}{x!\,(r-x)!} p^x q^{r-x}$$
$$= \sum_{x=0}^{r} \frac{r(r-1)(r-2)!}{(x-2)!\,(r-x)!} p^x q^{r-x} + rp.$$

Let $x - 2 = T$

$$r(r-1)p^2 \sum_{x=0}^{r} \frac{s!\, p^T q^{s-T}}{T!\,(s-T)!} + rp$$
$$E(x^2) = r(r-1)p^2 + rp = r^2 p^2 - rp^2 + rp$$
$$\sigma^2 = E(x^2) - \mu^2 = r^2 p^2 - rp^2 + rp - r^2 p^2$$
$$\sigma^2 = rp(1-p) = rpq.$$

The standard deviation (σ) is

$$\sigma = \sqrt{rpq}.$$

2.6.4 Extreme Value Probability

The relationship between standard deviation and the intuitive idea of concentration is emphasized by the fact that an upper bound (in standard deviation measure) may be set on extreme deviations from the mean. The following formula, known as Tchebysheff's[†] inequality and applicable to any distribution for which μ and σ are definable, gives such a relationship between μ and σ.

For any distribution with finite mean and variance,

$$p(|x - \mu| \geq t\sigma) \leq \frac{1}{t^2} \; (t > 0), \tag{2.45}$$

Example 1. Employ Eq. 2.45 to compute the probability of a deviation $\geq t = 2, 3, 4, 5$ standard deviations from the mean.

$$t = 2, \quad p(|x - \mu| \geq 2\sigma) \leq \frac{1}{2^2} = 0.2500,$$

$$t = 3, \quad p(|x - \mu| \geq 3\sigma) \leq \frac{1}{3^2} = 0.1111,$$

$$t = 4, \quad p(|x - \mu| \geq 4\sigma) \leq \frac{1}{4^2} = 0.0625,$$

$$t = 5, \quad p(|x - \mu| \geq 5\sigma) \leq \frac{1}{5^2} = 0.0400.$$

From the expression $(|x - \mu|)/\sigma$, it is seen that the probability of exceeding a stated magnitude of deviation from the mean tends to decrease as σ decreases.

Whereas the Tchebysheff inequality guarantees that no more than $\frac{1}{9}$ of any distribution with finite μ and σ likes beyond 3σ from its mean, the Camp-Meidall inequality

$$p \leq \frac{1}{2.25t^2},$$

guarantees that this region includes less than 5 percent of any distribution to which that inequality applies.

The probability p is indicated by the shaded area in Fig. 2.19.

Law of Large Numbers. The law of large numbers is the fundamental tool used in applying the theory of probability [18, p. 134]. The law states intuitively that the larger the sample, the greater the certainty that the sample mean is a good estimate of the population mean.

A sequence of jointly distributed random variables $X_1, X_2, \ldots, X_n$, with finite mean values, is said to obey the (classical) *law of large numbers* if

$$z_u = \frac{x_1 + x_2 + \cdots + x_n}{n} - \frac{E(x_1 + x_2 + \cdots + x_n)}{n} \to 0$$

in some mode of convergence as n tends to ∞.

† See [12].

2.7 MOMENTS OF RANDOM VARIABLES

Since the moments of a distribution correspond to theoretical mean value, variance, measure of skewness, etc., methods of finding moments, for variates and functional combination of variates, are very useful in solving design problems.

The following discussion of moments and moment generating functions accents continuous variates since it is largely with continuous variates that design applications are concerned. The procedure will be to discuss moments of a single random variable first, then moments of functions of a random variable, and moments of functions of more than one random variable.

The expected value is a theoretical average. It is not expected that x will equal its expected value in a given observation. However, it is reasonable to expect the average value of X (in a great many observations) to approach the expected value.

The moments of a distribution are the expected values of the powers of the random variable, which has the given distribution. The expected value[†] is therefore called the first moment of distribution, and the quantity $E(x^r)$ is called the rth moment, usually expressed [18]:

$$\mu'_r = E(x^r) = \int_{-\infty}^{\infty} x^r f(x)\, dx. \qquad (2.46)$$

The moments about an arbitrary point K are defined as

$$E[(x - K)^r] = \int_{-\infty}^{\infty} (x - K)^r f(x)\, dx. \qquad (2.47)$$

When K is set equal to the mean, the moments about the mean are commonly denoted by μ_r:

$$\mu_r = E[(x - \mu'_1)^r] = \int_{-\infty}^{\infty} (x - \mu'_1) f(x)\, dx.$$

For $r = 1$ we obtain the first moment about the mean:

$$\mu_1 = \int_{-\infty}^{\infty} x f(x)\, dx = \mu'_1 \int_{-\infty}^{\infty} f(x)\, dx = \mu'_1 - \mu'_1 = 0.$$

† The first moment μ' is called the mean of x.

The second moment about the mean is ($r = 2$):

$$\mu_2 = \int_{-\infty}^{\infty} (x - \mu_1')^2 f(x)\, dx$$

$$= \int_{-\infty}^{\infty} [x^2 - 2x\mu_1' + (\mu_1')^2] f(x)\, dx \qquad (2.48)$$

$$= \mu_2' - 2\mu_1'\mu_1' + (\mu_1')^2$$

$$= \mu_2' - (\mu_1')^2.$$

The second moment about the mean is called the variance of X (see Eq. 2.42).

2.7.1 Moment Generating Function

If all the moments of a distribution exist (are finite), it is possible to associate a moment generating function with the distribution. This, by definition, is $E(e^{xt})$, with x a random variable and t a continuous variable. The expected value is a function of t, which may be expressed by

$$m(t) = E(e^{xt}) = \int_{-\infty}^{\infty} e^{xt} f(x)\, dx. \qquad (2.49)$$

If the terms of Eq. 2.49 are differentiated r times with respect to t, the result is

$$\frac{d^r m(t)}{dt^r} = \int_{-\infty}^{\infty} x^r e^{xt} f(x)\, dx, \qquad (2.50)$$

recalling that

$$\frac{d(e^{xt})}{dt} = xe^{xt}. \qquad (2.50a)$$

Placing $t = 0$, we obtain

$$\frac{d^r m(0)}{dt^r} = E(x^r) = \mu_r'.$$

The term on the left is the rth derivative of $m(t)$ at $t = 0$. Hence the moments of a distribution may be obtained by differentiation.

Example 1. The technique may be illustrated by determining the moments that correspond to the mean and variance of a Poisson distribution (Eq. 2.18), the density of which is,

$$f(z) = \frac{e^{-a} a^z}{z!} \qquad z = 0, 1, 2, \ldots.$$

Substituting into the equivalent of Eq. 2.49 for discrete random variables,

$$m(t) = E(e^{zt}) = \sum_{z=0}^{\infty} \frac{e^{zt} e^{-a} a}{z!}$$

$$= e^{-a} \sum_{z=0}^{\infty} \frac{(ae^t)^z}{z!}$$

$$= e^{-a} \cdot e^a e^t.\dagger$$

Taking the first two derivatives with respect to t,

$$m'(t) = e^{-a} \cdot a \cdot e^t \cdot e^{ae^t},$$
$$m''(t) = e^{-a} \cdot a \cdot e^t \cdot e^{ae^t} \cdot (1 + ae^t),$$

and

$$\mu = m'(0) = a,$$
$$\mu_2' = m''(0) = a(1 + a),$$
$$\sigma^2 = a(1 + a) - a^2 = a.$$

Example 2. Determine $E(z)$, employing moment generating functions, given the density

$$f(z) = (1 - \alpha)\alpha^z \qquad (z = 0, 1, 2, \ldots, \infty;\ 0 < \alpha < \infty).$$

Solution. Substituting values into the equivalent of Eq. 2.49,

$$m(t) = \sum_{z=0}^{\infty} e^{tz} f(z) = (1 - \alpha) \sum_{z=0}^{\infty} (\alpha e^t)^z = \frac{1 - \alpha}{1 - \alpha e^t}.$$

$E(z)$ is obtained by differentiation with respect to t:

$$E(z) = \left[\frac{dm(t)}{dt}\right]_{t=0}$$

$$= \frac{(1 - \alpha)e^t}{(1 - \alpha e^t)^2} = \frac{\alpha}{1 - \alpha}.$$

It is often necessary to obtain the moments of a function of a random variable, such as the moments of $h(x)$, where x has the density $f(x)$. The rth moment of $h(x)$ is given by (see Eq. 2.37, with $h(x) = \phi(x)$ and $r = 1$)

$$E[h(x)]^r = \int_{-\infty}^{\infty} [h(x)]^r f(x)\, dx. \qquad (2.51)$$

The function that generates the moments of $h(x)$ is [18]

$$m(t) = E(e^{th(x)}) = \int_{-\infty}^{\infty} e^{th(x)} f(x)\, dx. \qquad (2.52)$$

$\dagger \sum_{z=0}^{\infty} \frac{(ae^t)^z}{z!} = \sum_{z=0}^{\infty} \frac{(y)^z}{z!} = e^y = e^{ae^t}$

2.7.2 Multivariate Moments

The results of Section 2.7.1 may be extended to distributions of several random variables. Consider three variates x, y, z with joint density function $f(x, y, z)$. The rth moment of z, for instance, is

$$E(z^r) = \int\!\!\int\!\!\int_{-\infty}^{\infty} z^r f(x, y, z)\, dz\, dy\, dx.$$

In addition to the moments of the individual variates, joint moments are defined,

$$E(x^q y^r z^s) = \int\!\!\int\!\!\int_{-\infty}^{\infty} x^q y^r z^s f(x, y, z)\, dz\, dy\, dx \qquad (2.53)$$

where q, r, and s are any positive integers, including zero.

It is useful to define a joint moment-generating function

$$m(t_1, t_2, t_3) = E[e^{(t_1 x + t_2 y + t_3 z)}];\dagger \qquad (2.54)$$

for instance, the rth moment of the variate z is obtained by differentiating the moment-generating function r times with respect to t_3 and then setting

$$t_1 = t_2 = t_3 = 0.$$

In a similar way the joint moment $E(x^q y^r z^s)$ is obtained by differentiating q times with respect to t_1, r times with respect to t_2, and s times with respect to t_3, and then setting

$$t_1 = t_2 = t_3 = 0.$$

Example 1. In later applications, the following density function, known as the bivariate normal density function [18] will be very useful. First, Eq. 2.55 is shown to be a density function.

$$f(x, y) = \frac{1}{2\pi\sigma_x\sigma_y\sqrt{1-\rho^2}} \cdot \exp\left\{-\frac{1}{2(1-\rho^2)}\left[\left(\frac{x-\mu_x}{\sigma_x}\right)^2 - 2\rho\frac{(x-\mu_x)(y-\mu_y)}{\sigma_x\sigma_y} + \left(\frac{y-\mu_y}{\sigma_y}\right)^2\right]\right\} \qquad (2.55)$$

Equation 2.55 is a density function if

$$\int\!\!\int_{-\infty}^{\infty} f(x, y)\, dy\, dx = 1.$$

† See trivariate moment-generating function, Eq. 2.65.

Introduce the changes of variable

$$u = \frac{x - \mu_x}{\sigma_x}, \quad du = \frac{dx}{\sigma_x},$$

$$v = \frac{y - \mu_y}{\sigma_y}, \quad dv = \frac{dy}{\sigma_y}.$$

Substitute these values for x, y, dx, and dy:

$$\int\int_{-\infty}^{\infty} \frac{1}{2\pi\sqrt{1-\rho^2}} \cdot \exp\{-[\tfrac{1}{2}(1-\rho^2)](u^2 - 2\rho uv + v^2)\} dv\, du.$$

Complete the square on u in the exponent

$$\int\int_{-\infty}^{\infty} \frac{1}{2\pi\sqrt{1-\rho^2}} \exp\{-[\tfrac{1}{2}(1-\rho^2)][(u-\rho v)^2 + (1-\rho^2)v^2]\} dv\, du.$$

Make a second change of variable:

$$w = \frac{u - \rho v}{\sqrt{1-\rho^2}}, \quad dw = \frac{du}{\sqrt{1-\rho^2}}.$$

The integral is expressed as the product of two simple integrals,

$$\int_{-\infty}^{\infty} \frac{1}{\sqrt{2\pi}} e^{-w^2/2}\, dw \cdot \int_{-\infty}^{\infty} \frac{1}{\sqrt{2\pi}} \cdot e^{(-v^2/2)}\, dv = 1.$$

Conclusion. $f(x, y)$ is a density function.

Now, the moment generating function for the bivariate normal distribution is derived. The moment generating function in random variables X and Y is [18]

$$m(t_1, t_2) = E[e^{(t_1 x + t_2 y)}] = \int\int e^{t_1 x + t_2 y} f(x, y)\, dy\, dx. \qquad (2.56)$$

The equivalent of $f(x, y)\, dy\, dx$ in terms (as above) of variables u and v, substituted into Eq. 2.56, is

$$m(t_1, t_2) = e^{t_1 \mu_x + t_2 \mu_y} \int\int e^{t_1 \sigma_x u + t_2 \sigma_y v}$$

$$\cdot \frac{1}{2\pi\sqrt{1-\rho^2}} \cdots \exp\{-[\tfrac{1}{2}(1-\rho^2)][u^2 - 2\rho uv + v^2]\}\, dv\, du.$$

Writing the combined exponents in the integrand as

$$-\frac{1}{2(1-\rho^2)} \cdot [u^2 - 2\rho uv + v^2 - 2(1-\rho^2)t_1 \sigma_x u - 2(1-\rho^2)t_2 \sigma_y v]$$

and completing the square on first u and then on v, the expression becomes

$$\frac{1}{2(1-\rho^2)}\{[u - \rho v - (1-\rho^2)t_1\sigma_x]^2 + (1-\rho^2)(v - \rho t_1\sigma_x - t_2\sigma_y)^2$$
$$-(1-\rho^2)[t_1^2\sigma_x^2 + 2\rho t_1 t_2 \sigma_x \sigma_y + t_2^2\sigma_y^2]\}.$$

Making the following changes of variable:

$$w = \frac{u - \rho v - (1-\rho^2)t_1\sigma_x}{\sqrt{1-\rho^2}},$$

$$z = v - \rho t_1\sigma_x - t_2\sigma_y$$

yields

$$-\tfrac{1}{2}w^2 - \tfrac{1}{2}z^2 + \tfrac{1}{2}(t_1^2\sigma_x^2 + 2\rho t_1 t_2 \sigma_x \sigma_y + t_2^2\sigma_y^2).$$

The integral is written

$$m(t_1, t_2) = e^{t_1\mu_x + t_2\mu_y} \cdot \exp\left[\tfrac{1}{2}(t_1^2\sigma_x^2 + 2\rho t_1 t_2 \sigma_x \sigma_y + t_2^2\sigma_y^2)\right] \cdots$$

$$\times \int_{-\infty}^{\infty}\int_{-\infty}^{\infty} \frac{1}{2\pi} \exp\left[-\left(\frac{w^2}{2}\right) - \left(\frac{z^2}{2}\right)\right] dw\, dz.$$

Since, by Eq. 2.29 and property c of density functions, Section 2.3.1,

$$\int_{-\infty}^{\infty}\int_{-\infty}^{\infty} \frac{1}{2\pi} \exp\left[-\left(\frac{w^2}{2}\right) - \left(\frac{z^2}{2}\right)\right] dw\, dz = 1,$$

the bivariate normal moment generating function is

$$m(t_1, t_2) = \exp\left[t_1\mu_x + t_2\mu_y + \tfrac{1}{2}(t_1^2\sigma_x^2 + 2\rho t_1 t_2 \sigma_x \sigma_y + t_2^2\sigma_y^2)\right] = e^R.\dagger$$
(2.57)

Moments are obtained by evaluating the derivatives of $m(t_1, t_2)$ at $t_1 = t_2 = 0$. For an example see Section 3.2.3. The mean and variance of the random variable X are computed by means of moment generating functions.

Two remaining problems must be considered: (1) that of finding the moments of variates and functions of variates, and (2) that of finding the distributions of variates and functions of variates. The first problem has been at least partially solved in the preceding discussion.

† For economy of notation, let

$$R = [t_1\mu_x + t_2\mu_y + \tfrac{1}{2}(t_1^2\sigma_x^2 + 2\rho t_1 t_2 \sigma_x \sigma_y + t_{2y}\sigma_y^2)].$$

2.8 DISTRIBUTION OF FUNCTIONS †

An important problem in the application of probability theory to design is finding the distribution of some function of one or more random variables.

Suppose that Y is a known function of another random variable Z. Consider the problem of finding the distribution of Y. If Y is a monotonic function of Z with derivative that exists, a general solution may be found in terms of density functions.

2.8.1 Simple Monotonic Types

If y increases whenever z increases, y is called a monotonically increasing function. For instance, $y = e^z$ is a monotonically increasing function, and $y = e^{-z}$ is a monotonically decreasing function. For $z > 0$, the function $y = z^2$ is monotonically increasing. For any monotonic function a unique relationship exists between the dependent and independent variables. Mathematically, the independent variables may be considered a function of the dependent variables. In addition, if the original function is differentiable with respect to z, then the inverse function is differentiable with respect to y.

Monotonically Increasing Functions. ‡ Assume a random variable Z with density function $f(z)$. Let $F(z)$ be the distribution function (cumulative) and $G(y)$ be the distribution function of Y. Further, let Y be a differentiable, monotonically increasing function of Z, such that $y \to \beta$ when $z \to \alpha$. Then $y \leq \beta$ when and only when $z \leq \alpha$, and

$$p(y \leq \beta) = p(z \leq \alpha)$$
$$G(\beta) = F(\alpha).$$

As stated, y is a function of z, and z may be considered a function of y; for example, $\phi(y)$, then

$$\alpha = \phi(\beta).$$

If the expression

$$G(\beta) = F(\alpha)$$

is differentiated with respect to β, the result is

$$\frac{dG(\beta)}{d\beta} = \frac{dF(\alpha)}{d\beta} = \frac{dF(\alpha)}{d\alpha} \cdot \frac{d\alpha}{d\beta} = \frac{dF(\alpha)}{d\alpha} \cdot \frac{d\phi(\beta)}{d\beta}.$$

† Portions adapted by permission from *Probability and Random Variables* by G. P. Wadsworth and J. G. Bryan, McGraw-Hill Book Co., N.Y., 1960.
‡ Recall that monotonic increasing implies a function such that for $z_2 > z_1; f(z_2) \geq f(z_1)$. See [17] for detailed discussion.

Recalling Eq. 2.22,

$$g(\beta) = f(\alpha) \cdot \frac{d\phi(\beta)}{d\beta} = f[\phi(\beta)] \frac{d\phi(\beta)}{d\beta},$$

and replacing β with y,

$$g(y) = f[\phi(y)] \frac{d\phi(y)}{dy}. \tag{2.58}$$

Example 1. Assume that the random variable Z is exponentially distributed (see Eq. 2.24 with $\beta = 1$) with density

$$f(z) = e^{-z} \quad (0 < z < \infty).$$

Derive the distribution of y, where (see Fig. 2.20)

$$y = (z)^{1/4}.$$

In this instance

$$\phi(y) = z = y^4,$$

where

$$f(z) = f[\phi(y)] = e^{-z} = e^{-y^4},$$

$$\frac{d\phi(y)}{dy} = \frac{dz}{dy} = 4y^3$$

and the density of y is given by

$$g(y) = 4y^3 e^{-y^4} \quad (0, \infty).$$

Monotonically Decreasing Functions.† When y is a monotonically decreasing function of z, and $y \to \beta$ as $z \to \alpha$, then

$$p(y \leq \beta) = p(z \geq \alpha)$$

with the notation and assumptions that led to Eq. 2.58:

$$G(\beta) = 1 - G(\alpha)$$

$$g(\beta) = f[\phi(\beta)] \frac{d\phi(\beta)}{d\beta}.$$

If y is a monotonically decreasing function of z, then $\phi(y)$ is also a monotonically decreasing function of y. Thus $d\phi(\beta)/d\beta$ is always negative and $-d\phi(\beta)/d\beta$ is always positive (and may be regarded as the absolute value of the derivative). The equation for $g(\beta)$ may be expressed as

$$g(\beta) = -f[\phi(\beta)] \frac{d\phi\beta}{d\beta}.$$

In more general form

$$g(y) = f[\phi(y)] \frac{d\phi(y)}{dy}. \tag{2.58a}$$

† It is recalled that monotonic decreasing implies a function such that for $z_2 > z_1$; $f(z_2) < f(z_1)$.

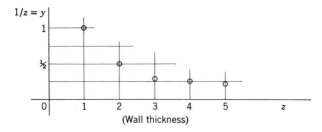

Figure 2.21

Example 2. Consider the expression $q = K_m A(t_1 - t_2)/z$ where K_m is the mean conductivity, A = area, ft^2, and $(t_1 - t_2)$ = temperature difference °F. Let $C = K_m A(t_1 - t_2)$, and $q/C = 1/z$, where z is wall thickness in feet. If it is assumed that Z is approximately described by the density function

$$f(z) = 6(1 + z)^{-4} \quad (0, \infty),$$

find the distribution of q/C, and for convenience let $q/C = y$.

From Fig. 2.21, $y = 1/z$, and $\phi(y) = z = 1/y$

$$\frac{d\phi(y)}{dy} = \left|\frac{dz}{dy}\right| = \frac{1}{y^2}.$$

By Eq. 2.58a

$$g\frac{q}{C} = g(y) = \frac{6/y}{(1 + 1/y)^4} \cdot \frac{1}{y^2} = \frac{6y}{(1 + y)^4} \quad (0, \infty).$$

Since $y \to 0$ as $z \to \infty$ and $y \to \infty$ as $z \to 0$, the range is as given above. See Fig. 2.22.

Example 3. This example provides a proof that is needed in Chapter 3. Assume that the random variable Z is normally distributed (μ_z, σ_z). Let y be a linear function of z, where

$$y = c_1 + c_2 z.$$

For nonzero values of c_2, the function is monotonically increasing if c_2 is positive, and monotonically decreasing if c_2 is negative. If $c_2 = 0$, the linear function degenerates to a

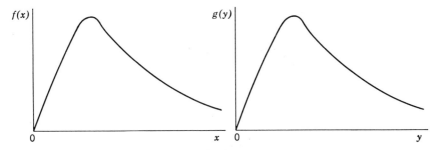

Figure 2.22 Distribution of Z and its inverse [12].

78 Foundation

constant. Thus, for either case,
$$z = \frac{y - c_1}{c_2},$$
and
$$\left|\frac{dy}{dz}\right| = \frac{1}{|c_2|}.$$

The density of Z is given by (Eq. 2.28)
$$f(z) = \frac{1}{\sigma_z \sqrt{2\pi}} \exp\left[-\frac{1}{2}\left(\frac{z - \mu_z}{\sigma_z}\right)^2\right] \quad (-\infty, \infty);$$

z must be expresssed in terms of y, in the expression above, then multiplied by $1/|c_2|$. Thus
$$z - \mu_z = \frac{y - c_1 - c_2\mu_z}{b}$$
$$\frac{(z - \mu_z)^2}{2\sigma_z^2} = \frac{(y - c_1 - c_2\mu_z)^2}{2\sigma_z^2 c_2^2} = \frac{(y - \mu)^2}{2\sigma^2},$$
where
$$\mu = c_1 + c_2\mu_z,$$
and
$$\sigma = |c_2|\,\sigma_z.$$
Thus
$$g(y) = \frac{1}{|c_2|} \cdot \frac{1}{\sigma\sqrt{2\pi}} \exp\left[-\frac{1}{2}\left(\frac{y-\mu}{\sigma}\right)^2\right] = \frac{1}{\sigma\sqrt{2\pi}} \exp\left[-\frac{1}{2}\left(\frac{y-\mu}{\sigma}\right)^2\right] \quad (-\infty, \infty).$$

The range of y is $(-\infty, \infty)$. The exercise proves that any nontrivial linear function of a normally distributed random variable is itself normally distributed.

If z is normally distributed (μ_z, σ_z) and y is any linear function of z,
$$y = c_1 + c_2 z \quad (c_2 \neq 0),$$
then y is normally distributed with corresponding parameters $\mu = c_1 + c_2\mu_z$ and $\sigma = |c_2|\,\sigma_z$ (see Section 3.1.1).

Example 1. Let $y = 1/x^2$, where the density function of x is
$$f(x) = \frac{6x}{(1+x)^4} \quad (0, \infty).$$

Determine the distribution of y. (See Fig. 2.22)
$$x = \sqrt{\frac{1}{y}} = \frac{1}{y^{1/2}}; \quad f[\phi(y)] = \frac{6y^{-1/2}}{(1 + y^{-1/2})^4},$$
so
$$x = \phi(y) = y^{-1/2}$$
and
$$\frac{d\phi(y)}{dy} = -\tfrac{1}{2}y^{-3/2}; \quad f[\phi(y)] = \frac{6y^{3/2}}{(y^{1/2} + 1)^4}$$
$$\frac{|d[\phi(y)]|}{dy} = \tfrac{1}{2}y^{-3/2}.$$

Then
$$g(y) = f[\phi(y)]\left|\frac{d\phi(y)}{dy}\right|$$
$$= \frac{6y^{3/2}}{(y^{1/2}+1)^4}(\tfrac{1}{2}y^{-3/2}) \quad (0 < y < \infty)$$
$$g(y) = \frac{3}{(y^{1/2}+1)^4} \quad (0 < y < \infty).$$

2.8.2 The Basic Distribution Transformation

This transformation, sometimes called the probability transformation, is a variable change of much importance. It is defined by the equation
$$y = F(z).$$
This transformation exists for all distributions, and in the set of distributions that possess probability density functions, it has the property of being rectangularly distributed. From the definition, the range of y is $(0, 1)$ and, if $f(z)$ exists,
$$\frac{dy}{dz} = f(z)$$
and
$$\frac{dz}{dy} = \frac{1}{f(z)},$$
from which
$$g(y)\frac{f(z)}{f(z)} = 1 \quad (0 \le y \le 1). \tag{2.59}$$

Example 1. Density function of a square. In Section 3.2.5, the moments of the square of a normally distributed random variable are derived. The square approximates normality for moderate σ. In this example,
$$y = x^2,$$
where x is exponentially distributed (Eq. 2.24) with density,
$$f(x) = \frac{1}{\beta}e^{-x/\beta} \quad (\beta > 0; 0 \le x < \infty).$$
Determine the distribution of y.
$$x = \pm\sqrt{y}$$
and since
$$g(y) = \frac{f(\sqrt{y}) + f(-\sqrt{y})}{2\sqrt{y}},$$

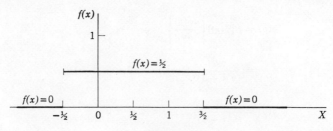

Figure A.

$$f(x) = \frac{1}{\beta} e^{-\sqrt{y}/\beta} = f(-\sqrt{y})$$

$$g(z) = \frac{\frac{1}{\beta} e^{-\sqrt{y}/\beta} + \frac{1}{\beta} e^{-\sqrt{y}/\beta}}{2\sqrt{y}}$$

we obtain

$$g(y) = \frac{1}{\beta} y^{-\frac{1}{2}} e^{-\sqrt{y}/\beta} \quad (0 \leq y < \infty).$$

Example 2. In Section 2.3.1 the rectangular distribution was discussed in Example 1, and in Section 2.3.5 the rectangular distribution was found to be a special case of the beta distribution. In this example the density $g(y)$ of the square of a rectangularly distributed random variable x is computed. Let the distribution of x be that shown in Fig. A.

$$y = x^2; \quad x = +\sqrt{y}.$$

The fact that $f(x) = 0$ when $x < -\frac{1}{2}$, leads to the conclusion that $f(-\sqrt{y}) = 0$ when $y > \frac{1}{4}$. Thus

$$g(y) = \frac{\frac{1}{2} + \frac{1}{2}}{2\sqrt{y}} = \frac{1}{2\sqrt{y}}, \quad (0 \leq y \leq \frac{1}{4}),$$

and

$$g(y) = \frac{0 + \frac{1}{2}}{2\sqrt{y}} = \frac{1}{4\sqrt{y}}, \quad (\tfrac{1}{4} < y \leq \tfrac{9}{4}).$$

The distribution of y displays a discontinuity at $y = \frac{1}{4}$, and thus requires the two equations given above.

Example 3. Determine the function of z with a distribution

$$u(y) = [3 - 3y^{1/2}], \quad (0 \leq y \leq 1),$$

assuming that

$$v(z) = [6z - 6z^2], \quad (0 \leq z = 1).$$

The two distribution functions are

$$V(z) = \int_0^z v(z)\,dz = 3z^2 - 2z^3$$

$$U(y) = \int_0^y u(y)\,dy = 3y - 2y^{3/2}.$$

Setting
$$U(y) = V(z)$$
gives
$$3y - 2y^{3/2} = 3z^2 - 2z^3;$$
hence
$$y = z^2.$$

2.8.3 Density Function of Powers

When y is not a monotonic function of z, Eqs. 2.58 and 2.59 do not hold. Consider
$$y = z^2,$$
then
$$z = \pm\sqrt{y}.$$

$$T(y) = \int_{-\sqrt{y}}^{\sqrt{y}} f(z)\, dz.$$

Differentiation of $T(y)$ yields Eq. 2.60:

$$t(y) = \frac{f(\sqrt{y}) + f(-\sqrt{y})}{2\sqrt{y}}. \tag{2.60}$$

Example 1. In this example the distribution of the square of a normally distributed random variable z is computed. The result may be compared with that obtained in Section 3.2.5. Let
$$y = z^2,$$
from which
$$z = \pm\sqrt{y},$$
and let the density of z be (Eq. 2.28)
$$f(z) = \frac{1}{\sqrt{2\pi}} e^{-(z^2/2)} \quad (-\infty, \infty).$$

Then
$$f(\sqrt{y}) = \frac{1}{\sqrt{2\pi}} e^{-(y/2)} = f(-\sqrt{y}).$$

Thus by Eq. 2.60
$$t(y) = \frac{1}{\sqrt{2\pi y}} \cdot e^{-(y/2)} \quad (0 \le y \le \infty).$$

2.8.4 Method of Convolutions[†]

In many applications the derived distributions involve more than a simple change of variable. There is often need for random variables that are derived mathematically as functions (sums, differences, products, etc.) of random

---

[†] See [12], [13], and [14]. *The Convolution Theorem*, I. I. Hirshman and D. V. Widder, Princeton University Press, New Jersey, 1955.

variables that are directly measurable. Suppose that X and Y are random variables with marginal density functions $f_1(x)$ and $f_2(y)$, respectively. With independence specified, the joint density $f(x, y)$ equals the product of the individual density functions. Thus by Eq. 2.31

$$f(x, y) = f_1(x) f_2(y).$$

Specify u as a monotonic function (of x and y), such that u is a monotonic function of x (with y constant) and a monotonic function of y (with x constant). Examples of such functions (see Chapter 3) are

$$u = x + y$$

$$u = xy$$

$$u = \frac{x}{y}, \text{ etc.}$$

To derive the distribution of u, (as in taking partial derivatives) one of the random variables x, y is held constant, usually at the mean value. The second variate varies over its specified range, according to its marginal distribution; for instance, let $f(x, y)$ be the density of two continuous random variables. Attention is focused on one variate, say x. The desired function of x is such that when it is integrated over an interval $\alpha < x < \beta$ the probability that x lies in that interval is obtained. Suppose that x is held constant with y free to vary. The probability distribution of u that results is, by Eq. 2.12, the conditional distribution of u, given x. Since u becomes a monotonic function of the single random variable y, the functional expression in u and y will contain parameters that involve x. Thus the conditional density function $\phi(u \mid x)$ may be derived from $f_2(y)$ by the methods of Sections 2.8.1 to 2.8.3. The joint density function of x and u may be written [e.g., $h(x, u)$] by using Eq. 2.12. From which

$$h(x, u) = f_1(x) f_2(u \mid x).$$

Finally, the marginal density of u is obtained by integrating this joint density function with respect to x. The method is clarified by examples.

Example 1. Given are two independent random variables X, Y with densities (see Fig. 2.23):

$$f_1(x) = \tfrac{1}{3}, \quad (1 \leq x \leq 4)$$
$$f_2(y) = e^{-(y-2)}, \quad (2 \leq y < \infty).$$

The distribution of the quotient random variable is required:

$$z = \frac{x}{y}.$$

Mathematical Considerations

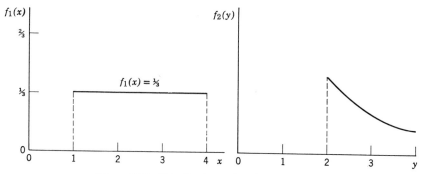

Figure 2.23 Distributions of variables x and y.

Since $f_1(x)$ is mathematically simpler to handle than $f_2(y)$, it is easier to determine $\phi(z \mid y)$. Holding y constant, the conditional distribution of $z = x/y$ is found, by regarding it as a function of x alone. Thus

$$x = yz$$

$$\frac{\partial x}{\partial z} = y,$$

then

$$[f_1(y \cdot z)] \cdot y = \tfrac{1}{3}(y).$$

When $x = 1$, $z = 1/y$, and when $x = 4$, $z = 4/y$. Thus the conditional distribution of z, given y, is

$$\phi(z \mid y) = \tfrac{1}{3}y \left(\frac{1}{y}, \frac{4}{y}\right)$$

from which the joint distribution of y and z is

$$g(y, z) = \frac{y}{3} e^{-(y-2)} \left(2 \leq y < \infty; \frac{1}{y} \leq z \leq \frac{4}{y}\right).$$

Of interest is the shaded part of the first quadrant bounded on the left by $y = 2$, above by $z = 4/y$, below by $z = 1/v$, and open to the right (Fig. 2.24). $z \to 0$ as $y \to \infty$. The greatest

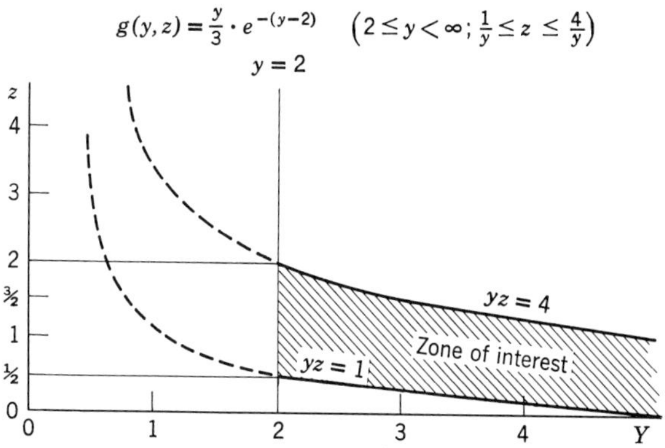

$$g(y,z) = \frac{y}{3} \cdot e^{-(y-2)} \quad \left(2 \leq y < \infty; \tfrac{1}{y} \leq z \leq \tfrac{4}{y}\right)$$

Figure 2.24 Quotient distribution [12].

value of z corresponds to the point of intersection of $z = 4/y$ and $y = 2$. The integration requires two steps. The lower hyperbola intersects $y = z$ at $z = \frac{1}{2}$. Below this level, ranges between the two hyperbolas, the limits being $y = 1/z$ to $y = 4/z$. Above the level $z = \frac{1}{2}$, the left boundary is the line $y = 2$, and the limits are $y = 2$ to $y = 4/z$. Hence the distribution of z must be expressed by two different functions, each of which applies to a stated range.

RANGE 1. $(0 \leq z \leq \frac{1}{2})$

$$h(z) = \int_{1/z}^{4/z} g(y, z)\, dy = \frac{-e^2}{3}\left[e^{-y}(y+1)\right]_{1/z}^{4/z}$$

$$= \frac{e^2}{3z}\left[e^{-1/z}(1+z) - e^{-4/z}(4+z)\right]$$

RANGE 2. $(\frac{1}{2} \leq z \leq 2)$

$$h(z) = \int_{2}^{4/z} g(y, z)\, dy = \frac{-e^2}{3}\left[e^{-y}(y+1)\right]_{2}^{4/z},$$

$$h(z) = 1 - \frac{e^2}{3z}(4+z)e^{-4/z}.$$

Example 2. The surface of a body in the open will exchange radiant energy with the sun and the earth. The net energy received by the surface (ignoring energy emitted) may be expressed as

$$q = q_{su} + q_e, \qquad q = \text{energy in Btu/Hour}$$

Assume that q_{su} and q_e are normally distributed independent random variables and set $x = q_{su}$ and $y = q_e$. Then $q = Z = X + Y$. Also, let X be normal (μ_1, σ_1) and Y be normal (μ_2, σ_2). Thus by Eq. 2.28 the densities of X and Y are

$$f_1(x) = \frac{1}{\sigma_1\sqrt{2\pi}} \exp\left[-\frac{1}{2}\left(\frac{x - \mu_1}{\sigma_2}\right)^2\right], \qquad (-\infty, \infty),$$

$$f_2(y) = \frac{1}{\sigma_2\sqrt{2\pi}} \exp\left[-\frac{1}{2}\left(\frac{y - \mu_2}{\sigma_2}\right)^2\right], \qquad (-\infty, \infty).$$

The distribution of the sum variate $q = z = X + Y$ is required. With x constant, the conditional distribution of z given x is obtained by changing variables in $f_2(y)$. Thus

$$y = z - x$$

and

$$\frac{\partial y}{\partial z} = 1.$$

Thus

$$\phi(z, x) = \frac{1}{\sigma_2\sqrt{2\pi}} \exp\left[-\frac{1}{2}\left(\frac{z - x - \mu_2}{\sigma_2}\right)^2\right], \qquad (-\infty, \infty).$$

Mathematical Considerations

The range of z is $(-\infty, \infty)$ since y has this range, and for fixed x the quantity $x + y$ will approach ∞ in either direction as $y \to \pm\infty$.

$$g(x, z) = f_1(x)\phi(z \mid x),$$

$$g(x, z) = \left\{\frac{1}{\sigma_1\sqrt{2\pi}} \cdot \exp\left[-\frac{1}{2}\left(\frac{x - \mu_1}{\sigma_1}\right)^2\right]\right\} \cdot \left\{\frac{1}{\sigma_2\sqrt{2\pi}} \cdot \exp\left[-\frac{1}{2}\left(\frac{z - x - \mu_2}{\sigma_2}\right)^2\right]\right\},$$

$$g(x, z) = \frac{1}{\sigma_1\sigma_2 \cdot 2\pi} \exp\left\{-\frac{1}{2}\left[\left(\frac{x - \mu_1}{\sigma_1}\right)^2 + \left(\frac{z - x - \mu_2}{\sigma_2}\right)^2\right]\right\}.$$

Next, let

$$K = \frac{1}{2\pi\sigma_1\sigma_1}, \qquad Q = \frac{(x - \mu_1)^2}{2\sigma_1^2} + \frac{(z - x - \mu_2)^2}{2\sigma_2^2};$$

$$g(x, z) = Ke^{-Q}, \qquad (-\infty, \infty)$$

each and the marginal density function of z will be given by

$$h(z) = K \int_{-\infty}^{\infty} e^{-Q}\, dx.$$

In order to affect the integration, the square on x is completed:

$$Q = \frac{(\sigma_1^2 + \sigma_2^2)}{2\sigma_1^2\sigma_2^2}\left[x - \frac{(\sigma_2^2\mu_1 + \sigma_1^2 z - \sigma_1^2\mu_2)}{\sigma_1^2 + \sigma_2^2}\right]^2 + \frac{(z - \mu_1 - \mu_2)^2}{2(\sigma_1^2 + \sigma_2^2)}.$$

Let

$$v = x - \frac{(\sigma_2^2\mu_1 + \sigma_1^2 z - \sigma_1^2\mu_2)}{\sigma_1^2 + \sigma_2^2},$$

and

$$dv = dx.$$

Set

$$\mu = (\mu_1 + \mu_2)$$

and

$$\sigma^2 = (\sigma_1^2 + \sigma_2^2).$$

Then

$$h(z) = K \exp\left[-\frac{1}{2}\left(\frac{z - \mu}{\sigma}\right)^2\right] \int_{-\infty}^{\infty} e^{-\sigma^2 v^2/2\sigma_1\sigma_2}\, dv.$$

By comparing with a suitable gamma density function, the integral reduces to

$$\frac{\sigma_1\sigma_2\sqrt{2\pi}}{\sigma}.$$

Multiplying by K

$$h(z) = \frac{1}{\sigma\sqrt{2\pi}} \exp\left[-\frac{1}{2}\left(\frac{z - \mu}{\sigma}\right)^2\right].$$

Note 1. This derivation proves that the sum of two normal independent random variables is a normally distributed random variable (closure under addition).

Note 2. By induction, summing may be extended to n independent normally distributed variates.

Example 3. Under the initial assumptions of Example 2, derive the distribution of the variate Z, where

$$Z = \frac{Y}{X}.\dagger$$

The densities of x and y are (Eq. 2.28)

$$f_1(x) = \frac{1}{\sigma_1\sqrt{2\pi}} \exp\left[-\frac{(x-\mu_1)^2}{2\sigma_1^2}\right], \quad (-\infty, \infty)$$

and

$$f_2(y) = \frac{1}{\sigma_2\sqrt{2\pi}} \exp\left[-\frac{(y-\mu_2)^2}{2\sigma_2^2}\right], \quad (-\infty, \infty)$$

$$y = xz \quad \text{and} \quad \frac{dy}{dz} = x.$$

Writing the conditional density of z, given x,

$$\phi(z\mid x) = f_2(xz)\frac{dy}{dz} = \frac{x}{\sigma_2\sqrt{2\pi}} \exp\left[-\frac{(xz-\mu_2)^2}{2\sigma_2^2}\right],$$

$$g(x, z) = f_1(x)[\phi(zx)],$$

$$g(x, z) = \frac{1}{\sigma_1\sqrt{2\pi}} \exp\left[-\frac{(x-\mu_1)^2}{2\sigma_1^2}\right] \frac{x}{\sigma_2\sqrt{2\pi}} \exp\left[-\frac{(xz-\mu_2)^2}{2\sigma_2^2}\right].$$

The marginal density of z is obtained by integrating $g(x, z)$ with respect to x:

$$h(z) = \int_{-\infty}^{\infty} g(x, z)\, dx = \frac{1}{2\pi\sigma_1\sigma_2} \int_{-\infty}^{\infty} \exp\left\{-\left[\frac{(x-\mu_1)^2}{2\sigma_1^2} + \frac{(xz-\mu_2)^2}{2\sigma_2^2}\right]\right\} dx.$$

Let

$$Q = \left[\frac{(x-\mu_1)^2}{\sigma_1^2} + \frac{(xz-\mu_2)^2}{\sigma_2^2}\right],$$

$$Q = \frac{\sigma_2^2 + \sigma_1^2 z^2}{\sigma_1^2 \sigma_2^2}\left[x^2 - 2\frac{(\sigma_2^2\mu_1 + \sigma_1^2\mu_2 z)}{\sigma_2^2 + \sigma_1^2 z^2}x + \left(\frac{\sigma_2^2\mu_1 + \sigma_1^2\mu_2 z}{\sigma_2^2 + \sigma_1^2 z^2}\right)\right.$$

$$\left.+ \frac{\sigma_2^2\mu_1^2 + \sigma_1^2\mu_2^2}{\sigma_2^2 + \sigma_1^2 z^2} - \left(\frac{\sigma_2^2\mu_1 + \sigma_1^2\mu_2 z}{\sigma_2^2 + \sigma_1^2 z^2}\right)^2\right].$$

† See also [22].

Let
$$A = \sigma_2^2 + \sigma_1^2 z^2,$$

$$\frac{v^2}{A} = x^2 - 2\frac{(\sigma_2^2 \mu_1 + \sigma_1^2 \mu_2 z)}{\sigma_2^2 + \sigma_1^2 z^2} x + \left(\frac{\sigma_2^2 \mu_1 + \sigma_1^2 \mu_2 z}{\sigma_2^2 + \sigma_1^2 z^2}\right)^2$$

$$\frac{2v}{A} dv = 2x\, dx - 2\frac{\sigma_2^2 \mu_1 + \sigma_1^2 \mu_2 z}{\sigma_2^2 + \sigma_1^2 z^2} dx,$$

$$h(z) = \frac{1}{2\pi\sigma_1\sigma_2} \int_{-\infty}^{\infty} \left(x - \frac{\sigma_{21} + \sigma_1 \mu_2 z}{\sigma_2^2 + \sigma_1^2 z^2} + \frac{\sigma_2 \mu_1 + \sigma_1 \mu_2 z}{\sigma_2^2 + \sigma_1^2 z^2}\right)$$

$$\exp\left\{-\frac{1}{2}\left[\frac{(x-\mu_1)^2}{\sigma_1^2} + \frac{xz - \mu_2)^2}{\sigma_2^2}\right]\right\} dx.$$

Now, write the first term and denote it by R

$$R = \frac{1}{2\pi\sigma_1\sigma_2} \int_{-\infty}^{\infty} \left(x - \frac{\sigma_2\mu_1 + \sigma_1\mu_2 z}{\sigma_2^2 + \sigma_1^2 z^2}\right) \exp\left\{-\frac{1}{2}\left[\frac{(x-\mu_1)^2}{\sigma_1^2} + \frac{(xz-\mu_2)^2}{\sigma_2^2}\right]\right\} dx.$$

Next write the second term and denote it by S:

$$S = \frac{1}{2\pi\sigma_1\sigma_2} \int_{-\infty}^{\infty} \left(\frac{\sigma_2\mu_1 + \sigma_1\mu_2 z}{\sigma_2^2 + \sigma_1^2 z^2}\right) \exp\left[-\frac{1}{2}\frac{(x-\mu_1)^2}{\sigma_1^2} + \frac{(xz-\mu_2)^2}{\sigma_2^2}\right] dx,$$

$$R = \frac{1}{2\pi\sigma_1\sigma_2(\sigma_2^2 + \sigma_1^2 z^2)} \exp\left[-\frac{1}{2}\frac{(\mu_2 + \mu_1 z)^2}{\sigma_1^2 \sigma_2^2}\right] \int_{-\infty}^{\infty} v \exp\left(-\frac{v^2}{2\sigma_1^2\sigma_2^2}\right) dv.$$

Changing variables and simplifying,

$$S = \frac{1}{2\pi\sigma_1\sigma_2} \int_{-\infty}^{\infty} \left(\frac{\sigma_2\mu_1 + \sigma_1\mu_2 z}{\sigma_2^2 + \sigma_1^2 z^2}\right) \exp\left\{-\frac{1}{2}\left[\frac{(x-\mu_1)^2}{\sigma_1^2} + \frac{(xz-\mu_2)^2}{\sigma_2^2}\right]\right\} dx,$$

$$S = \frac{\sigma_2\mu_1 + \sigma_1\mu_2 z}{2\pi\sigma_1\sigma_2(\sigma_2^2 + \sigma_1^2 z^2)^{3/2}} \exp\left[-\frac{(\mu_2 + \mu_1 z)^2}{2(\sigma_2^2 + \sigma_1^2 z^2)}\right] \int_{-\infty}^{\infty} \exp\left(-\frac{v^2}{2\sigma_1^2\sigma_2^2}\right) dv.$$

GAMMA FUNCTION:

$$A = \int_{-0}^{\infty} \frac{1}{\beta(\alpha+1)} x^\alpha e^{-x/\beta}\, dx = \int_{-\infty}^{\infty} v \exp\left(-\frac{v^2}{2\sigma_1^2\sigma_2^2}\right) dv,$$

$$h(z) = \frac{1}{2\pi\sigma_1\sigma_2(\sigma_2^2 + \sigma_1^2 z^2)} \exp\left[-\frac{(\mu_2 + \mu_1 z)^2}{2(\sigma_2^2 + \sigma_1^2 z^2)}\right] \int_{-\infty}^{\infty} v \exp\left(-\frac{v^2}{2\sigma_1^2\sigma_2^2}\right) dv + \cdots$$

$$+ \frac{\sigma_2\mu_1 + \sigma_1\mu_2 z}{2\pi\sigma_1\sigma_2(\sigma_2^2 + \sigma_1^2 z^2)^{3/2}} \exp\left\{-\left[\frac{(\mu_2 + \mu_1 z)^2}{2(\sigma_2^2 + \sigma_1^2 z^2)}\right]\right\} \int_{-\infty}^{\infty} \exp\left(-\frac{v^2}{2\sigma_1^2\sigma_2^2}\right) dv.$$

Utilizing the gamma function [3, p. 112]:

$$\int_{-\infty}^{\infty} v \exp\left[-\left(\frac{v^2}{2\sigma_1^2\sigma_2^2}\right)\right] dv = \int_{-\infty}^{\infty} \sigma_1^2\sigma_2^2 e^{-z} dz$$

$$z = \frac{v^2}{2\sigma_1^2\sigma_2^2}$$

$$dz = \frac{2v\,dv}{2\sigma_1^2\sigma_2^2}, \qquad v\,dv = \sigma_1^2\sigma_2^2\,dz,$$

$$\int_{-\infty}^{\infty} \exp\left[-\left(\frac{v^2}{2\sigma_1^2\sigma_2^2}\right)\right] dv = \sqrt{2}\,\sigma_1\sigma_2 \int_{-\infty}^{\infty} e^{-z^2} dz$$

$$z^2 = \frac{v^2}{2\sigma_2^2\sigma_1^2}$$

$$z = \frac{v}{\sqrt{2}\cdot\sigma_1\sigma_2}$$

$$dv = \sqrt{2}\cdot\sigma_1\sigma_2\,dz,$$

$$h(z) = \frac{1}{\pi(\sigma_2^2\sigma_1^2 z^2)}\left[\sigma_1\sigma_2 + \frac{1}{\sqrt{\sigma_2^2 + \sigma_1^2 z^2}}\right] \exp\left\{-\left[\frac{(\mu_2 + \mu_1 z)^2}{2(\sigma_2^2 + \sigma_1^2 z^2)}\right]\right\}.$$

Setting

$$A = (\mu_2 + \mu_1 z),$$

and

$$B^2 = (\sigma_2^2 + \sigma_1^2 z^2),$$

$$h(z) = \frac{1}{\pi B^2}\left[\sigma_1\sigma_2 + \frac{1}{B}\right] \exp\left[-\left(\frac{A^2}{2B^2}\right)\right].$$

2.9 MOMENTS OF FUNCTIONS OF RANDOM VARIABLES

Having discussed methods of deriving distributions of functions in Section 2.8.4, the next task is that of computing moments of functional combinations of random variables.

Because a distribution is uniquely defined by its moments (if they are finite), the moments of a function of random variables uniquely define the distribution of the function. In particular, the normal distribution is uniquely defined by its first and second moments about the mean (variance). Disregarding constraints, the distribution of a function of normally distributed (independent) random variables is defined by the mean of the function and the variance of the function.

Limiting consideration to combinations of normal random variables, the following are approaches to the problem of computing moments of functions.

1. Maximum likelihood methods may be extended to functional combinations.
2. Partial derivative methods yield good approximations to variances of functional combinations.
3. Moment generating functions may be employed, provided that the moments of the component distributions are known.

In Sections 2.9.1, 2.9.2, and 2.9.3, each of the approaches above is discussed in turn.

2.9.1 Maximum Likelihood Estimators

Maximum likelihood estimators were introduced in Section 2.5 and used to compute estimators of mean values, variances, and standard deviations of univariates. The method may be extended to functional combinations of independent random variables, such as sums, products, etc. (see Section 3.1). Visualize random variables X and Y as described by populations of values:

$$x_1, x_2, x_3, \ldots, x_{n-2}, x_{n-1}, x_n$$

$$y_1, y_2, y_3, \ldots, y_{r-2}, y_{r-1}, y_r.$$

A set of couples may be generated by random pairing of a member from the X population with a member from the Y population, thus

$$(x_1, y_1), (x_1, y_2), \ldots, (x_1, y_r),$$
$$\cdots (x_2, y_1), (x_2, y_2), \ldots, (x_2, y_r),$$
$$\cdots (x_n, y_1), (x_n, y_2), \ldots, (x_n, y_r).$$

If a combinatorial operation is applied to each couple in the population (of $n \cdot r$ couples), such as summing or taking differences, the result is a new set which defines the random variable Z; for example,

$$Z = X + Y,$$
$$Z = X - Y,$$
$$Z = XY,$$
$$Z = X/Y.$$

To compute the mean value estimator of the sum variate ($\bar{z}$), the required algebraic formula is

$$\bar{z} = \frac{\sum_{i=1}^{n} \sum_{j=1}^{r} (x_i + y_j)}{n \cdot r}$$

$$\bar{z} = \frac{1}{n \cdot r} [(x_1 + y_1)] + [(x_1 + y_2) + \cdots + (x_1 + y_r)]$$
$$+ \cdots + (x_2 + y_1) + (x_2 + y_2) + \cdots + (x_2 + y_r)$$
$$+ \cdots + (x_n + y_1) + (x_n + y_2) + \cdots + (x_n + y_r)].$$

Similarly, the standard deviation estimator of the sum variate (s_z) is computed as follows:

$$s_z = \left\{ \sum_{i=1}^{n} \sum_{j=1}^{r} \frac{[\bar{z} - (x_i + y_j)]^2}{n \cdot r} \right\}^{1/2}$$

$$s_z = \left\{ \frac{1}{n \cdot r} [[\bar{z} - (x_1 + y_1)]^2 + [\bar{z} - (x_1 + y_2)]^2 + \cdots + [\bar{z} - (x_1 + y_r)]^2 \right.$$
$$+ \cdots [\bar{z} - (x_2 + y_1)]^2 + [\bar{z} - (x_2 + y_2)]^2 + \cdots + [\bar{z} - (x_2 + y_r)]^2$$
$$\left. + \cdots [\bar{z} - (x_n + y_1)]^2 + [\bar{z} - (x_n + y_2)]^2 + \cdots + [\bar{z} - (x_n + y_r)]^2] \right\}^{1/2}.$$

Applications of maximum likelihood estimators appear in Chapter 3 (The Algebra of Normal Functions), where the moment estimators of sums, differences, products, quotients, and squares of random variables are given.

2.9.2 Partial Derivative Methods [20]

Let the random variable ψ be functionally related to normally distributed random variables X and Y. Then

$$\psi = f(x, y),$$
$$\psi_i = f(x_i, y_i),$$

and

$$\psi_0 = f(\bar{x}, \bar{y}).$$

The function is assumed to be continuous and differentiable. Let X and Y be statistically independent with deviations

$$\delta x_i = x_i - \bar{x}$$

and

$$\delta y_i = y_1 - \bar{y}$$

which are assumed to be relatively small. The deviation is

$$\delta\psi_i = \psi_i - \bar{\psi} \approx \frac{\partial \psi}{\partial x} \delta x_i + \frac{\partial \psi}{\partial y} \delta y_i, \qquad (a)$$

which follows from the definition of the partial differential when written for small increments† or from the Taylor expansion,

$$\psi_i = f([\bar{x} + \delta x_i], [\bar{y} + \delta y_i]) = f(\bar{x}, \bar{y}) + \frac{\partial \psi}{\partial x} x_i \delta + \frac{\partial \psi}{\partial y} \delta y_i \qquad (a')$$

$$\delta_{\psi_i} = \psi_i - \bar{\psi} = \frac{\partial \psi}{\partial x} \delta x_i + \frac{\partial \psi}{\partial y} \delta y_i.$$

Equation a', for $\delta\psi_i$, is the same as Eq. a if second order (and higher) terms are neglected. Notice that $\bar{\psi}$ is taken as $f(\bar{x}, \bar{y})$ rather than as

$$\bar{\psi} = \sum_{i=1}^{n} \frac{\psi_i}{n}, \qquad \text{(Eq. 2.38)}. \qquad (b)$$

These two definitions of $\bar{\psi}$ are essentially the same, if the deviations are of moderate size. The partial derivates are taken at

$$x = \bar{x},$$
$$y = \bar{y},$$

and thus are constants.

The expressions for $\delta\psi_i$ show the way in which individual deviations arise to produce a deviation in ψ. The effect in ψ of a deviation in x is multiplied by $\partial\psi/\partial x$, hence, if x appears in ψ with an exponent much greater than one, the effect may be large. The variance estimator σ_x^2 (see Section 2.5, Example 2) is given by

$$\sigma_\psi^2 = \frac{\sum_{i=1}^{n} (\delta\psi_i)^2}{n}. \qquad (c)$$

Squaring Eq. a for $\delta\psi_i$ gives

$$(\delta\psi_i)^2 = \left(\frac{\partial \psi}{\partial x}\right)^2 (\partial x_i)^2 + 2 \frac{\partial \psi}{\partial x} \frac{\partial \psi}{\partial y} \delta x_i \, \delta y_i + \left(\frac{\partial \psi}{\partial y}\right)^2 (\delta y_i)^2. \qquad (d)$$

Substituting Eq. d into Eq. c yields

$$\sigma_\psi^2 = \frac{\left(\frac{\partial \psi}{\partial x}\right)^2 \sum (\delta x_i)^2 + 2 \frac{\partial \psi}{\partial x} \frac{\partial \psi}{\partial y} \sum (\delta x_i \, \delta y_i) + \left(\frac{\partial \psi}{\partial y}\right)^2 \sum (\delta y_i)^2}{n}.$$

† See [17], p. 372.

The sum $\sum (\delta x_i\, \delta y_i) \to 0$ if x_i and y_i are independent. Then, since

$$\sigma_x^2 = \frac{\sum (\delta x_i)^2}{n}$$

and

$$\sigma_y^2 = \frac{\sum (\delta x_i)^2}{n},$$

the expression for s_ψ is

$$\sigma_\psi \approx \left[\left(\frac{\partial \psi}{\partial x}\right)^2 s_x^2 + \left(\frac{\partial \psi}{\partial y}\right)^2 s_y^2\right]^{\frac{1}{2}}.$$

The standard deviation of a function (approximate) is

$$\sigma_\psi \approx \left[\sum_{j=1}^{j}\left(\frac{\partial \psi}{\partial x_j}\right)^2 s_{x_j}^2\right]^{\frac{1}{2}}. \tag{2.61}$$

Example 1. Let the functional relationship be

$$\psi = Ay \log x,$$

where X and Y are normally distributed random variables. Compute the standard deviation of ψ.

$$\frac{\partial \psi}{\partial x} = \frac{Ay}{x}$$

and

$$\frac{\partial \psi}{\partial y} = A \log x.$$

By Eq. 2.61, the standard deviation of ψ is

$$\sigma_\psi \approx \left[\frac{A^2 \bar{y}^2}{\bar{x}^2}(\sigma_x)^2 + A^2(\log x)^2 \sigma_y^2\right]^{\frac{1}{2}}.$$

Had the function been $\psi = A \log x$, the result for s_ψ would be

$$\sigma_\psi \approx \left(\frac{A^2}{\bar{x}^2} s_x^2\right)^{\frac{1}{2}} = \frac{A s_x}{\bar{x}}.$$

Example 2. Let the functional relationship be

$$\psi = Cy \sin x.$$

X and Y are normally distributed random variables. Compute the standard deviation of ψ [*Note.* The units of x are expressed in radians (x is not near $\pi/2$).] Taking partial deviatives,

$$\frac{\partial \psi}{\partial x} = Cy \cos x$$

and

$$\frac{\partial \psi}{\partial y} = C \sin x.$$

Then
$$\sigma_\psi \approx [C^2\bar{y}^2 \cos^2 x (s_x)^2 + C^2 \sin^2 x (s_y)^2]^{1/2}.$$
With $\psi = C \sin x$,
$$\sigma_\psi \approx [C^2 \cos^2 x (s_x)^2]^{1/2} = Cs_x \cos \bar{x}.$$

Note. For a function whose first derivative (in the region of interest) is very small, the higher terms in the Taylor expansion cannot be ignored.

Additional examples of this method are found in the algebra of normal functions (see Chapter 3).

2.9.3 Moment Generating Function Methods

In this section, procedures are discussed for utilizing moment generating functions. Applications include computing the first moment (mean value) and the second moment about the mean of univariates, functions of one random variable, and functions of several random variables.

Mean or Expected Value. In Section 2.6, expected values are introduced. The equation for expected value of a discrete random variable (Eq. 2.36) is

$$E[\phi(x)] = \sum_{i=1}^{r} \phi(x_i) f(x_i),$$

and, for a continuous random variable (Eq. 2.37) it is

$$E[\phi(x)] = \int_a^b \phi(x) f(x)\, dx.$$

Setting $\phi(x) = x$ in Eq. 2.37 and integrating yields the expected value of X given by (Eq. 2.40)

$$E(x) = \int_a^b x f(x)\, dx.$$

Consider Eq. 2.49:

$$m(t) = E(e^{xt}) = \int_{-\infty}^{-\infty} e^{xt} f(x)\, dx.$$

If Eq. 2.49 is differentiated once with respect to t,

$$\frac{dm(t)}{dt} = \int_{-\infty}^{-\infty} x e^{xt} f(x)\, dx$$

and t is set equal to zero,

$$\frac{dm(t)}{dt} = \int_{-\infty}^{-\infty} x f(x)\, dx.$$

Referring to Eq. 2.40, it is seen that the

$$\text{average of } x = E(x) = \left[\frac{dm(t)}{dt}\right]_{t=0}.$$

Thus the moment generating function may be utilized to compute mean values of single variates.

The logical extension is to show that moment generating functions may be utilized to compute mean values of functions of one random variable. With Eq. 2.37,

$$E[\phi(x)] = \int_a^b \phi(x) f(x) \, dx.$$

By the change of variable

$$z = \phi(x),$$

$$E[\phi(x)] = E(z) = \int_a^b z f(z) \, dz$$

Consider Eq. 2.51 with $r = 1$ and $\phi(x) = h(x)$:

$$m(t) = E[\phi(x)] = \int_{-\infty}^{\infty} e^{t\phi(x)} f(x) \, dx \cdot = E(e^{t\phi(x)}),$$

differentiating once with respect to t,

$$\frac{dm(t)}{dt} = \int_{-\infty}^{\infty} \phi(x) e^{t\phi(x)} f(x) \, dx$$

and setting $t = 0$,

$$\frac{dm(t)}{dt} = \int_{-\infty}^{\infty} \phi(x) f(x) \, dx.$$

Referring to Eq. 2.37,

$$\text{average of } \phi(x) = E[\phi(x)] = \left[\frac{dm(t)}{dt}\right]_{t=0}.$$

Thus moment generating functions may be used to compute mean values of functions of one random variable.

For n continuous random variables, let the definition (Eq. 2.37) be extended.

$$E[\phi(x_1, x_2, \ldots, x_n)] = \int\int_{-\infty}^{\infty} \cdots \int_{-\infty}^{\infty} \phi(x_1, x_2, \ldots, x_n)$$
$$\times f(x_1, x_2, \ldots, x_n) \, dx_1 \, dx_2 \ldots dx_n).$$

Thus it is concluded that the

$$\text{average of } \phi(x_1, x_2, \ldots, x_n) = E[\phi(x_1, x_2, \ldots, x_n)]$$
$$= \frac{\partial^n m(t)}{\partial t_1 \, \partial t_2 \cdots \partial t_n}\bigg|_{t_1 = t_2 = \cdots = t_n = 0}.$$

Moment generating functions may be utilized to compute expected values of functions of two or more random variables. For examples involving two random variables see Section 3.2. (An example involving three random variables is given later in this section.) To make use readily and economically of moment generating functions two properties of E are needed. Let the random variable X be distributed by $f(x)$. Let k be any constant and, if $g(x)$ and $h(x)$ are any functions of X,

and
$$E[kg(x)] = kE[g(x)] \qquad (2.62)$$

$$E[g(x) + h(x)] = E[g(x)] + E[h(x)]. \qquad (2.63)$$

The two properties above follow from

and
$$\int kg(x)f(x)\,dx = k\int g(x)f(x)\,dx$$

$$\int [g(x) + h(x)]f(x)\,dx = \int g(x)f(x)\,dx + \int h(x)f(x)\,dx.$$

Example 1. In Section 2.7.1 the derivation of the moment generating function of the Poisson distribution appears. The equations are then employed to determine the expected value.

In Section 2.7.2 the moment generating function of the bivariate normal distribution is derived. This expression is used to prove that the mean value of X is μ_x.

Example 2. Consider the function

$$y = 3x^3 + 4x^2 + 5x + 1.$$

Let X be a normally distributed random variable. Then

$$E(y) = E(3x^3 + 4x^2 + 5x + 1).$$

Applying Eqs. 2.62 and 2.63,

$$E(y) = E(3x^3) + E(4x^2) + E(5x) + E(1).$$
$$E(y) = 3E(x^3) + 4E(x^2) + 5E(x) + 1.$$

The expected value of a constant is a constant, thus

$$E(1) = 1.$$

It remains to determine the expected values of powers of X, $E(x^3)$, $E(x^2)$, and $E(x)$; for instance, $E(x^3)$ is determined by taking the derivative of the univariate moment generating function (Eq. 2.64) three times with respect to t, then setting $t = 0$, and simplifying the resulting expression. In Section 2.7.2 it is shown that

$$E(x) = \mu_x$$

and
$$E(x^2) = \mu_x^2 + \sigma_x^2.$$

96 Foundation

Example 3. Consider the function

$$w = x^2y + xyz + y^3z^2.$$

Let X, Y, and Z be independent normally distributed random variables

$$E(w) = E(x^2y + xyz + y^3z^2)$$
$$E(w) = E(x^2y) + E(xyz) + E(y^3z^2)$$

To compute $E(w)$, the trivariate moment generating function (Eq. 2.65) is needed. Equation 2.53,

$$E(x^q y^r z^s) = \int\int\int_{-\infty}^{\infty} x^q y^r z^s f(x, y, z) \, dz \, dy \, dx$$

and Eq. 2.54,

$$M(t_1, t_2, t_3) = E[e^{(t_1 x + t_2 y + t_3 z)}].$$

In the computation of $E(w)$, each term on the right is determined separately. $E(x^2y)$ is obtained from the bivariate moment generating function by differentiating twice with respect to t_1 and then once with respect to t_2. Then t_1 and t_2 are set equal to zero and the result is obtained, Similarly, $E(xyz)$ is obtained from the trivariate moment generating function by differentiating once with respect to t_1, t_2, and t_3 in turn, and then setting $t_1 = t_2 = t_3 = 0$ to obtain the needed result.

Variance and Standard Deviation. Variance is discussed in Section 2.6.3 and defined by Eq. 2.42.

$$\text{Variance} \equiv \sigma^2 \equiv E[(x - \mu)^2].$$

Equation 2.47 with $K = \mu$ and $r = 2$ is

$$\sigma^2 = E[(x - \mu)^2] = \int_{-\infty}^{\infty} (x - \mu)^2 f(x) \, dx$$

(see Eq. 2.51) with $h(x) = x - \mu$ and $r = 2$. Then by Eq. 2.52 $h(x) = (x - \mu)$,

$$m(t) = E(e^{t(x-\mu)}) = \int_{-\infty}^{\infty} e^{t(x-\mu)} f(x) \, dx.$$

Differentiating $m(t)$ twice with respect to t and setting $t = 0$,

$$\left. \frac{d^2 m(t)}{dt_2} \right]_{t=0} = \int_{-\infty}^{\infty} (x - \mu)^2 f(x) \, dx.$$

Since, by Eq. 2.44, $\sigma^2 = E(x^2) - \mu^2$, the variance of a normal random variable X may be computed by moment generating functions.

Examples of the variance computations for univariates appear in Sections 2.7.1 and 2.7.2 and in Chapter 3.

It is now shown that moment generating functions may be used to compute the variance of a function $\phi(x)$ of the normal random variable X. Equation 2.51, with $h(x) = \phi(x) - \mu_{\phi(x)}$ and $r = 2$, is

$$E[(\phi(x) - \mu_{\phi(x)})^2] = \int_{-\infty}^{\infty} [\phi(x) - \mu_{\phi(x)}]^2 f(x)\, dx.$$

Apply Eq. 2.52, with
$$h(x) = [\phi(x) - \mu_{\phi(x)}],$$

$$m(t) = E[e^{[t(\phi(x) - \mu\phi(x))]}] = \int_{-\infty}^{\infty} \exp\,[t(\phi(x) - \mu_{\phi(x)})] f(x)\, dx.$$

Differentiating $m(t)$ twice with respect to t, then setting $t = 0$,

$$\left[\frac{d^2 m}{dt}\right]_{t=0} = \int_{-\infty}^{\infty} [\phi(x) - \mu_{\phi(x)}]^2 f(x)\, dx.$$

Thus moment generating functions may be used to compute the variance of functions of a univariate. An example, the variance of a quadratic, is computed in Section 3.4.8.

The variance of a function $h(x_1, x_2, \ldots, x_n)$ of n random variables $x_1, x_2, \ldots, x_n$ may be computed by means of moment generating functions. Let

$$z = h(x_1, x_2, \ldots, x_n)$$
$$\phi(z) = \phi[h(x_1, x_2, \ldots, x_n)]$$
$$dz = dx_1\, dx_2 \cdots dx_n.$$

Equation 2.47, with $K = \mu_z$ and $r = 2$, is

$$E[(z - \mu_z)^2] = \int_{-\infty}^{\infty} (z - \mu_z)^2 f(z)\, dz$$

$$= \int_{-\infty}^{\infty} \cdots \int_{-\infty}^{\infty} [h(x_1, x_2, \ldots, x_n) - \mu_z]^2 \cdots$$
$$\times f(x_1, x_2 \cdots x_n)\, dx_1\, dx_2, \ldots dx_n.$$

By extension of Eq. 2.52,

$$m(t_1, t_2, \ldots, t_n) = E[\exp\,(t_1 x_1 + t_2 x_2 + \cdots + t_n x_n)]$$

$$= \iiint_{-\infty}^{\infty} \exp\,[h(x_1, x_2, \ldots, x_n) - \mu_z]$$
$$\times f(x_1, x_2, \ldots, x_n)\, dx_1\, dx_2 \cdots dx_n.$$

Differentiating twice each with respect to $t_1, t_2, \ldots, t_n$ and setting $t_1 = t_2 = \cdots = t_n = 0$, gives

$$\frac{\partial^{2n} m(t_1, t_2, \ldots, t_n)}{\partial^2 t_1 \, \partial^2 t_2 \cdots \partial^2 t_n}\bigg]_{t_1=t_2=\cdots=t_n=0} = \int_{-\infty}^{\infty} \cdots \int_{-\infty}^{\infty} [h(x_1, x_2, \ldots, x_n) - \mu_z]^2 \cdots$$
$$\times f(x_1, x_2, \ldots, x_n) \, dx_1 \, dx_2 \cdots dx_n.$$

Moment generating functions may be employed to compute the variance of multivariate functions. An example appears later in this section (also see Section 2.7.2). In the remainder of this section, the univariate, bivariate, and trivariate moment generating functions of normally distributed random variables are discussed.

The Standard Normal Distribution; Univariate.

$$f(x; 0, 1) = \frac{1}{\sqrt{2\pi}} e^{-(x^2/2)} \quad \text{(Eq. 2.28)}$$

and

$$F(x; 0, 1) = \int_{-\infty}^{x} n(t; 0, 1) \, dt.$$

First, compute the moments of

$$f(x; \mu, \sigma^2)$$

from its moment generating function

$$m(t) = E(e^{tx}) = e^{t\mu} E(e^{t(x-\mu)})$$
$$= e^{t\mu} \int_{-\infty}^{\infty} \sigma\sqrt{2\pi} \, e^{t(x-\mu)} \exp\left[-\frac{1}{2}\left(\frac{x-\mu}{\sigma}\right)^2\right] dx$$
$$= e^{t\mu} \frac{1}{\sigma\sqrt{2\pi}} \int_{-\infty}^{\infty} \exp\left\{-\left(\frac{1}{2\sigma^2}\right)[(x-\mu)^2 - 2\sigma^2(x-\mu)t]\right\} dx.$$

Completing the square inside the bracket of the exponent,

$$(x-\mu)^2 - 2\sigma^2 t(x-\mu) = (x-\mu)^2 - 2\sigma^2 t(x-\mu) + \sigma^4 t^2 - \sigma^4 t^2$$
$$= (x - \mu - \sigma^2 t)^2 - \sigma^4 t^2$$

and

$$m(t) = e^{t\mu} \exp\left(\frac{\sigma^2 t^2}{2}\right) \frac{1}{\sigma\sqrt{2\pi}} \int_{-\infty}^{\infty} \exp\left[-\frac{(x - \mu - \sigma^2 t)^2}{2\sigma^2}\right] dx.$$

Univariate normal, moment generating function:

$$m(t) = \exp\left(t\mu + \frac{\sigma^2 t^2}{2}\right) \tag{2.64}$$

Bivariate Normal (see Section 2.7.2). To find the moments of X and Y in functional combinations requires the joint moment generating function, Eq. 2.56:

$$m(t_1, t_2) = E(e^{t_1 x + t_2 y}) = \iint e^{t_1 x + t_2 y} f(x, y) \, dx \, dy.$$

and Eq. 2.57:

$$m(t_1, t_2) = e^{t_1 \mu_x + t_2 \mu_y} \exp\left[\tfrac{1}{2}(t_1^2 \sigma_x^2 + 2\rho t_1 t_2 \sigma_x \sigma_y + t_2^2 \sigma_y^2)\right].$$

Examples of moment computations of bivariates in functional combinations appear in the algebra of normal functions (Chapter 3).

Trivariate Normal Distribution. (statistically independent). Let X, Y, and Z be mutually independent and normally distributed (correlation coefficient, $\rho_1 = \rho_2 = \rho_3 = 0$). The joint density function of X, Y, and Z is $f(x, y, z) = f(x) g(y) h(z)$. Thus

$$f(x, y, z) = \frac{1}{\sqrt{2\pi}\,\sigma_x} \exp\left[-\frac{1}{2}\left(\frac{x - \mu_x}{\sigma_x}\right)^2\right] \cdot \frac{1}{\sqrt{2\pi}\,\sigma_y} \exp\left[-\frac{1}{2}\left(\frac{y - \mu_y}{\sigma_y}\right)^2\right]$$

$$\cdot \frac{1}{\sqrt{2\pi}\,\sigma_z} \exp\left[-\frac{1}{2}\left(\frac{z - \mu_z}{\sigma_z}\right)^2\right]$$

$$= \frac{1}{(2\pi)^{3/2} \sigma_x \sigma_y \sigma_z} \cdot \exp\left\{-\frac{1}{2}\left[\left(\frac{x - \mu_x}{\sigma_x}\right)^2 + \left(\frac{y - \mu_y}{\sigma_y}\right)^2 + \left(\frac{z - \mu_z}{\sigma_z}\right)^2\right]\right\}. \quad (a)$$

Trivariate normal moment generating function. By Eq. 2.54,

$$m(t_1, t_2, t_3) = E(e^{t_1 x + t_2 y + t_3 z})$$

$$= \iiint e^{t_1 x + t_2 y + t_3 z} f(x, y, z) \, dz \, dy \, dx. \quad (b)$$

Changing variables. Let

$$u = \frac{x - \mu_x}{\sigma_x}, \qquad \sigma_x \, du = dx,$$

$$v = \frac{y - \mu_y}{\sigma_y}, \qquad \sigma_y \, dv = dy,$$

$$w = \frac{z - \mu_z}{\sigma_z}, \qquad \sigma_z \, dw = dz,$$

$$x = \sigma_x u + \mu_x,$$

$$y = \sigma_y v + \mu_y,$$

$$z = \sigma_z w + \mu_z.$$

Substituting for x, y, and z in Eq. b,

$$m(t_1, t_2, t_3) = \iiint e^{t_1(\sigma_x u + \mu_x) + t_2(\sigma_y v + \mu_y) + t_3(\sigma_z w + \mu_z)} \cdot f(x, y, z) \, dx \, dy \, dz$$

$$m(t_1, t_2, t_3) = e^{t_1\mu_x + t_2\mu_y + t_3\mu_z} \iiint e^{(t_1\sigma_x u + t_2\sigma_y v + t_3\sigma_z w)} \cdots f(x, y, z) \, dx \, dy \, dz. \quad (c)$$

Substituting Eq. a for $f(x, y, z)$ in Eq. c and making the changes of variable,

$$m(t_1, t_2, t_3) = e^{(t_1\mu_x + t_2\mu_y + t_3\mu_z)} \iiint e^{t_1\sigma_x u + t_2\sigma_y v + t_3\sigma_z w}$$

$$\cdot \frac{1}{(2\pi)^{3/2}} \cdots \exp[-\tfrac{1}{2}(u^2 + v^2 + w^2)] \cdot dx \, dy \, dz. \quad (d)$$

Let
$$k = e^{t_1\mu_x + t_2\mu_y + t_3\mu_z}$$

and substitute in Eq. d,

$$m(t_1, t_2, t_3) = \frac{k}{(2\pi)^{3/2}} \iiint \exp[-\tfrac{1}{2}(u^2 + v^2 + w^2 - 2t_1\sigma_x u - 2t_2\sigma_y v - 2t_3\sigma_z w)] \cdots du \, dv \, dw. \quad (e)$$

Completing the squares on the terms in the exponent of Eq. e,

$$u^2 = 2t_1\sigma_x u + t_1^2\sigma_x^2 - t_1^2\sigma_x^2$$
$$v^2 = 2t_2\sigma_y v + t_2^2\sigma_y^2 - t_2^2\sigma_y^2$$
$$w^2 = 2t_3\sigma_z w + t_3^2\sigma_z^2 - t_3^2\sigma_z^2$$

$$m(t_1, t_2, t_3) = \frac{k}{(2\pi)^{3/2}} \iiint \exp\{-\tfrac{1}{2}[u - t_1\sigma_x)^2 + (v - t_2\sigma_y)^2 \cdots$$
$$+ (w - t_3\sigma_z)^2 \cdots - t_1^2\sigma_x^2 - t_2^2\sigma_y^2 - t_3^2\sigma_z^2]\} \, du \, dv \, dw. \quad (f)$$

Since the expression in Eq. f that remains under the integral sign equals one, after the terms $t_1^2\sigma_x^2 + t_2^2\sigma_y^2 + t_3^2\sigma_z^2$ are removed, we obtain the following function.

Trivariate Normal Moment Generating Function

$$\dagger m(t_1, t_2, t_3) = \exp(t_1\mu_x + t_2\mu_y + t_3\mu_z + \tfrac{1}{2}t_1^2\sigma_x^2 + \tfrac{1}{2}t_2^2\sigma_y^2 + \tfrac{1}{2}t_3^2\sigma_z^2).$$
(2.65)

Example 1. Test the validity of the distributive law (Section 3.5.7) in the algebra of normal functions. Assume normal, mutually independent variates A, B, and C. Test the equality
$$Z = A(B + C) = AB + AC.$$

† This result may also be obtained from the univariate moment generating function and theorem E, p. 215, [7].

Since the mean value relationship is valid (being relations among real numbers), the proof is confined to computing the variance of each side of the expression above.

Left side:
$$Z = A(B + C)$$

$$E(z - \mu_z)^2 = E[a(b + c) - \mu_a(\mu_b + \mu_c)]^2$$
$$= E[a^2(b + c)^2 - 2a(b + c)\mu_a(\mu_b + \mu_c) + \mu_a{}^2(\mu_b + \mu_c)^2]$$
$$= E[a^2(b + c)^2] - 2E[a(b + c)\mu_a(\mu_b + \mu_c)]$$
$$\quad + \mu_a{}^2(\mu_b{}^2 + 2\mu_b\mu_c + \mu_c{}^2)$$
$$= E[a^2b^2 + 2a^2bc + a^2c^2] - 2\mu_a(\mu_b + \mu_c)E[a(b + c)]$$
$$\quad + \mu_a{}^2\mu_b{}^2 + 2\mu_a{}^2\mu_b\mu_c + \mu_a{}^2\mu_c{}^2.$$

Employing the bivariate moment generating function (Eq. 2.57).

$$E(a^2b^2) = \sigma_a{}^2\sigma_b{}^2 + \sigma_a{}^2\mu_b{}^2 + \mu_a{}^2\sigma_b{}^2 + \mu_a{}^2\mu_b{}^2$$

$$\sigma^2_{a(b+c)} = E(z - \mu_z)^2,$$

the expectancy of $2E(a^2bc)$ is computed by utilizing Eq. 2.65.

$$E(x^2yz)$$

From Eq. 2.65, setting the exponent of e equal to R,

$$R = t_1\mu_x + t_2\mu_y + t_3\mu_z + \tfrac{1}{2}t_1{}^2\sigma_x{}^2 + \tfrac{1}{2}t_2{}^2\sigma_y{}^2 + \tfrac{1}{2}t_3{}^2\sigma_z{}^2.$$

Taking the partial derivative of m with respect to t_1 in Eq. 2.65,

$$\frac{\partial m}{\partial t_1} = (\mu_x + t_1\sigma_x{}^2)e^R.$$

The second partial derivative with respect to t_1 is

$$\frac{\partial^2 m}{\partial t_1{}^2} = \sigma_x{}^2 e^R + (\mu_x + t_1\sigma_x{}^2)e^R.$$

Containing, as in Examples 2 and 3 of Section 2.9.3,

$$\frac{\partial^3 m}{\partial t_1{}^2 \partial t_2} = (\sigma_x{}^2 + \mu_x{}^2 + t_1\mu_x\sigma_x{}^2 + t_1{}^2\sigma_x{}^2)(\mu_y + t_2\sigma_y{}^2)e^R,$$

$$\frac{\partial^4 m}{\partial t_1{}^2 \partial t_2 \partial t_3} = (\sigma_x{}^2 + \mu_x{}^2 + t_1\mu_x\sigma_x{}^2 + t_1{}^2\sigma_x{}^2)(\mu_y + t_2\sigma_y{}^2\mu_z + t_3\sigma_z{}^2)e^R,$$

@$t_1, t_2, t_3 = 0$

$$= (-\sigma_x{}^2 + \mu_x{}^2)\mu_y\mu_z = \sigma_x{}^2\mu_y\mu_z + \mu_x{}^2\mu_y\mu_z,$$

$$\sigma^2_{a(b+c)} = \sigma_a{}^2\sigma_b{}^2 + \sigma_a{}^2\mu_b{}^2 + \mu_a{}^2\sigma_b{}^2 + \mu_a{}^2\mu_b{}^2 + 2\sigma_a{}^2\mu_b\mu_c + 2\mu_a{}^2\mu_b\mu_c$$
$$\quad + \sigma_a{}^2\sigma_c{}^2 + \sigma_a{}^2\mu_c{}^2 + \mu_a{}^2\sigma_c{}^2 + \mu_a{}^2\mu_c{}^2 - 2\mu_a{}^2\mu_b{}^2 - 4\mu_a{}^2\mu_b\mu_c$$
$$\quad - 2\mu_a{}^2\mu_c{}^2 + \mu_a{}^2\mu_b{}^2 + 2\mu_a{}^2\mu_b\mu_c + \mu_a{}^2\mu_c{}^2,$$

$$\sigma^2_{a(b+c)} = \mu_a{}^2\sigma_b{}^2 + \mu_a{}^2\sigma_c{}^2 + \mu_b{}^2\sigma_a{}^2 + 2\mu_b\mu_c\sigma_a{}^2 + \mu_c{}^2\sigma_a{}^2 + \sigma_a{}^2\sigma_b{}^2 + \sigma_a{}^2\sigma_c{}^2.$$

Right side:
$$Z = AB + AC$$

$$E(z - \mu_2)^2 = E[(ab + ac) - (\mu_a\mu_b + \mu_a\mu_c)]^2$$
$$= E[(ab + ac)^2 - 2(ab + ac)(\mu_a\mu_b + \mu_a\mu_c) + (\mu_a\mu_b + \mu_a\mu_c)^2]$$
$$= E[a^2b^2 + 2a^2bc + a^2c^2 - 2ab(\mu_a\mu_b + \mu_a\mu_c) - 2ac(\mu_a\mu_b + \mu_a\mu_c)$$
$$+ \mu_a^2\mu_b^2 + 2\mu_a^2\mu_b\mu_c + \mu_a^2\mu_c^2],$$

$$\sigma^2_{(ab+ac)} = \mu_a^2\sigma_b^2 + \mu_a^2\sigma_c^2 + \mu_b^2\sigma_a^2 + 2\mu_b\mu_c\sigma_a^2 + \mu_c^2\sigma_a^2 + \sigma_a^2\sigma_b^2 + \sigma_a^2\sigma_c^2.$$

Comparing these two expansions,

$$\sigma^2_{a(b+c)} = \sigma^2_{ab+ac} = \mu_a^2\sigma_b^2 + \mu_a^2\sigma_c^2 + \mu_b^2\sigma_a^2 + 2\mu_b\mu_c\sigma_a^2 + \mu_c^2\sigma_a^2 + \sigma_a^2\sigma_b^2 + \sigma_a^2\sigma_c^2.$$

PROBLEMS

1. Describe a procedure for estimating the volume of a body of arbitrary shape and size utilizing the process (relative frequency) described in the example, Section 2.1.2. Discuss the example as a process of integration.
2. (a) What is the probability of tossing a penny three heads in a row (the three events being independent)?
   (b) What is the probability of tossing two heads in a row then a tail?
   (c) What is the probability of tossing two heads and a tail in three tosses if the particular sequence of events is not specified? In this case we have a combination of mutually exclusive and independent events.
3. If two pennies are tossed together three times, what is the probability of two matches and one mismatch in any (unspecified) sequence? Since the outcome of the first penny is never specified but in each of the three tosses the outcome of the second penny is specified (with $p = \frac{1}{2}$) and there are three independent ways (different possible sequences in which the winning outcomes may appear. The three different ways refer to the fact that the mismatch may follow two matches, come between them, or proceed them.
4. Another example of combined probabilities of independent events may be stated as follows. An incident is witnessed by one person, who describes it to another who transmits it on. If 20 people are in the chain before the incident is related to you, and if the component probability for truth is 0.9 per person,
   (a) What is the probability that you are told the truth?
   (b) What is the number n of such people in the chain before the combined probability has dropped to 0.5?
5. Overlapping compound events are not entirely mutually exclusive. If one card is drawn from each of two (well shuffled) decks of 52 cards, what is the probability that *at least* one of them (either or both) is the ace of spades?
6. A *compound event* consists of two or more single events. As in the following problem: two balls are drawn, one at a time, from an urn containing two black, three white, and four red balls. What is the probability that the first is red and the second is white? (The first is not replaced before the second is drawn.)
7. The general multiplication theorem involves the ideas of partially dependent events, leading to conditional probability. Consider the following: three white balls and one black ball are placed in a jar and one white ball and two black balls are placed in an identical jar. If one of two jars is selected at random and one ball withdrawn from it, what is the probability that the ball will be white?

8. Six cards are drawn with replacement from an ordinary well shuffled deck. What is the probability that each of the four suits will be represented at least once among the six cards? We solve this problem by finding first the probability that all suits do not appear (that one or more is absent). Let A symbolize the appearance of all suits and B, the *nonappearance* of at least one of the suits.
9. Two dice cast together. Let A be the event that the sum of the faces are odd: B the event that at least one is a one. What is the probability that
   (a) both A and B will occur,
   (b) either A or B or both will occur,
   (c) A and not B will occur,
   (d) B and not A will occur?
10. Five cards are dealt from an ordinary deck with replacement. What is the density function for the distribution of spades?
11. Ten balls are tossed into four boxes so that each ball is equally likely to fall into any box. What is the density function for the number of balls in the first box?
12. A coin is tossed until a head appears. What is the density function for the number of tosses?
13. Ten dice are cast. What is the density of the number of ones and twos?
14. A machine makes nails with an average of one percent defective. What is the density of the number of defectives in a sample of 50 nails?
15. In a town of 5000 adults a sample of 100 is asked opinions of a proposed municipal project; 60 are found to favor it and 40 to oppose it. If, in fact, the adults of the town were equally divided on the proposal, what would be the probability of obtaining a majority of 60 or more favoring it in a sample of 100?
16. If 20 percent of the values produced by a certain supplier are found to be defective, determine the probability that out of four valves selected at random, from a typical lot, at most two valves will be defective.
17. A manual circuit breaker trips as expected four out of five times the handle is struck. What is the probability that it will trip
    (a) exactly four times in four attempts to actuate it,
    (b) exactly four times in five attempts,
    (c) the first four times are successes and the fifth fails?
18. The distribution of life (hours of service prior to failure) of a certain type of vacuum tube is exponential with density,

$$f(x) = \frac{1}{\beta} \cdot e^{-x/\beta} \quad (0 \leq x < \infty)$$

with β approximately 180 hours. Compute the probability that (a) a tube will fail in less than 90 hours, (b) a tube will fail after more than 360 hours, and (c), if three tubes are selected at random, one will fail in less than 90 hours, another will fail between 90 and 360 hours and a third in more than 360 hours?
Note. Consider the permutations of three tubes in computing the answer to part 3.
19. The distribution of the length of telephone conversations (in minutes) has been found to be exponentially distributed,

$$f(x) = \frac{1}{\beta} \cdot e^{-x/\beta},$$

with β approximately 2.26 (between a certain pair of towns). A man going to a public telephone booth notes that five persons are ahead of him, including the person who just entered the booth. Neglecting dialing time and time to change occupants, what is the average time in minutes that he must expect to wait?

20. Approximately, the individual values of the average monthly temperature for a given region are normally distributed about the climatological mean temperature at that facility for the month in question. Toward the end of December, a meteorologist predicts that at two localities A and B, the average temperatures for the following January will be at least 5°F above normal. These forecasts prove to be correct. If σ for the distribution of the average January temperature is 2°F at A and 4°F at B, what are the respective probabilities of obtaining average temperatures within the range of forecast? Which forecast is the more indicative of skill?

21. Method of maximum likelihood. Assume that t is the time interval between counts in a Geiger counter measurement of the intensity of cosmic rays, and assume that the frequency function for t takes the form

$$f(t, \theta) = \theta \cdot e^{-\theta t},$$

where θ is some unknown instrumental parameter whose value is to be determined in this problem. Assume that a sample of n measurements of t have been made, $(t_1, t_2, \ldots, t_n)$. Use the method of maximum likelihood to derive an expression for θ_e.

22. A certain type of commercially manufactured resistor is known to have normally distributed values with mean $\bar{R} = 150\,\Omega$ and standard deviation $S_R = 6\,\Omega$. Find (1) $P(140 < R < 160)$, (2) $P(R < 135)$, and (3) $P(R > 162)$.

23. The finished shaft diameter in a certain machining operation is normally distributed about a mean value $\bar{D} = 1.1560$ in. with standard deviation 0.0052. The design dimensions were shown as 1.560 ± 0.0030. What percentage of the parts will be out of tolerance?

24. In Problem 23 what must $\bar{D}$ be if the probability is to be $P(D = 1.1600) = 0.001$?

25. Consider a screw with threads of pitch diameter, normally distributed about a mean, $\overline{PD} = 0.3160$ in. and standard deviations $S_{PD} = 0.0005$ in. The design requirements specify $PD = 0.361 \pm 0.0001$ in. What must S_{PD} be in order that no more than one in 250 screws is rejected (because of faulty PD)?

26. Values $x_1, x_2, \ldots, x_r$ are drawn at random from a population having the functional form

$$f(x) = ke^{-kx} \quad (0 < x < \infty).$$

Determine the maximum likelihood estimate of k.

27. The parameter m in the density function

$$f(x) = [1 + m(x - \tfrac{1}{2})] \quad (0 < x < \infty)$$

is to be estimated from two observations x_1 and x_2 drawn at random. What is the maximum likelihood estimate of m?

28. Determine the maximum likelihood estimate of P^* as in Example 1, Section 2.5 except that two samples of size r each show success w_1 times and w_2 times.

29. Given are r numbers selected at random from the internal $(0, A)$. A is positive. Let $x_1, x_2, \ldots, x_r$ denote the numbers given. What is the maximum likelihood estimate of A?

30. Consider a Poisson distribution with unknown parameter μ, and density

$$f(x) = e^{-\mu}\mu^x/x!$$

values $x_1, x_2, \ldots, x_r$ are obtained from a series of r experiments. Each x_i represents the number of successes observed during experiment i. Determine the maximum likelihood estimator of μ.

CHAPTER 3

Algebra of Normal Functions[†]

Since it is usually necessary to evaluate functions of random variables when applying probabilistic methods, a suitable statistical mathematical system is needed.

Initially, consider the set of all ordered pairs, (μ, σ), that can be formed from the real numbers. μ and σ are each real numbers (or real valued). The set described contains an uncountable number of couples, since the set of real numbers is uncountable. If it is specified that μ and σ be finite valued, the set of couples (μ, σ) will still be an uncountable set, because any finite real interval contains uncountably many numbers. Of interest here is the set S (of normally distributed random variables) restricted to those couples: 1. having finite valued μ and σ, and 2. satisfying the closure restrictions described in Section 3.5. The definition of closure is given in Section 3.5.6.

The statistical mathematical system, developed in Chapter 3, reflects the properties of the elements in the set S. Common algebra, for example, reflects the properties of the set R of real numbers. The set S contains an infinite number of elements, each of which is a normal random variable.

Specifically, the properties of real numbers that shape the mathematical system are primarily those that define an integral domain. "Let D be a set of elements $a, b, c, \ldots$, for which the sum $a + b$ and the product ab of any two elements a and b (distinct or not) of D are defined. Then D is called an integral domain if the following postulates (1 to 9) hold: 1. closure, 2. uniqueness, 3. commutative laws, 4. associative laws, 5. distributive law, 6. zero, 7. unity, 8. additive inverse, and 9. cancellation law.[‡] The real numbers also satisfy the requirements of a number field F, defined in Section 3.6.1.

[†] See [22] and [23].
[‡] See [16].

106 Foundation

The properties of the set S (as constrained above) are stated in the following sections.

As stated earlier, the normal distribution is bell-shaped and symmetrical. It is uniquely described by two parameters, the mean value μ_x (or $\bar{x}$) and the standard deviation σ_x (or s_x).

In this next section, we shall employ the method of maximum likelihood to obtain estimates of the mean value and standard deviation of sum, difference, product, and quotient variates resulting from the random combinations of statistically independent pairs of normally distributed random variables. Results involving correlated random variables are also given.

3.1 INDEPENDENT BINARY OPERATIONS

In Section 2.5 the method of maximum likelihood is discussed. It is stated that the method is efficient for large n and satisfactorily consistent for the estimation of most parameters. The way in which the method of maximum likelihood may be employed to derive parameter estimators of functionally related pairs of statistically independent random variables is discussed in Section 2.9.1.

In deriving the estimators for moments of a variate Z resulting from the binary combination (sum, difference, etc.) of any pair X, Y of normally distributed random variables, the methods of Section 2.9.1 are now employed. Recall that the estimator

$$\bar{x} = \frac{x_1 + x_2 + \cdots + x_{n-1} + x_n}{n},$$

where (see Section 2.5):

$$\lim_{n \to \infty} \bar{x} = \mu_x.$$

Estimator of the standard deviation is

$$s_x = \left[\frac{1}{n} \sum_{i=1}^{n} (\mu_x - x_i)^2 \right]^{1/2},$$

where

$$\lim_{n \to \infty} s_x = \sigma_x.$$

The sum relationship of two normally distributed random variables is shown in Fig. 3.1.

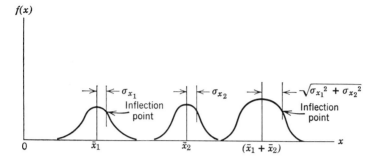

Figure 3.1 Sum of two normal variates (statistically independent).

3.1.1. Sums (Maximum Likelihood Estimators) $Z = X + Y$

The *mean* $\bar{z}$, with X and Y statistically independent is as follows.

$$\bar{z} = \frac{1}{n}\left[\frac{(x_1 + y_1) + (x_1 + y_2) + \cdots + (x_1 + y_w)}{w}\right.$$

$$+ \frac{(x_2 + y_1) + (x_2 + y_2) + \cdots + (x_2 + y_w)}{w}$$

$$+ \cdots + \frac{+ \cdots + (x_n - y_1) + (x_n + y_2) + \cdots + (x_n + y_w)}{w}\bigg],$$

$$\bar{z} = \frac{1}{n}\left[\frac{wx_1 + (y_1 + y_2 + \cdots + y_w)}{w} + \frac{wx_2 + (y_1 + y_2 + \cdots + y_w)}{w}\right.$$

$$+ \cdots + \frac{wx_n + (y_1 + y_2 + \cdots + y_w)}{w}\bigg],$$

$$\bar{z} = \left[\frac{wx_1 + wx_2 + \cdots + wx_n}{nw} + \frac{n(y_1 + y_2 + \cdots + y_w)}{nw}\right],$$

$$\bar{z} = \left(\frac{x_1 + x_2 + \cdots + x_n}{n}\right) + \left(\frac{y_1 + y_2 + \cdots + y_w}{w}\right),$$

and from Section 2.5, Example 2,

$$\bar{z} = \bar{x} + \bar{y}. \tag{3.1}$$

The standard deviation estimator s_z, with X and Y statistically independent is as follows:

$$[(x_i + y_j) - \bar{z}] = (x_i + y_j - \bar{x} - \bar{y}) = [(x_i - \bar{x}) + (y_j - \bar{y})].$$

$$s_z = \left[\frac{1}{n \cdot w} \sum_{i=1}^{n} \sum_{j=1}^{w} [(x_i - \mu_x) + (y_j - \mu_y)]^2\right]^{1/2}.$$

$$s_z = \left\{\frac{1}{n \cdot w} \sum_{j=1}^{w} [(x_1 - \bar{x}) + (y_j - \bar{y})]^2 + [(x_2 - \bar{x}) + (y_j - \bar{y})]^2 \right.$$
$$\left. + \cdots + [(x_n - \bar{x}) + (y_j - \bar{y})]^2\right\}^{1/2}.$$

$$s_z = \left\{\frac{1}{nw} \sum_{j=1}^{w} (x_1 - \bar{x})^2 - 2(x_1 - \bar{x})(y_j - \bar{y}) + \cdots + (y_j - \bar{y})^2 \right.$$
$$+ \cdots + (x_2 - \bar{x})^2 - 2(x_2 - \bar{x})(y_j - \bar{y}) + \cdots + (y_j - \bar{y})^2$$
$$\left. + \cdots + (x_n - \bar{x})^2 - 2(x_n - \bar{x})(y_j - \bar{y}) + \cdots + (y_j - \bar{y})^2\right\}^{1/2}.$$

$$s_z = \left\{\sum_{j=1}^{w} \frac{[(x_1 - \bar{x})^2 + (x_2 - \bar{x}) + \cdots + (x_n - \bar{x})^2]}{nw}\right.$$
$$\times \frac{-2[(x_1 - \bar{x}) + (x_2 - \bar{x}) + \cdots + (x_n - \bar{x})](y_j - \bar{y})}{nw}$$
$$\left. + \frac{n(y_j - \bar{z})^2}{nw}\right\}^{1/2}.$$

The middle term disappears because of the symmetry of the deviation in a normal population. Summing from $j = 1$ to $j = w$,

$$s_z = \left[\frac{(x_1 - \bar{x})^2 + (x_2 - \bar{x})^2 + \cdots + (x_n - \bar{x})^2}{n} \right.$$
$$\left. + \frac{(y_1 - \bar{y})^2 + (y_2 - \bar{y})^2 + \cdots + (y_w - \bar{z})^2}{w}\right]^{1/2}$$

and from Section 2.5, Example 2,

$$s_z = \sqrt{s^{72} + s_y^2}.\dagger \qquad (3.2)$$

Example 1. Given two resistors X_1 and X_2 with the resistance values of each normally distributed and described by

$$(\bar{x}_1, s_1) = (500, 30) \text{ ohms},$$
$$(\bar{x}_2, s_2) = (300, 25) \text{ ohms}.$$

† For check and verification see Sections 3.2.2 and 3.3.1.

If resistors X_1 and X_2 are coupled in series, compute the sum random variable R.

$$\bar{R} = \bar{x}_1 + \bar{x}_2 = (500 + 300) \text{ ohms} = 800 \text{ ohms},$$
$$s_R = \sqrt{s_1^2 + s_2^2} = \sqrt{(30)^2 + (25)^2} \text{ ohms} = 39 \text{ ohms}.$$

Thus

$$(\bar{R}, s_R) = (\bar{x}_1, s_1) + (\bar{x}_2, s_2) = (800, 39) \text{ ohms}.$$

A proof that addition of independent normally distributed variates is closed (i.e., is a normal variate) is given in Example 2 of Section 2.8.4. As shown later, the associative law for addition (see Section 3.5.3) holds for sums of independent normal variates, thus summing may be extended to n variates.

MEAN AND VARIANCE OF LINEAR COMBINATIONS. Consider the linear combination

$$W = aX + bY + cZ,$$

where X, Y and Z are independent normal variates and a, b, and c are arbitrary constants. The *mean* of the linear combination is (see Fig. 3.2)

$$\bar{w} = a\bar{x} + b\bar{y} + c\bar{z}. \tag{3.3}$$

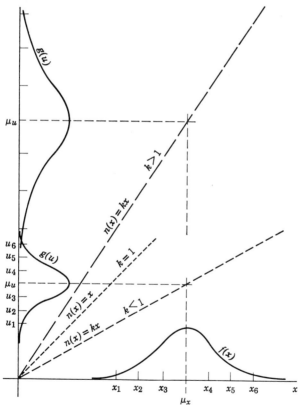

Figure 3.2 Linear transformation: normal variate.

The variance of the linear combination is

$$s_w^2 = a^2 s_x^2 + b^2 s_y^2 + c^2 s_z^2. \tag{3.4}$$

Because of the results of Section 3.2.2, μ and σ may be used rather than sample moments (in the limit) on sums.

Example 2. Consider a linear combination of random variables C, W_1, W_2, B_1, and B_2 which are independent and normal. A common problem in design is to obtain overall tolerance of a stack, consisting of a number of components. Consider an assembly which consists of a crank (C), two washers (W_1 and W_2), and two side bars (B_1 and B_2). The nominal dimensions (averages) are

$$\bar{c} = 0.750 \text{ in.}$$
$$\bar{w}_1 = 0.060 \text{ in.}$$
$$\bar{w}_2 = 0.060 \text{ in.}$$
$$\bar{b}_1 = 0.450 \text{ in.}$$
$$\bar{b}_2 = 0.450 \text{ in.}$$

$$\text{stack average} = \bar{s} = 1.770 \text{ in.}$$

Assume that the dimensional requirements are 1.770 ± 0.025 in. The components have the same variance, s^2. No more than 0.0027 of the assemblies are outside tolerance. The variance of the stack is $s_s^2 = 5s^2$

$$R(z) = \frac{0.0027}{2} = 0.00135$$

from Table 2.4, $z = 3.0$. Thus

$$\frac{1.795 - 1.770}{s_s} = 3, \quad \text{and} \quad s_s = 0.00833,$$

$$s_s^2 = 0.000069,$$

$$s^2 = \frac{0.000069}{5} = 0.000014,$$

$$s = 3.72 \cdot 10^{-3}.$$

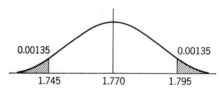

Thus the component tolerance at $\pm 3s$ is ± 0.0112 in.

ADDITION WITH ONE VARIATE NEGATIVE.

$$(\mu_a, \sigma_a) - (\mu_b, \sigma_b) = (\mu_a, \sigma_a) + (-1)(\mu_b, \sigma_b)$$
$$= (\mu_a, \sigma_a) + (-\mu_b, -\sigma_b)$$
$$= (\mu_a + (-\mu_b), \sqrt{\sigma_a^2 + (-\sigma_a)^2})$$
$$= (\mu_a - \mu_b, \sqrt{\sigma_a^2 + \sigma_b^2}).$$

BOTH VARIATES DEGENERATE.

$$(\mu_x, 0) + (\mu_y, 0) = (\mu_x + \mu_y, 0) = \text{constant}.$$

ONE VARIATE DEGENERATE.

$$(\mu_x, \sigma_x) + (\mu_y, 0) = (\mu_x + \mu_z, \sqrt{\sigma_x + 0}) = (\mu_x + \mu_y, \sigma_x).$$

MEANS OF BOTH VARIATES ZERO.

$$(0, s_x) + (0, s_y) = (0, \sqrt{s_x^2 + s_y^2}).$$

MEAN OF ONE VARIATE DEGENERATE.

$$(\mu_x, \sigma_x) + (0, \sigma_y) = (\mu_x, \sqrt{\sigma_x^2 + \sigma_y^2}).$$

Correlated Sums. † The standard deviation σ_z with X and Y correlated is

$$\sigma_z = \sigma_{x+y} = \sqrt{\sigma^{\prime 2} + \sigma_y^2 + 2r\sigma_x\sigma_y}, \qquad (3.5)$$

where r is the correlation coefficient.

When $r = 0$ (independence) Eq. 3.5 yields

$$\sigma_z = \sigma_{x+y} = \sqrt{\sigma_x^2 + \sigma_y^2}.$$

This result agrees with Eq. 3.2.

When $r = +1$ and variates x and y are identical

$$\sigma_{x+y} = \sqrt{\sigma_x^2 + 2\sigma_x\sigma_y + \sigma_y^2} = \sigma_x + \sigma_y = 2\sigma_x = 2\sigma_y,$$

which agrees with the results in Section 3.4 for the sum of two identical variates.

When $r = -1$ and variates X and Y are identical

$$\sigma_x = \sigma_y$$

and

$$\sigma_{x+y} = \sqrt{\sigma_x^2 - 2\sigma_x\sigma_y + \sigma_y^2} = 0,$$

which is the result expected when two variates are equal and absolutely (negatively) correlated (see Section 3.4 for the difference of two identical variates.)

---

† Correlation in the sense of physical variates functionally related, may be explained by example. Consider any function of random variables, such as $W = TV + T^2V^3$. T and V are random variables; hence W and each term on the right are also random variables. Thus $X = TV$ and $Y = T^2V^3$; hence $W = X + Y$. X and Y are not statistically independent as defined in Section 2.1.2, Theorem 3 for independent events, or in Section 2.4 Eqs. 2.29 and 2.31 for independent variates.

If $p^2 = 1$ (p = correlation coefficient), the random variables X and Y are linearly dependent; that is, each is a linear function of the other. If $p = 1$, there is absolute positive correlation between X and Y. If $p = -1$, there is absolute negative correlation between them. (See [7] p. 196.)

A common problem is as follows. Given the normally distributed random variable Z, where $Z = X - Y$, and the normally distributed random variable Y, how can the random variable X be reclaimed from Z and Y?

Combining X and Y as a difference involves the random combination of statistically independent random variables. The reverse process of obtaining $X = Z + Y$ involves correlated random variables Z and Y.

Thus when $Z = X - Y$ (by Eq. 3.7)

$$\sigma_z = \sqrt{\sigma_x^2 + \sigma_y^2}. \qquad (a)$$

When $X = Z + Y$ (by Eq. 3.5)

$$\sigma_x = \sqrt{\sigma_z^2 + \sigma_y^2 + 2r\sigma_z\sigma_y}, \qquad (b)$$

and the correlation coefficient r must be known.

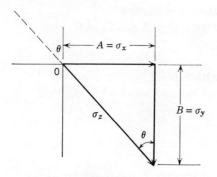

Figure 3.3

Noting the strong similarity in form between Eq. b and the cosine law,

$$A = \sqrt{C^2 + B^2 - 2CB \cos \theta}, \qquad (c)$$

a vector analogy is illuminating (see Fig. 3.3).

Then

$$r = \cos \theta = -\frac{\sigma_y}{\sqrt{\sigma_x^2 + \sigma_y^2}} = -\left(\frac{\sigma_y}{\sigma_z}\right). \qquad (d)$$

Substitution of Eq. d into Eq. b and $\sigma_x^2 + \sigma_y^2$ for σ_z^2 yields

$$\sigma_x = \left[(\sigma_x^2 + \sigma_y^2) + \sigma_y^2 + \sigma_z\sigma_y - \left(\frac{\sigma_x}{\sigma_z}\right)\right]^{1/2} = \sqrt{\sigma_x^2}.$$

The needed correlation coefficient is

$$r = \left(-\frac{\sigma_y}{\sigma_z}\right).$$

Algebra of Normal Functions 113

3.1.2 Differences (Maximum Likelihood Estimators) $Z = X - Y$

The difference relationship of two normally distributed statistically independent variates is shown in Fig. 3.4. The mean $\bar{z}$, with X and Y statistically independent, is

$$\bar{z} = \frac{1}{n}\left[\frac{(x_1 - y_1) + (x_1 - y_2) + \cdots + (x_1 - y_w)}{w}\right.$$
$$+ \cdots + \frac{(x_2 - y_1) + (x_2 - y_2) + \cdots + (x_2 - y_w)}{w}$$
$$+ \cdots + \left.\frac{(x_n - y_1) + (x_n - y_2) + \cdots + (x_n - y_w)}{w}\right],$$
$$\mu_{x-y} = \mu_x - \mu_y,$$
$$\bar{z} = \bar{x} - \bar{y}. \tag{3.6}$$

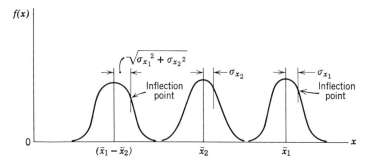

Figure 3.4 Difference of two normal variates statistically independent.

The standard deviation s_z, with X and Y statistically independent, is

$$s_z = \left\{\frac{1}{nw}\sum_{i=1}^{n}\sum_{j+1}^{w}[(x_i - \bar{X}) - (y_j - \bar{Y})]^2\right\}^{1/2}$$
$$s_z = \sqrt{s_x^2 + s_y^2}. \tag{3.7}$$

Details of the derivation are left as an exercise.

Example 1. Assume two normally distributed independent load components X and X_2, aligned but acting in opposite directions. Given $(\bar{x}_1, s_1) = (200, 45)$ lb and $(\bar{x}_2, s_2) = (100, 15)$ lb compute the resultant variate x:

$$\bar{x} = \bar{x}_1 - \bar{x}_2 = 200 - 100 = 100 \text{ lb}$$
$$s_x = \sqrt{s_1^2 + s_2^2} = \sqrt{(45)^2 + (15)^2} = 47.4 \text{ lb}$$
$$(\bar{x}, s_x) = (\bar{x}_1, s_1) - (x_2, s_2) = (100, 47.4) \text{ lb}$$

Correlated Differences. The standard deviation σ_z (with X and Y correlated) is

$$\sigma_z = \sigma_{x-y} = \sqrt{\sigma_x^2 + \sigma_y^2 - 2r\sigma_x\sigma_y}, \qquad (3.8)$$

where r is the correlation coefficient. Notice that when $r = 0$ (independence) the variance is

$$\sigma_{x-y}^2 = \sigma_x^2 + \sigma_y^2,$$

which agrees with the result obtained in Eq. 3.7.

When $r = +1$ and variates X and Y are equal, then

$$\sigma_{x-y} = \sqrt{\sigma_x^2 - 2\sigma_x\sigma_y + \sigma_y^2} = 0,\dagger$$

since $\sigma_y = \sigma_x$, which agrees with the results in Section 3.4 for the difference of two identical variates.

When $r = -1$ and variates x and y are equal,

$$\sigma_{x-y} = \sqrt{\sigma_x^2 - (-1)2\sigma_x\sigma_y + \sigma_y^2} = \sigma_x + \sigma_y = 2\sigma_x = 2\sigma_y,$$

which is the result expected when the difference is taken between two identical variates completely negatively correlated (see Section 3.4 for the sum of two identical variates).

A problem that often occurs is as follows. Given the normally distributed random variable Z, where $Z = X + Y$, and the normally distributed random variable X, how can the random variable Y be reclaimed from X and Z?

Combining X and Y as a sum involves random combination of statistically independent random variables. The reverse process of obtaining $Y = Z - X$ involves correlated random variables Z and X.

Thus when $Z = X + Y$ (by Eq. 3.2):

$$\sigma_z = \sqrt{\sigma_x^2 + \sigma_y^2}. \qquad (a)$$

When $Y = Z - X$ (by Eq. 3.8)

$$\sigma_y = \sqrt{\sigma_z^2 + \sigma_x^2 - 2r\sigma_z\sigma_x}, \qquad (b)$$

and the correlation coefficient r must be known.

Noting the identity in the form of Eq. b with the cosine law, Eq. c,

$$B = \sqrt{C^2 + A^2 - 2CA \cos \theta}. \qquad (c)$$

Again a vector interpretation is helpful (see Fig. 3.5).

Now

$$r = \cos \theta = \frac{\sigma_x}{\sqrt{\sigma_y^2 + \sigma_x^2}} = \left(\frac{\sigma_x}{\sigma_z}\right). \qquad (d)$$

† This result may be interpreted as demonstrating the existence of the additive inverse in the algebra of normal functions.

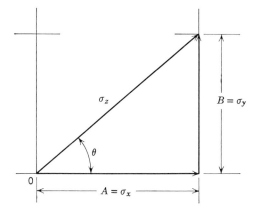

Figure 3.5

Substituting Eq. *d* into Eq. *b* and $(\sigma_x^2 + \sigma_y^2)$ for σ_z^2, the identity

$$\sigma_y = \left[(\sigma_x^2 + \sigma_y^2) + \sigma_x^2 - 2\sigma_x\sigma_z \left(\frac{\sigma_x}{\sigma_z}\right) \right]^{1/2} = \sqrt{\sigma_y^2}$$

is shown.

Hence the correlation coefficient is

$$r = \left(\frac{\sigma_x}{\sigma_z}\right).$$

The proof that subtraction of normally distributed random variables is closed (i.e., a normal variate) is given in the second derivation of Section 4.2.

Example 2. In mechanical design, a common problem is the mating of bearings and shafts. This problem involves the difference between two statistically independent variates. For

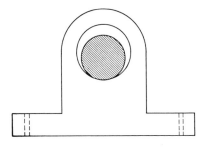

example, given the diameter (*OD*) variate of the shaft, $(\bar{D}_s, s_s) = (1.048, 0.0013)$ in., and the diameter (*ID*) variate of the bearing is $(\bar{D}_B, s_B) = (1.059, 0.0017)$ in. What is the clearance to be expected on random pairing of such bearings and shafts?

We define clearance as a value C, such that P (clearance $< C$) $= 0.00135$ (3σ value), (see $R(z)$, Table 2.4). We are interested in examining the clearance variate, between the shaft OD variate and the bearing ID variate. Thus the clearance variate (Z) is $Z = D_B - D_S$, from which $\bar{z} = \bar{D}_B - \bar{D}_S = 0.011$ in., and the standard deviation of Z is

$$s_z = \sqrt{s_B^2 - s_S^2} = \sqrt{(0.0017)^2 + (0013)^2}$$

$$= 0.00214 \text{ in.}$$

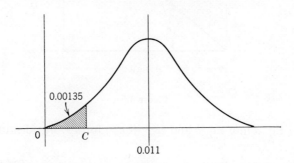

On the sketch, the measure in standard deviations of the joint distribution from C to 0.011 is $s_z = 3$. Thus

$$\frac{C - 0.011}{0.00214} = -3$$

and

$$C = 0.00458$$

Interference exists bearing ID is less than shaft OD. Thus the probability of interference is $p(z < 0)$.

$$p(z < 0) = p\left(z < \frac{0 - 0.011}{0.00214}\right) = p(z < -5.1),$$

and since -5.1 is negative,

$$p(z < -5.1) = p(z > 5.1) = 0.0000005.$$

3.1.3 Products (Maximum Likelihood Estimators) $Z = XY$

The mean $\bar{z}$, with X and Y statistically independent is

$$\bar{z} = \frac{1}{n}\left[\frac{x_1 y_1 + x_1 y_2 + \cdots + x_1 y_w}{w} + \frac{x_2 y_1 + x_2 y_2 + \cdots + x_2 y_w}{w} + \cdots + \frac{x_n y_1 + x_n y_2 + \cdots + x_n y_w}{w}\right]$$

$$\bar{z} = \bar{x}\bar{y}. \tag{3.9}$$

Algebra of Normal Functions 117

The standard deviation s_z, with X and Y statistically independent is

$$s_z = \left[\frac{1}{nw}\sum_{i=1}^{n}\sum_{j=1}^{w}\bar{z}(-x_iy_j)^2\right]^{1/2},$$

$$s_z^2 = \frac{1}{nw}[(\bar{z}-x_1y_1)^2 + (\bar{z}-x_1y_2)^2 + \cdots + (\bar{z}-x_1y_w)^2$$

$$+\cdots+(\bar{z}-x_2y_1)^2+(\bar{z}-x_2y_2)^2+\cdots+(\bar{z}-x_2y_w)^2$$

$$+\cdots+(\bar{z}-x_ny_1)^2+(\bar{z}-x_ny_2)^2+\cdots+(\bar{z}-x_ny_w)^2]$$

$$s_z = \sqrt{\bar{x}^2 s_y^2 + {}^{-2}s_x^2 + s_x^2 s_y^2}\dagger \qquad (3.10)$$

Example 1. The tensile strength variate P determined by normally distributed random variables A (cross section area) and f_{ty} (tensile yield strength)? Given values

$$(\bar{A}, \sigma_A) = (10, 0.4) \text{ in.}^2$$

$$(\bar{f}, \sigma_f) = (5\cdot 10^4, 3\cdot 10^3) \text{ psi.}$$

Applying Eq. 3.9,

$$\bar{P} = \bar{A}\cdot\bar{f} = 5\cdot 10^5 \text{ lb.}$$

Applying Eq. 3.10,

$$s_P = \sqrt{s_A^2\bar{f}^2 + s_f^2\bar{A}^2 + s_A^2\cdot s_f^2},$$

$$s_P = \sqrt{(0.4)^2(5\cdot 10^4)^2 + (3\cdot 10^3)^2(10)^2 + (0.4)^2(3\cdot 10^3)^2},$$

$$s_P = 3.61\cdot 10^4 \text{ lb.}$$

Thus

$$(\bar{P}, s_P) = (\bar{A}, s_A)(\bar{f}, s_f) = (5\cdot 10^5, 3.61\cdot 10^4) \text{ lb.}$$

LIMIT CONDITIONS ON PRODUCTS
(a) Both Variates Degenerate.

$$(\mu_x, 0)(\mu_y, 0) = \mu_x\mu_y, \sqrt{\mu_x^2\cdot 0 + \mu_y^2\cdot 0 + 0\cdot 0}$$

$$= (\mu_x\mu_2, 0) = \text{constant.}$$

(b) One variate Degenerate. (See Fig 3.2.)

$$(\mu_x, 0)(\mu_y, \sigma_y) = (\mu_x\mu_y, \sqrt{\mu_x^2\sigma_y^2 + \mu_y^2\cdot 0 + 0\cdot \sigma_y^2})$$

$$= (\mu_x\mu_y, \mu_x\sigma_y).$$

Note. $(\mu_x, 0)$ is a constant. The result therefore, is in agreement with that of a linear function (see Section 3.1.1).

† For check and verification see Sections 3.2.4 and 3.3.2.

(c) Both variates—standard deviation small.

$$(\mu_x, \sigma_x)(\mu_y, \sigma_y) = (\mu_x\mu_y, \sqrt{\mu_x^2\sigma_y^2 + \mu_y^2\sigma_x^2 + \sigma_x^2\sigma_y^2}),$$

and, for $\sigma_x^2\sigma_y^2$ very small,

$$(\mu_x, \sigma_x)(\mu_y, \sigma_y) \approx (\mu_x\mu_y, \sqrt{\mu_x^2\sigma_y^2 + \mu_y^2\sigma_x^2}).$$

(d) One Variate Negative,

$$(\mu_a, \sigma_a) \cdot [-(\mu_b, \sigma_b)] = (\mu_a, \sigma_a)(-\mu_b, -\sigma_b)$$
$$= (-\mu_a\mu_b), \sqrt{\mu_a^2(-\sigma_a)^2 + (-\mu_b)^2\sigma_a^2 + \sigma_a^2(-\sigma_b)^2}$$
$$= (-\mu_a\mu_b, \sqrt{\mu_a^2\sigma_b^2 + \mu_b^2\sigma_a^2 + \sigma_a^2\sigma_b^2})$$

(e) Mean of Both Variates Zero,

$$(0, \sigma_x)(0, \sigma_y) = (0, \sqrt{0 \cdot \sigma_y^2 + 0\sigma_x^2 + \sigma_x^2 + \sigma_x^2\sigma_y^2}) = (0, \sigma_x\sigma_y).$$

(f) Mean of One Variate Zero.

$$(\mu_x, \sigma_x)(0, \sigma_y) = (\mu_x \cdot 0, \sqrt{\mu_x^2\sigma_y^2 + 0\sigma_x^2 + \sigma_x^2 + \sigma_x^2\sigma_y^2})$$
$$= (0, \sigma_y\sqrt{\mu_x^2 + \sigma_x^2}).$$

Multiplication of normally distributed random variables X and Y is not closed, but for variates of moderate variance, products approximate closure (see Table 3.1).

Correlated Products. Standard deviation σ_z, with X and Y correlated is

$$\sigma_z = \sigma_{xy} = \mu_x^2\mu_y^2\left(\frac{\sigma_x^2}{\mu_x^2} + 2r\frac{\sigma_x\sigma_y}{\mu_x\mu_y} + \frac{\sigma_y^2}{\mu_y^2}\right). \quad (3.11)$$

Another expression for σ_{xy}^2 is given by

$$\sigma_z^2 = \sigma_{xy}^2 = \mu_x^2\mu_y^2\left[\frac{\sigma_x^2}{\mu_x^2} + \frac{\sigma_y^2}{\mu_x^2} + \frac{\sigma_x^2\sigma_y^2}{\mu_x^2\mu_y^2}\right](1 + r^2).\dagger \quad (3.12)$$

Consider Eq. 3.12 when $r = 0$ and $r = \pm 1$.

When $r = 0$:

$$\sigma_{xy}^2 = [\sigma_x^2\mu_y^2 + \sigma_y^2\mu_x^2 + \sigma_x^2\sigma_y^2].$$

This is in agreement with the results obtained by maximum likelihood methods and by moment generating function methods. When $r = \pm 1$ (absolute correlation) and variates X and Y are identical

$$\sigma_{xy}^2 = [2\mu_y^2\sigma_x^2 + 2\mu_x^2\sigma_y^2 + 2\sigma_x^2\sigma_y^2],$$

† Madeline J. Alexander, *Variance of Products and Quotients of Random Variables*, NAA-Rocketdyne, SORUM 63-1.

and since $\mu_x = \mu_y$ and $\sigma_x = \sigma_y$

$$\sigma_{xy}^2 = [4\mu_x^2\sigma_x^2 + 2\sigma_x^4] = [4\mu_y^2\sigma_y^2 + 2\sigma_y^4],$$

and

$$\sigma_{x^2} = \sigma_{y^2} = \sqrt{4\mu_x^2\sigma_x^2 + 2\sigma_x^4}.$$

This is in agreement with the results for the variance of a square of a normal variate, given in Section 3.2, and for the product of two identical and absolutely correlated normal variates given in Section 3.4.

The mean value μ_z with X and Y correlated is

$$\mu_z = \mu_{xy} = (\mu_x\mu_y + r\sigma_x\sigma_y),$$

which for $r = 0$, yields

$$\mu_{xy} = \mu_x\mu_y.$$

For $r = 1$, with $\mu_x = \mu_y$ and $\sigma_x = \sigma_y$,

$$\mu_{xy} = \mu_{x^2} = \mu_x^2 + \sigma_x^2 = \mu_y^2 + \sigma_y^2.\dagger$$

Product of a Variate and Its Inverse. Assume a normally distributed variate X described by (μ_x, σ_x), with inverse (see Eqs. 3.13 and 3.14)

$$(\mu_{1/x}, \sigma_{1/x}) \approx \left(\frac{1}{\mu_x}, \frac{\sigma_x}{\mu_x^2}\right)^{\ddagger}$$

computed by the algebra of normal functions. The product required is $(\mu_x, \sigma_x)(1/\mu_x, \sigma_x/\mu_x^2)$. The mean of the product is

$$\mu_{x(1/x)} = \mu_x\left(\frac{1}{\mu_x}\right) = 1.$$

By considering the populations of values of the variate X and the variate $1/x$, it is clear that the correlation of x_i and $1/x_i$ is negative. Thus, for $r = -1$, we obtain, by applying Eq. 3.11 above

$$\sigma_{x(1/x)}^2 = \mu_x^2\left(\frac{1}{\mu_x}\right)^2\left[\frac{\sigma_x^2}{\mu_x^2} - 2\frac{\sigma_x}{\mu_x} \cdot \frac{\sigma_x}{\mu_x^2} \cdot \frac{1}{(1/\mu_x)} + \frac{\sigma_x}{\mu_x^2} \cdot \frac{1}{(1/\mu_x^2)}\right]$$

$$= \left[\frac{\sigma_x^2}{\mu_x^2} - 2\frac{\sigma_x^2}{\mu_x^2} + \frac{\sigma_x^2}{\mu_x^2}\right] = 0,$$

which is a result $(1, 0)$ in agreement with that of Section 3.4.6.

† Refer to A. S. Merrill, Frequency Distribution of an Index when Both the Components Follow the Normal Law, Biometrics, XX^A, 1928, 55-63.
‡ A note on the expectation of a reciprocal and square root of a random variable, V. N. Murthy and C. S. Pillai, Central Statistical Organization, New Delhi, The American Statistician, December 1966.

4.1.4 Quotients (Maximum Likelihood Estimators) $Z = X/Y$

(*Note.* As with real numbers, division by zero is excluded.) The mean $\bar{z}$ with X and Y statistically independent is

$$\bar{z} = \frac{1}{nw}[x_{1/y_1} + x_{1/y_2} + \cdots + x_{1/y_w} + x_{2/y_1} + x_{2/y_2}$$
$$+ \cdots + x_{2/y_w} + x_{n/y_1} + x_{n/y_2} + \cdots + x_{n/y_w}]$$
$$\bar{Z} = \bar{X}/\bar{Y} \qquad (3.13)$$

Standard deviation s_z with X and Y statistically independent is

$$s_z^2 = \sum_{j=1}^{w}\sum_{i=1}^{n}\left(\frac{1}{n \cdot w}\right)\left(\frac{\mu_x}{\mu_y} - \frac{x_i}{y_j}\right)^2 = \left(\frac{1}{n \cdot w}\right)\sum_{j=1}^{w}\sum_{i=1}^{n}\left(\frac{\mu_x y_j - \mu_y x_i}{\mu_y y_j}\right)^2$$

$$s_z = \frac{1}{\mu_y}\left(\frac{\mu_y^2 \sigma_x^2 + \mu_x^2 \sigma_y^2}{\mu_y^2 + \sigma_y^2}\right)^{1/2}. \qquad (3.14)$$

Example 2. Given a normally distributed moment M, what is the opposing load variate P (at the end of a lever) required for statistical balance? Denote the lever variate by l.
Given
$$(\bar{M}, s_M) = (12{,}000;\ 980\ \text{in-lb})$$
$$(\bar{l}, s_l) = (100,\ 1.5)\ \text{in.,}$$
then
$$\bar{P} = \frac{\bar{M}}{\bar{l}} = \frac{12{,}000\ \text{in-lb}}{100\ \text{in.}} = 120\ \text{lb}$$

$$s_P = \frac{1}{100}\left(\frac{(12{,}000)^2(2.5)^2 + (100)^2(980)^2}{(100)^2 + (2.5)^2}\right)^{1/2}$$
$$= 10.24\ \text{lb}.$$

Thus
$$(\bar{P}, s_P) = \frac{(\bar{M}, s_M)}{(\bar{l}, s_l)} = (120,\ 10.24)\ \text{lb}.$$

LIMIT CONDITIONS ON QUOTIENTS
(a) Both Variates Degenerate.

$$\frac{(\mu_x, 0)}{(\mu_y, 0)} = \left(\frac{\mu_x}{\mu_y}, 0\right) = \text{constant}.$$

(b) Denominator (only) Degenerate (division of a normal variate by a constant).

$$\frac{(\mu_x, \sigma_x)}{(\mu_y, 0)} = \left(\frac{\mu_x}{\mu_y}, \frac{1}{\mu_y}\sqrt{\frac{\mu_y^2 \sigma_x^2 + \mu_x^2 0}{\mu_y^2 + 0}}\right) = \left(\frac{\mu_x}{\mu_y}, \frac{\sigma_x}{\mu_y}\right).$$

(c) Numerator (only) Degenerate (division of a constant by a normal variate).

$$\frac{(\mu_x, 0)}{(\mu_y, \sigma_y)} = \left(\frac{\mu_x}{\mu_y}, \frac{1}{\mu_y}\sqrt{\frac{\mu_y^2 0 + \mu_x^2 \sigma_y^2}{\mu_y^2 + \sigma_y^2}}\right) = \left(\frac{\mu_x}{\mu_y}, \frac{\mu_x \sigma_y}{\mu_y \sqrt{\mu_y^2 + \sigma_y^2}}\right).$$

(d) Mean of Numerator is Zero.

$$\frac{(0, \sigma_x)}{(\mu_y, \sigma_y)} = \left(\frac{0}{\mu_y}, \frac{1}{\mu_y}\sqrt{\frac{\mu_y^2 \sigma_y^2 + 0\sigma_y^2}{\mu_y^2 + \sigma_y^2}}\right) = \left(0, \frac{\sigma_x}{\sqrt{\mu_y^2 + \sigma_y^2}}\right).$$

(e) Mean of Numerator and Denominator—Zero.

$$\frac{(0, \sigma_x)}{(0, \sigma_y)} = \left(\frac{0}{0}, \frac{1}{0} \cdot \sqrt{\frac{0 \cdot \sigma_x^2 + 0 \cdot \sigma_y^2}{0 + \sigma_y^2}}\right) = \left(\frac{0}{0}, \frac{0}{0}\right).$$

Indeterminate

(f) Mean of Denominator—Zero.

$$\frac{(\mu_x, \sigma_x)}{(0, \sigma_y)} = \left(\frac{\mu_x}{0}, \frac{1}{0}\sqrt{\frac{0^2\sigma_x^2 + \mu_x^2\sigma_y^2}{0 + \sigma_y^2}}\right) = \left(\infty, \frac{\mu_x \sigma_y}{0}\right).$$

Undefined

Division x/y of normally distributed random variables X and Y is not closed (see derivation of the distribution of a quotient in Section 2.8.4, Example 3), but for random variables of reasonably small variance, division approximates closure.

Standard deviation s_z with X and Y correlated is

$$\sigma_z = \sigma_{y/x} = \frac{\mu_y}{\mu_x}\left[\frac{\sigma_x^2}{\mu_x^2} + \frac{\sigma_y^2}{\mu_y^2} - 2r\frac{\sigma_x \sigma_y}{\mu_x \mu_y}\right], \qquad (3.15)$$

and a more exact expression for $\sigma_{y/x}$ is given by

$$\sigma_z = \sigma_{y/x} = \frac{\mu_y}{\mu_x}\left[\frac{\sigma_x^2}{\mu_x^2} - 2r\frac{\sigma_x \sigma_y}{\mu_x \mu_y} + \frac{\sigma_y^2}{\mu_y^2} + 8\frac{\sigma_x^4}{\mu_x^4} - 16r\frac{\sigma_x^3 \sigma_y}{\mu_x^3 \mu_y}\right.$$

$$\left. + 3\frac{\sigma_x^2 \sigma_y^2}{\mu_x^2 \mu_y^2} + 5r^2\frac{\sigma_x^2 \sigma_y^2}{\mu_y^2 \mu_y^2} + \cdots\right]^{1/2}. \qquad (3.16)$$

In either Eqs. 3.15 or 3.16, when $r = 0$,

$$\sigma_{y/x}^2 = \frac{\mu_y^2}{\mu_x^2}\left[\frac{\sigma_x^2}{\mu_x^2} + \frac{\sigma_y^2}{\mu_y^2}\right] = \frac{1}{\mu_x^2}\left[\frac{\sigma_x^2 \mu_y^2 + \sigma_y^2 \mu_x^2}{\mu_x^2}\right]$$

$$\sigma_{y/x} = \frac{1}{\mu_x}\sqrt{\frac{\sigma_x^2 \mu_y^2 + \sigma_y^2 \mu_x^2}{\mu_x^2}}.$$

This result is in agreement with that shown in Section 3.3. *Note.* The terms in Eq. 3.16 not considered involve 4th or higher powers of ratios of σ/μ.

When $r = +1$ and variates X and Y are identical, we obtain, by applying Eq. 3.15,

$$\sigma_{y/x}^2 = \frac{\mu_y^2}{\mu_x^2}\left[\frac{\sigma_x^2}{\mu_x^2} - 2\frac{\sigma_x\sigma_y}{\mu_x\mu_y} + \frac{\sigma_y^2}{\mu_y^2}\right],$$

and since $\mu_x = \mu_y$ and $\sigma_x = \sigma_y$,

$$\sigma_{y/x}^2 = 0; \qquad \sigma_{y/x} = 0.$$

This yields $(1, 0)$. These results are in agreement with those given in Section 3.4. It should be noted that division yields the unit element $(1, 0)$ when identical positively and absolutely correlated variates appear in both numerator and denominator.

When $r = -1$ and variates X and Y are identical,

$$\sigma_{y/x}^2 = \frac{\mu_y^2}{\mu_x^2}\left[\frac{\sigma_x^2}{\mu_x^2} - (-1)2\frac{\sigma_x\sigma_y}{\mu_x\mu_y} + \frac{\sigma_y^2}{\mu_x^2}\right]$$

$$= \frac{\mu_y^2}{\mu_x^2}\left[\frac{\sigma_x^2}{\mu_x^2} + \frac{2\sigma_x\sigma_y}{\mu_x\mu_y} + \frac{\sigma_y^2}{\mu_y^2}\right] = \frac{4\sigma_x^2}{\mu_x^2} = \frac{4\sigma_y^2}{\mu_x^2}$$

$$\sigma_z = \sigma_{y/x} = \frac{2\sigma_x}{\mu_x}.$$

The mean value μ_z, with X and Y correlated is

$$\mu_z = \mu_{y/x} = \frac{\mu_y}{\mu_x}\left[1 + \frac{\sigma_x}{\mu_x}\left(\frac{\sigma_x}{\mu_x} - r\frac{\sigma_y}{\mu_y}\right)\left(1 + \frac{\sigma_x^2}{\mu_x^2} + \cdots\right)\right].$$

If the second and higher powers in the last term on the right are ignored

$$\mu_{y/x} = \frac{\mu_y}{\mu_x} + \frac{\sigma_x\mu_y}{\mu_x}\left(\frac{\sigma_x}{\mu_x} - \frac{r\sigma_y}{\mu_y}\right).$$

For $r = 0$

$$\mu_{y/x} = \frac{\mu_y}{\mu_x} + \frac{\sigma_x\mu_y}{\mu_x^3}.$$

3.1.5 Summary: Binary Operations

Independent   *Correlated*

Addition $(x + y)$:

$$\mu_{x+y} = \mu_x + \mu_y,$$

$$\sigma_{x+y} = \sqrt{\sigma_x^2 + \sigma_y^2}. \qquad \sigma_{x+y} = \sqrt{\sigma_x^2 + \sigma_y^2 + 2r\sigma_x\sigma_y},$$

Subtraction $(x - y)$:

$$\mu_{x-y} = \mu_x - \mu_y,$$

$$\sigma_{x-y} = \sqrt{\sigma_x^2 + \sigma_y^2}. \qquad \sigma_{x-y} = \sqrt{\sigma_x^2 + \sigma_y^2 - 2r\sigma_x\sigma_y}$$

Multiplication (xy):

$$\mu_{xy} = \mu_x\mu_y,$$

$$\sigma_{xy} = \sqrt{\mu_x^2\sigma_y^2 + \mu_y^2\sigma_x^2 + \sigma_x^2\sigma_y^2}. \qquad \sigma_{xy} = \mu_x\mu_y\left\{\left[\frac{\sigma_x^2}{\mu_x^2} + \frac{\sigma_y^2}{\mu_y^2} + \frac{\sigma_x^2\sigma_y^2}{\mu_x^2\mu_y^2}\right](1 + r^2)\right\}^{1/2}.$$

Division $(x \div y)$:

$$\mu_{(x/y)} = \mu_x/\mu_y,$$

$$\sigma_{x/y} = \frac{1}{\mu_y}\left(\frac{\mu_x^2\sigma_y^2 + \mu_y^2\sigma_x^2}{\mu_y^2 + \sigma_y^2}\right)^{1/2}. \qquad \sigma_{x/y} = \frac{\mu_x^2}{\mu_y^2}\left(\frac{\sigma_x^2}{\mu_x^2} + \frac{\sigma_y^2}{\mu_y^2} - 2r\frac{\sigma_x\sigma_y}{\mu_x\mu_y}\right)^{1/2}.$$

3.2 MOMENT GENERATING FUNCTIONS

The powerful methods of moment generating functions are now applied to verify some of the results given in Sections 3.1.1, 3.1.2, 3.1.3, and 3.1.4. In addition, expressions for the mean value and standard deviation of the square and the square root of a variate X are derived, as well as the quadratic form.

Moments and moment generating functions were introduced in Section 2.7, and applications were developed in Section 2.9.3. In this section, the univariate normal and the bivariate normal moment generating functions are employed. Moments of univariates can be computed utilizing the bivariate moment generating function.

3.2.1 Moments of the Sum $Z = X + Y$

In an example in Section 3.2.3 it is shown that, given $E(x) = \mu_x$ and $E(y) = \mu_y$, from the properties of E given by Eqs. 2.62 and 2.63,

$$E(z) = E(x + y) = E(x) + E(y) = \mu_x + \mu_y = \mu_z.$$

Similarly, for the variance of a sum σ_z^2, by Eq. 2.42,

$$E[(z - \mu_z)^2] = E\{[(x + y) - (\mu_x + \mu_y)]^2\}$$
$$= E[(x + y)^2] \cdots - E[2(x + y)(\mu_x + \mu_y)] + E[(\mu_x + \mu_y)^2]$$
$$= E(x^2) + E(2xy) + E(y^2) - 2\mu_x E(x) - 2\mu_x E(y)$$
$$\quad - 2\mu_y E(x) - 2\mu_y E(y) + E(\mu_x^2) + 2E(\mu_x \mu_y) + E(\mu_y^2).$$

Variance of $Z = \sigma_x^2 + \sigma_y^2 = \sigma_z^2$ (see Eq. 3.2).

3.2.2 Moments of the Difference $Z = X - Y$

$$E(z) = E(x - y) = E(x) - E(y) = \mu_x - \mu_y = \mu_z,$$

and, by a computation similar to that for sums,

$$E[(x - y) - (\mu_x - \mu_y)]^2 = \sigma_x^2 + \sigma_y^2 = \sigma_z^2 \quad \text{(see Eq. 3.7)}.$$

Moments may be obtained directly by evaluating the derivatives of $m(t_1, t_2)$ at $t_1 = 0$, $t_2 = 0$ (Eq. 2.57).

The mean of a normal variate X is

$$E(x) = \frac{\partial m}{\partial t_1}\bigg]_{t_1 = t_2 = 0} = (\mu_x + t_1 \cdot \sigma_x^2 + 2\rho t_2 \sigma_x \sigma_y)e^{(R)},\dagger$$

and, since

$$e^{(R)} = 1 \text{ at } t_1, t_2 = 0,$$
$$E(x) = \mu_x.$$

The variance of x is (see Eq. 2.42),

$$E[(x - \mu_x)^2] = E(x^2) - E(2x\mu_x) + E(\mu_x^2)$$
$$= (\mu_x^2 + \sigma_x^2) - 2\mu_x^2 + \mu_x^2 = \sigma_x^2,$$

since

$$E(x^2) = \frac{\partial^2 m}{\partial t_1^2}\bigg]_{t_1, t_2 = 0}.$$

The partial derivative of m with respect to t_1 is

$$\frac{\partial m}{\partial t_1} = (\mu_x + t_1 \cdot \sigma_x^2 + \rho \cdot t_2 \cdot \sigma_x \cdot \sigma_y)e^{(R)}.$$

† See Section 2.8.4 (p. 81 footnote).

The second partial derivative of m with respect to t_1 is

$$\frac{\partial^2 m}{\partial t_1^2} = (\sigma_x^2)e^{(R)} + (\mu_x + t \cdot \sigma_x^2 + \rho t_2 \sigma_x \sigma_y)^2 e^{(R)}.$$

Setting t_1 and t_2 equal to zero,

$$E(x^2) = \frac{\partial^2 m}{\partial t_1^2}\bigg]_{t_1=t_2=0} = \sigma_x^2 + \mu_x^2.$$

The variance of x, by Eq. 2.44, is

$$E[(x - \mu_x)^2] = E(x^2) - \mu_x^2 = \sigma_x^2.$$

3.2.3 Moments of the Product $Z = XY$

$$E(z) = \frac{\partial^2 m}{dt_1\, dt_2}\bigg]_{t_1=t_2=0}.$$

The first partial derivative of m with respect to t_2 is

$$\frac{\partial m}{dt_2} = (\mu_x + t_1 \cdot \sigma_x^2 + \rho t_2 \cdot \sigma_x \cdot \sigma_y)e^{(R)},$$

and the second partial derivative of m with respect to t_1 is

$$\frac{\partial^2 m}{dt_1\, dt_2} = (\rho \sigma_x \sigma_y)e^{(R)} + (\mu_y + \rho t_1 \sigma_x \sigma_y + t_2 \sigma_y^2) \cdots$$

$$\cdot (\mu_x + t_1 \sigma_x^2 + \rho t_2 \sigma_x \sigma_y) \cdot e^R.$$

For independent variates ($\rho = 0$),

$$\mu_z = \frac{\partial^2 m}{dt_1\, dt_2}\bigg|_{t_1=t_2=0} = 0 + \mu_x \mu_y = \mu_x \mu_y$$

The variance of Z is computed as follows:

$$\sigma_z^2 = E(xy - \mu_{xy})^2 = E(x^2 y^2 - 2xy\mu_{xy} + \mu_{xy}^2)$$

$$= E(x^2 y^2) - 2\mu_x \mu_y E(xy) + \mu_x^2 \mu_y^2$$

$$= \sigma_x^2 \sigma_y^2 + \sigma_x^2 \mu_y^2 + \sigma_y^2 \mu_x^2 + \mu_x^2 \mu_y^2 - 2\mu_x^2 \mu_y^2 + \mu_x^2 \mu_y^2$$

$$\sigma_z^2 = \sigma_{xy}^2 = E(xy - \mu_{xy})^2 = \mu_x^2 \sigma_y^2 + \mu_y^2 \sigma_x^2 + \sigma_x^2 \sigma_y^2$$

$$\sigma_z = \sigma_{xy} = \sqrt{\mu_x^2 \sigma_y^2 + \mu_y^2 \sigma_x^2 + \sigma_x^2 \sigma_y^2} \qquad \text{(see Eq. 3.10).}$$

3.2.4 Moments of a Square, $Z = X^2$

It has been shown that Eq. 2.57 yields

$$\mu_z = \mu_{x^2} = E(x^2) = \mu_x^2 + \sigma_x^2, \tag{3.17}$$

and, from Section 3.2.3, we have the following result

$$E(x^2) = \frac{\partial^2 m}{\partial t_1^2} = \sigma_x^2 e^R + (\mu_x + t_1 \sigma_x^2 + \rho t_2 \sigma_x \sigma_y)^2 e^R,$$

$$E(x^2) = \frac{\partial^2 m}{\partial t_1^2}\bigg|_{t_1=t_2=0} = \mu_x^2 + \sigma_x^2.$$

The variance of $Z = X^2$ is (by Eq. 2.42), expanding and applying the properties of E (Eqs. 2.62 and 2.63),

$$\sigma_{x^2}^2 = E[(x^2 - \mu_x^2)^2] = E(x^4) - E(2x^2 \mu_x^2) + E(\mu_x^4)$$
$$= E(x^4) - \mu_x^4 - 2\mu_x^2 \sigma_x^2 - \sigma_x^4.$$

$E(x^4)$ is computed. Taking the derivative of $\partial^2 m/\partial t_1^2$ (the third derivative of m) with respect to t_1 yields

$$E(x^3) = \frac{\partial^3 m}{\partial t_1^3} = 0 + \sigma_x^2(\mu_x + t_1 \sigma_x^2 + \rho t_2 \sigma_x \sigma_y) e^R \dagger + \cdots$$
$$+ 2(\mu_x + t_1 \sigma_x^2 + \rho t_2 \sigma_x \sigma_y)\sigma_x^2 e^R + \cdots$$
$$+ (\mu_x + t_1 \sigma_x^2 + \rho t_2 \sigma_x \sigma_y)^3 e^R,$$

$$E(x^3) = \frac{\partial^3 m}{\partial t_1^3}\bigg|_{t_1=t_2=0} = \mu_x^3 + 3\mu_x \sigma_x^2, \tag{3.18}$$

Taking the derivative of $\partial^3 m/\partial t_1^3$ with respect to t_1,

$$\frac{\partial^4 m}{\partial t_1^4} = 3\sigma_x^2 \sigma_x^2 e^R + 3\sigma_x^2(\mu_x + t_1 \sigma_x^2 + \rho t_2 \sigma_x \sigma_y)^2 e^R + \cdots$$
$$+ 3\sigma_x^2(\mu_x + t_1 \sigma_x^2 + \rho t_2 \sigma_x \sigma_y)^2 e^R$$
$$+ (\mu_x + t_{1x}^2 + \rho t_2 \sigma_x \sigma_y)^4 e^R.$$

$$E(x^4) = \frac{\partial^4 m}{\partial t_1^4}\bigg|_{t_1=t_2=0} = \mu_x^4 + 6\mu_x^2 \sigma_x^2 + 3\sigma_x^4, \tag{3.19}$$

† See Section 2.8.4.

Substituting for $E(x^4)$ in the equation for $\sigma_{x^2}^2$,

$$\sigma_z^2 = \sigma_{x^2}^2 = \mu_x^4 + 6\mu_x^2\sigma_x^2 + 3\sigma_x^4 - \mu_x^4 - 2\mu_x^2\sigma_x^2 - \sigma_x^4,$$

$$\sigma_{x^2}^2 = 4\mu_x^2\sigma_x^2 + 2\sigma_x^4, \tag{3.20}$$

and the standard deviation of x^2 is

$$\sigma_{x^2} = \sqrt{4\mu_x^2\sigma_x^2 + 2\sigma_x^4}. \tag{3.21}$$

Example 1. Given that the radius is a normally distributed random variable, R, determine the area (random variable) A.

$$(\bar{R}, \sigma_R) = (1.5, 0.10) \text{ in.}$$

Applying Eq. 3.17,

$$\mu_{R^2} = \bar{R}^2 + \sigma_R^2 = (1.5)^2 + (0.10)^2 = 2.26 \text{ in.}^2$$

Applying Eq. 3.21,

$$\sigma_{R^2} = \sqrt{4(1.5)^2(0.10)^2 + 2(0.10)^4} = 0.316 \text{ in.}^2$$

Thus

$$(\bar{A}, \sigma_A) = 3.14(2.26, 0.316) = (7.096, 0.992) \text{ in.}^2$$

3.2.5 Moments of a Root, $Z = X^{\frac{1}{2}}$†

In the derivation the following results are used:

$$\mu_{x^2} = \mu_x^2 + \sigma_x^2, \quad (3.17) \tag{1}$$

$$\sigma_{x^2}^2 = 4\mu_x^2\sigma_x^2 + 2\sigma_x^4, \quad (3.20) \tag{2}$$

$$\mu_x = \sqrt{\mu_{x^2} - \sigma_x^2}. \quad (3.7) \tag{3}$$

Then

$$\sigma_{x^2}^2 = 4(\sigma_x^2 - \sigma_x^2)\sigma_x^2 + 2\sigma_x^4,$$

$$\sigma_{x^2}^2 = 4\mu_x^2\sigma_x^2 - 2\sigma_x^4,$$

$$0 = 2\sigma_x^4 - 4\mu_{x^2}\sigma_x^2 + \sigma_{x^2}^2.$$

Solving the quadratic in σ_x^2,

$$\sigma_x^2 = \mu_{x^2} \pm \tfrac{1}{2}\sqrt{4\mu_{x^2}^2 - 2\sigma_{x^2}^2}.$$

Note that when $\sigma_{x^2} = 0$, $\sigma_x = 0$. Returning to the evaluation of the mean,

$$\mu_x^2 = \mu_{x^2} + \sigma_x^2.$$

† See footnote, p. 119.

Substituting for σ_x^2 into the expression above,

$$\mu_x^2 = \mu_{x^2} - \mu_{x^2} \pm \sqrt{\frac{4\mu_{x^2}^2 - 2\sigma_{x^2}^2}{2}},$$

$$\mu_x^2 = \pm\tfrac{1}{2}\sqrt{4\mu_{x^2}^2 - 2\sigma_{x^2}^2}.$$

Changing the expressions to a more convenient form,

$$\mu_{\sqrt{x}} = [\tfrac{1}{2}\sqrt{4\mu_x^2 - 2\sigma_x^2}]^{1/2}, \tag{3.22}$$

$$\sigma_{\sqrt{x}} = [\mu_x - \tfrac{1}{2}\sqrt{4\mu_x^2 - 2\sigma_x^2}]^{1/2}. \tag{3.23}$$

3.2.6 Moments of the Quadratic Form

Let

$$Z = h(x) = ax^2 + bx + c,$$

where X is a normally distributed random variable and a, b, and c are arbitrary constants. By utilizing Eq. 2.51 (Section 2.9.3) we obtain

$$E[h(x)]^r = \int_{-\infty}^{\infty} e^{t[h(x)]} f(x)\, dx$$

and

$$E(z) = E[ax^2 + bx + c] = \int_{-\infty}^{\infty} e^{t[ax^2+bx+c]} f(x)\, dx.$$

Then by properties of E (Eqs. 2.62 and 2.63)

$$E(z) = E[ax^2] + E[bx] + E(c) = aE(x^2) + bE(x) + E(c),$$

$$E(z) = a(\mu_x^2 + \sigma_x^2) + b\mu_x + c = \mu_z. \tag{3.24}$$

To compute the variance of z, the results of Section 2.9.3 are utilized:

$$\sigma_z^2 = E[(z - \mu_z)^2] = E[(h(x) - \mu_{h(x)})^2].$$

Expanding the term on the right and using the properties of E

$$\sigma_z^2 = \sigma_{h(x)}^2 = E[(h(x))^2] - 2\mu_{h(x)} E[h(x)] + (\mu_{h(x)})^2.$$

By expanding and simplifying, the variance of the quadratic form is

$$\sigma_z^2 = \sigma_{h(x)}^2 = \sigma_x^2(2a\mu_x + b)^2 + 2a^2\sigma_x^4. \tag{3.25}$$

3.3 METHOD OF PARTIAL DERIVATIVES

In Section 2.9.2 the method of partial derivatives is developed.

In this section, partial derivative methods are applied to check results in Sections 3.1 and 3.2, and to demonstrate the consistency of results regardless of method. The standard deviation of a function of random variables is given by Eq. 2.61:

$$S_z = \left[\sum_{j=1}^{j}\left(\frac{\partial z}{\partial x_j}\right)^2 (S_{x_j})^2\right]^{1/2}.$$

3.3.1 Sums and Differences

Assume that $Z = X \pm Y$, where X and Y are normally distributed random variables. Then the variance of Z is obtained by first obtaining the partial derivatives

$$\frac{\partial z}{\partial x} = 1 \quad \text{and} \quad \frac{\partial z}{\partial y} = \pm 1.$$

Then employing Eq. 2.61

$$\sigma_z^2 = \sigma_x^2 + \sigma_y^2.$$

The standard deviation of Z is (by Eq. 2.43)

$$\sigma_z = \sqrt{\sigma_x^2 + \sigma_y^2} \quad \text{(see Eqs. 3.2 and 3.7)}.$$

3.3.2 Products

Assume that $Z = XY$, where X and Y are normally distributed random variables. The variance of Z is obtained by applying Eq. 2.61, after obtaining the partial derivatives

$$\frac{\partial z}{\partial x} = y \quad \text{and} \quad \frac{\partial z}{\partial y} = x,$$

and

$$\sigma_z^2 = \mu_y^2 \sigma_x^2 + \mu_x^2 \sigma_y^2.$$

Approximate standard deviation is

$$\sigma_z = \sqrt{\mu_y^2 \sigma_x^2 + \mu_x^2 \sigma_y^2} \quad \text{(see Eq. 3.10)}.$$

3.3.3 Quotients

Assume that $Z = X/Y$, where X and Y are normally distributed random variables. First obtain the partial derivatives of X and Y with respect to Z:

$$\frac{\partial z}{\partial x} = \frac{1}{y} \quad \text{and} \quad \frac{\partial z}{\partial y} = -\frac{x}{y^2}.$$

Then the approximate variance in Z is

$$\sigma_z^2 = \frac{1}{\mu_y^2}\sigma_x^2 + \left(-\frac{\mu_x}{\mu_y^2}\right)^2 \sigma_y^2,$$

$$\sigma_z^2 = \left(\frac{\mu_y^2 \sigma_x^2 + \mu_x^2 \sigma_y^2}{\mu_y^4}\right)^{1/2}.$$

The standard deviation (approximate) of Z is

$$\sigma_z = \left(\frac{\mu_y^2 \sigma_x^2 + \mu_x^2 \sigma_y^2}{\mu_y^4}\right)^{1/2} \quad \text{(see Eq. 3.14)}$$

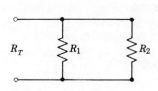

Example 1. Given the circuit as shown, and R_1 and R_2 statistically independent resistors, each normally distributed and described as follows,

$(R_1, s_{R_1}) = (10,000; 300)$ ohm

$(R_2, s_{R_2}) = (20,000; 600)$ ohms,

determine the resistance variate $(\bar{R}_T, s_{R_T})$.

Solution.

$$\frac{1}{R_T} = \frac{1}{R_1} + \frac{1}{R_2} \quad \text{and} \quad R_T = \frac{R_1 R_2}{R_1 + R_2}.$$

Solving for the mean value $\bar{R}_T$,

$$\bar{R}_T = \frac{\bar{R}_1 \bar{R}_2}{\bar{R}_1 + \bar{R}_2} = \frac{(10,000)(20,000)}{10,000 + 20,000} = 6.66 \cdot 10^3 \text{ ohms}.$$

Solving for s_{R_T}, the standard deviation of R_T involves applying Eq. 2.61:

$$s_{R_T}^2 = \sum_{i=1}^{2}\left(\frac{\partial R}{\partial R_i}T\right)^2 \cdot s_{R_i}^2.$$

First, taking the required partial derivatives,

$$\frac{\partial R_T}{\partial R_1} = \frac{-\bar{R}_2^2}{(\bar{R}_1 + \bar{R}_2)^2} = 0.444,$$

$$\frac{\partial R_T}{\partial R_2} = \frac{-\bar{R}^2}{(\bar{R}_1 + \bar{R}_2)^2} = 0.111,$$

$$s_{R_T}^2 = (0.444)^2(300)^2 + (0.111)^2(600)^2 = 2.212 \cdot 10^4 \text{ ohms}^2,$$

$$s_{R_T} = \sqrt{2.212 \cdot (10^4) \text{ ohms}^2} = 148.6 \text{ ohms}.$$

Thus

$$(\bar{R}_T, s_{R_T}) = (6.66 \cdot 10^3, 148.6) \text{ ohms}.$$

Note that dispersion, reflected by the standard deviation, is decreasing at a faster rate than is the mean value.

3.4 SPECIAL CORRELATED† COMBINATIONS

In this section certain combinations of variates that are correlated absolutely are considered. Included are binary combinations of identical variates such as powers and differences. The way in which identical correlated variate pairs behave in binary combinations is of importance in engineering. This reason justifies treating these operations separately.

3.4.1 Binary Operations—Correlated Variates

Keeping in mind the possible consequences of confusing a random combination with a correlated combination, certain important special cases are now considered. Maximum likelihood estimators are used for the derivations. Assume that S is a normally distributed random variable, $(\bar{s}, s_s)$, characterized by the population of values $s_1, s_2, \ldots, s_n$. With values arranged as shown, two identical, correlated random variables pair as shown:

$$
\begin{array}{cccccc}
s_1, & s_2, & s_3 & \cdots & s_{n-2}, & s_{n-1}, & s_n \\
\updownarrow & \updownarrow & \updownarrow & & \updownarrow & \updownarrow & \updownarrow \\
s_1, & s_2, & s_3 & \cdots & s_{n-2}, & s_{n-1}, & s_n
\end{array}
$$

3.4.2 Sum, $Z = S + S$

$$\bar{z} = \frac{(s_1 + s_1) + (s_2 + s_2) + \cdots + (s_{n-1} + s_{n-1}) + (s_n + s_n)}{n},$$

$$\bar{z} = \frac{2s_1 + 2s_2 + \cdots 2s_{n-1} + 2s_n}{n} = 2\bar{s}, \qquad (3.26)$$

$$s_z = \left(\frac{1}{n}\sum_{i=1}^{n}(2z - 2s_i)^2\right)^{1/2} = 2\left(\frac{1}{n}\sum_{i=1}^{n}(z - s_i)^2\right)^{1/2} = 2s_s. \qquad (3.27)$$

3.4.3 Difference, $Z = S - S$

$$\bar{z} = \frac{(s_1 - s_1) + (s_2 - s_2) + \cdots + (s_n - s_n)}{n} = 0, \qquad (3.28)$$

$$s_z = \left\{\frac{1}{n}\sum[\bar{z} - (s_i - s_i)]^2\right\}^{1/2} = \left[\frac{1}{n}\sum_{i=1}^{n}(0 - (0)^2\right]^{1/2} = 0. \qquad (3.29)$$

† See footnote, p. 75.

3.4.4 Quotient, $Z = S/S$

$$\bar{z} = \frac{s_1/s_1 + s_2/s_2 + \cdots + s_{n-1}/s_{n-1} + s_n/s_n}{n},$$

$$\bar{z} = \frac{1 + 1 + \cdots + 1 + 1}{n} = \frac{n}{n} = 1,$$

$$s_z = \left[\frac{\sum_{i=1}^{n}(\bar{z} - s_i/s_i)^2}{n}\right]^{1/2} = \left[\frac{\sum_{i=1}^{n}(1-1)^2}{n}\right]^{1/2} = 0. \qquad (3.30)$$

3.4.5 Product, $Z = SS$

$$\bar{z} = \frac{s_1 s_1 + s_2 s_2 + \cdots + s_{n-1} s_{n-1} + s_n s_n}{n},$$

$$\bar{z} = \mu_s^2 + \sigma_s^2, \qquad (3.31)$$

$$s_z = \left[\frac{\sum_{i=1}^{n}(\bar{z} - s_i^2)^2}{n}\right]^{1/2}$$

$$= \left[\frac{(\bar{z} - s_1^2)^2 + (\bar{z} - s_2^2)^2 + (\bar{z} - s_{n-1}^2)^2 + (\bar{z} - s_n^2)^2}{n}\right]^{1/2}$$

$$= \left[\bar{z}^2 - 2\bar{z}\frac{(s_1^2 + s_2^2 + \cdots + s_{n-1}^2 + s_n^2)}{n}\right.$$

$$\left. + \frac{s_1^4 + s_2^4 + \cdots s_{n-1}^4 + s_n^4}{n}\right]^{1/2}$$

$$s_z = \left[\frac{s_1^4 + s_2^4 + \cdots + s_{n-1}^4 + s_n^4}{n} - \bar{z}^2 - 2\bar{z}s_s^2 - s_s^4\right]^{1/2}.$$

Since

$$E(x^4) = \frac{x_1^4 + x_2^4 + \cdots + x_n^4}{n} = \bar{x}^4 + 6\bar{x}^2 s_x^2 + 3s_x^4,$$

(see Eq. 3.19), substituting s for x in the equation above,

$$s_z = (\bar{s}^4 + 6\bar{s}^2 s_s^2 + 3s_s^4 - \bar{s}^4 - 2\bar{s}^2 s_s^2 - s_s^4)^{1/2},$$

$$s_z = \sqrt{4\bar{s}^2 s_s^2 + 2s_s^4}. \qquad (3.32)$$

The result for products of identical correlated variates agrees with that obtained by moment generating functions for the square of a normal variate.

3.4.6 Product of a Variate (S) and Its Inverse (1/S), $Z = S(1/S)$

$$(\bar{s}, s_s)[(1/s), s_{(1/s)}].$$

The product of a real number and its inverse equals one. Recalling that a variate (S) and its inverse variate (1/S) are not statistically independent, we may picture the pairing of a variate and its inverse as

$$s_1, s_2, s_3, \ldots, s_{n-2}, s_{n-1}, s_n$$
$$\updownarrow \updownarrow \updownarrow \quad \updownarrow \updownarrow \updownarrow$$
$$\frac{1}{s_1}, \frac{1}{s_2}, \frac{1}{s_3}, \ldots, \frac{1}{s_{n-2}}, \frac{1}{s_{n-1}}, \frac{1}{s_n}.$$

Mean of $(S)(1/S)$ is

$$\bar{z} = \frac{1}{n}\sum_{i=1}^{n}(s_i)\left(\frac{1}{s_i}\right) = \frac{s_1(1/s_1) + s_2(1/s_2) + \cdots + s_n(1/s_n)}{n},$$

$$\bar{z} = \frac{n}{n} = 1 \tag{3.33}$$

Standard deviation of $(S)(1/S)$ is

$$s_z = \left[\frac{1}{n}\sum_{i=1}^{n}\left(s_i\left(\frac{1}{s_i}\right) - 1\right)^2\right]^{1/2} = 0 \tag{3.34}$$

3.4.7 Summary (Nonindependent Combinations) Binary Operations

1. Addition, $s + s$:
$$\mu_{s+s} = 2\mu_s,$$
$$\sigma_{s+s} = 2\sigma_s.$$

2. Subtraction, $s - s$:
$$\mu_{s-s} = 0,$$
$$\sigma_{s-s} = 0.$$

3. Division, s/s:
$$\mu_{s/s} = 1,$$
$$\sigma_{s/s} = 0.$$

4. Multiplications, $s \cdot s$:
$$\mu_{s \cdot s} = \mu_s^2 + \sigma_s^2,$$
$$\sigma_{s \cdot s} = \sqrt{4\mu_s^2\sigma_s^2 + 2\sigma_s^4}.$$

5. Product, (variate s) $\cdot$ (variate $1/s$):
$$\mu_{s(1/s)} = 1,$$
$$\sigma_{s(1/s)} = 0.$$

3.4.8 Correlated Functions: General

The solutions of many problems require series expressions involving a normal variate. Such series expressions are not termwise statistically independent; for instance, a Fourier series of the form

$$z = \sum_{n=1}^{n} a_n \sin \frac{n\pi x}{e}$$

or of the form

$$z = e^x = 1 + x + \frac{x^2}{2!} + \frac{x^3}{3} + \cdots,$$

involve statistical dependency. The standard deviations of such expressions,

Table 3.1 Constraints on Squares, Products, Quotients

Operation	μ_a	σ_a	μ_b	σ_b	σ_a/μ_a	σ_b/μ_b
Squaring	10.0	1.5			15.0%	
	100.0	15.0			15.0%	
Products	100.0	20.0	10	0.50	20.0%	5.0%
	100.0	5.0	10	0.75	5.0%	7.5%
	100.0	10.0	50	2.50	10.0%	5.0%
	100.0	20.0	50	5.0	20.0%	10.0%
	100.0	7.5	50	5.0	7.5%	10.0%
Quotients						
$x/y < 1$	5.0	0.5	10	0.5	10.0%	5.0%
	5.0	0.25	10	0.5	5.0%	5.0%
$x/y = 1$	5.0	0.25	5	0.50	5.0%	10.0%
$x/y > 1$	10.0	1.0	5	0.25	10.0%	5.0%
	10.0	0.50	5	0.50	5.0%	10.0%

where X is a normal variate, are computed by Eq. 2.61,

$$s_z^2 = \sum_{k=1}^{k} \left(\frac{\partial z}{dx_x}\right)^2 \sigma_{x_k}^2.$$

As was stated previously, Eq. 2.61 gives satisfactory approximations provided the variances of the random variables are moderate.

3.4.9 Constraints

Table 3.1 is a summary of constraints; for instance, the square of a variate remains normally distributed according to the chi-square test for coefficients of variation (σ_x/μ_x) to about 15 percent (see Section 6.2.1, Example 2).

3.4.10 Coefficient of Variation

The coefficient of variation of a random variable X, with mean μ_x and standard deviation σ_x, is defined as the ratio

$$\text{coeff var}_x = \frac{\sigma_x}{\mu_x} = V_x.$$

It may be recalled that, for sums and differences of normally distributed random variables, $Z = X \pm Y$, variance of Z is given by

$$\sigma_z^2 = \sigma_x^2 + \sigma_y^2 \quad \text{(see Eqs. 3.2 and 3.7)}.$$

Now consider the product $Z = XY$, from 3.12, with variates X and Y statistically independent and $r = 0$. We have

$$\sigma_z^2 = \mu_x^2 \mu_y^2 \left(\frac{\sigma_x^2}{\mu_x^2} + \frac{\sigma_z^2}{\mu_y^2} + \frac{\sigma_x^2 \sigma_y^2}{\mu_x^2 \mu_y^2} \right).$$

Dividing by $\mu_x^2 \mu_y^2$,

$$V_z^2 = V_x^2 + V_y^2 + V_x^2 V_y^2.$$

For V_x and V_y small, $V_x^2 V_y^2 \ll V_x^2 + V_y^2$, and

$$V_z^2 \approx V_x^2 + V_y^2. \tag{3.35}$$

Next consider the quotient $Z = X/Y$, from Eq. 3.14, with variates X and Y statistically independent:

$$\sigma_z^2 = \frac{1}{\mu_y^2} \left(\frac{\sigma_x^2 \mu_y^2 + \sigma_y^2 \mu_x^2}{\mu_y^2 + \sigma_y^2} \right),$$

$$\sigma_z^2 = \frac{\mu_x^2 (\sigma_y^2 / \mu_y^2) + \sigma_x^2}{\mu_y^2 (1 + \sigma_y^2 / \mu_y^2)} = \frac{\mu_x^2 V_y^2 + \sigma_x^2}{\mu_y^2 (1 + V_y^2)},$$

$$\left(\frac{\mu_x}{\mu_y} \right)^2 = \mu_z^2 \quad \text{and} \quad V_z^2 \mu_x^2 = \sigma_z^2 \mu_y^2,$$

$$\sigma_z^2 \mu_y^2 = V_z^2 \mu_x^2 = \frac{\mu_x^2 V_y^2 + \sigma_x^2}{(1 + V_y^2)},$$

$$V_z^2 = \frac{V_y^2 + V_x^2}{1 + V_y^2},$$

and, for V_y small, $V_y^2 \ll 1$, and as an approximation,

$$V_z^2 \approx V_x^2 + V_y^2. \tag{3.36}$$

It is interesting to note that the expressions for variance σ_z^2 of sums and differences of normally distributed random variables are formally the same as for coefficients of variation, V_z^2, of products and quotients.

3.5 LAWS OF COMBINATION

The set S of normally distributed random variables for which the laws of combination are valid is that for which closure (see Section 3.5.6) constraints are not exceeded.

Since μ_x and σ_x are real valued, the validity of several of the laws (Sections 3.5.1 to 3.5.5, and 3.5.7) follow from the properties of real numbers.

3.5.1 Characteristics

Equality in the Set S.

$$(\mu_a, \sigma_a) = (\mu_b, \sigma_b)$$

if and only if

$$\mu_a = \mu_b \quad \text{and} \quad \sigma_a = \sigma_b.$$

Otherwise,

$$(\mu_a, \sigma_a) \neq (\mu_b, \sigma_b).$$

Order Relations in the Set S. Does $(\mu_a, \sigma_a) \neq (\mu_b, \sigma_b)$ imply

$$(\mu_a, \sigma_a) > (\mu_b, \sigma_b) \quad \text{or} \quad (\mu_a, \sigma_a) < (\mu_b, \sigma_b)?$$

The answer is no.† If the mean value μ_a of a normal variate is plotted as an abscissa value and standard deviation σ_a is plotted as an ordinate value, (μ_a, σ_a) represents a point in a plane (see Fig. 3.6). Thus normal variates, in this respect, are analogues of the complex numbers.

In the sense of serial ordering, the expressions "greater than" or "less than" are meaningless in the algebra of normally distributed random variables.

Equivalence Relations. Reflexness, symmetry, and transitiveness are valid in this system. Consider any elements A, B, C in S.

Reflexive: $(\mu_a, \sigma_a) = (\mu_a, \sigma_a).$

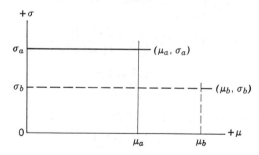

Figure 3.6 Normal variate representation in a plane.

† If the random variables A and B are both degenerate, that is, are real numbers, then they may be ordered.

Symmetric: if $(\mu_a, \sigma_a) = (\mu_b, \sigma_b)$, then

$$(\mu_b, \sigma_b) = (\mu_a, \sigma_a).$$

Transitive: if $(\mu_a, \sigma_a) = (\mu_b, \sigma_b)$ and $(\mu_b, \sigma_b) = (\mu_c, \sigma_c)$, then

$$(\mu_a, \sigma_a) = (\mu_c, \sigma_c).$$

3.5.2 Commutative Law: Addition and Multiplication

Addition:

$$\mu_a + \mu_b = \mu_b + \mu_a; \qquad \sqrt{\sigma_a^2 + \sigma_b^2} = \sqrt{\sigma_b^2 + \sigma_a^2}.$$

Products:

$$\mu_a \mu_b = \mu_b \mu_a;$$

$$\mu_a^2 \sigma_b^2 + \mu_b^2 \sigma_a^2 + \sigma_a^2 \sigma_b^2 = \mu_b^2 \sigma_a^2 + \mu_a^2 \sigma_b^2 + \sigma_b^2 \sigma_a^2.$$

3.5.3 Associative Law

The associative law for addition holds that

$$(\mu_x + \mu_y) + \mu_z = \mu_x + (\mu_y + \mu_z),$$

and that

$$(\sigma_x^2 + \sigma_y^2) + \sigma_z^2 = \sigma_x^2 + (\sigma_y^2 + \sigma_z^2).$$

The associative law for multiplication is

$$(\mu_x \mu_y)\mu_z = \mu_x(\mu_y \mu_z),$$
$$(\mu_{xy})\mu_z = \mu_x(\mu_{yz}).$$

Expanding the left side expression for variance,

$$\mu_{xy}^2 \sigma_z^2 + \mu_z^2 \sigma_{xy}^2 + \sigma_{xy}^2 \sigma_z^2$$

$$= \mu_x^2 \mu_y^2 \sigma_z^2 + \mu_z^2[\mu_x^2 \sigma_y^2 + \mu_y^2 \sigma_x^2 + \sigma_x^2 \sigma_y^2] + \cdots$$
$$+ \sigma_z^2[\mu_x^2 \sigma_y^2 + \mu_y^2 \sigma_x^2 + \sigma_x^2 \sigma_y^2]$$

$$= \mu_x^2 \mu_y^2 \sigma_z^2 + \mu_x^2 \mu_z^2 \sigma_y^2 + \mu_y^2 \mu_z^2 \sigma_x^2 + \mu_z^2 \sigma_x^2 \sigma_y^2 + \cdots$$
$$+ \mu_x^2 \sigma_y^2 \sigma_z^2 + \mu_y^2 \sigma_x^2 \sigma_z^2 + \sigma_x^2 \sigma_y^2 \sigma_z^2.$$

Now expanding the right side expression for variance,

$$\mu_x^2 \sigma_{yz}^2 + \mu_{yz}^2 \sigma_x^2 + \sigma_x^2 \sigma_{yz}^2$$

$$= \mu_x^2(\mu_y^2 \sigma_z^2 + \mu_z^2 \sigma_y^2 + \sigma_y^2 \sigma_z^2) + \mu_y^2 \mu_z^2 \sigma_x^2$$
$$+ \sigma_x^2 (\mu_y^2 \sigma_z^2 + \mu_z^2 \sigma_y^2 + \sigma_y^2 \sigma_z^2)$$

$$= \mu_x^2 \mu_y^2 \sigma_z^2 + \mu_x^2 \mu_z^2 \sigma_y^2 + \mu_x^2 \sigma_y^2 \sigma_z^2 + \mu_y^2 \mu_z^2 \sigma_x^2 + \cdots$$
$$+ \mu_y^2 \sigma_x^2 \sigma_z^2 + \mu_z^2 \sigma_x^2 \sigma_y^2 + \sigma_x^2 \sigma_y^2 \sigma_z^2.$$

Comparing the two expansions term by term shows the identity.

3.5.4 Existence of the Zero [(0, 0) is in the set S]

Addition of and multiplication by zero or the zero random variable $(0, 0)$ are valid. Applying Eqs. 3.1 and 3.2, it is sufficient to show that adding $(0, 0)$ to (μ_x, σ_x) leaves (μ_x, σ_x) unchanged. Thus

$$\mu_x + 0 = \mu_x \quad \text{and} \quad \sigma_x^2 + 0 = \sigma_x^2.$$

Applying Eqs. 3.9 and 3.10, it is sufficient to show that the product of (μ_x, σ_x) and $(0, 0)$ equals $(0, 0)$. Thus

$$\mu_x(0) = 0,$$

$$\sigma_x^2(0) + (0)\mu_x^2 + \sigma_x^2(0) = 0.$$

3.5.5 Existence of the Unity [(1, 0) is in the set S]

It is sufficient to show that the product of (μ_x, σ_x) and $(1, 0)$ equals (μ_x, σ_x). Thus

$$\mu_x(1) = \mu_x.$$

Applying Eq. 3.10,

$$\mu_x^2(0) + (1)\sigma_x^2 + \sigma_x^2(0) = \sigma_x^2.$$

3.5.6 Closure

Given any two normal variates (μ_x, σ_x) and (μ_y, σ_y), from the set S, we may associate with them other elements, such as $(\mu_{x+y}, \sigma_{x+y})$ and (μ_{xy}, σ_{xy}) of the set S. A set S is closed under a binary operation, and the operation is uniquely defined over the set S if, for all elements (in this case normal variates) X and Y in S, the element $(X + Y)$ or (XY) is a unique element in S.†

The set S is closed under addition (proved in Example 2, Section 2.4.8).

Under the operation of multiplication, closure holds approximately, provided the variances are moderate.

Let $G = g(x_1, x_2, \ldots, x_n)$ be any function of normal independent variates $(x_1, x_2, \ldots, x_n)$. G (approximately) follows the normal law if the variances of random variables $X_1, X_2, \ldots, X_n$ are moderate. The dispersion in G results from the dispersion in the variates X_i according to the following relation:

$$G + \Delta G = g(x_1 + \Delta x_1, x_2 + \Delta x_2, \ldots, x_n + \Delta x_n). \quad (a)$$

Expanding the right side of Eq. a by Taylor's theorem,

$$G + \Delta G = g(x_1, x_2, \ldots, x_n) + \cdots$$

$$+ \frac{\partial g}{\partial x_1} \Delta x_1 + \frac{\partial g}{\partial x_2} \Delta x_2 + \cdots + \frac{\partial g}{\partial x_n} \cdot \Delta x_n. \quad (b)$$

† An example of the absence of closure occurs in vector algebra. The dot product of two vectors is a scalar.

From Eq. b subtract $G = g(x_1, x_2, \ldots, x_n)$. Then

$$\Delta G = \frac{\partial g}{\partial x_1} \Delta x_1 + \frac{\partial g}{\partial x_2} \cdot \Delta x_2 + \cdots + \frac{\partial g}{\partial x_n} \Delta x_n. \tag{c}$$

ΔG is a linear relationship in $\Delta x_1, \Delta x_2, \ldots, \Delta x_n$; hence by Eq. c and a simple development (see Eq. 2.61)

$$\sigma_G^2 \approx \left(\frac{\partial g}{\partial x_1}\right)^2 \cdot \sigma_{x_1}^2 + \left(\frac{\partial g}{\partial x_2}\right)^2 \cdot \sigma_{x_2}^2 + \cdots + \left(\frac{\partial g}{\partial x_n}\right)^2 \cdot \sigma_{x_n}^2.$$

3.5.7 Distributive Law

The distributive law implies the following relation:

$$A(B + C) = AB + AC.$$

Validity of the distributive law is shown by both partial derivative methods and moment generating functions (trivariate). For demonstration of the validity of the distributive law in the set S, see the example in Section 2.7.4.

$$\sigma_{a(b+c)}^2 = \sigma_{ab+ac}^2 = \mu_b^2 \sigma_a^2 + 2\mu_b \mu_c \sigma_a^2$$
$$+ \mu_c^2 \sigma_a^2 + \mu_a^2 \sigma_b^2 + \mu_a^2 \sigma_c^2 + \sigma_a^2 \sigma_b^2 + \sigma_a^2 \sigma_c^2.$$

The mean values are trivially equal, for the reasons given in Section 3.5.

3.6 MATHEMATICAL STRUCTURE

3.6.1 Field

A field F is an integral domain† which contains for each element $a \neq 0$ an inverse element a^{-1} satisfying the equation

$$aa^{-1} = 1.$$

By the definition above, it is clear that within the constraints established (see Section 3.5), the set S behaves as a number field.‡

---

† See *A Survey of Modern Algebra*. G. Birkhoff and S. MacLane, The Macmillan Company, New York, 1959.
‡ The set S of normally distributed random variables may be considered a two parameter analogue of the real (decimal) numbers. As with the complex numbers, the real numbers may be regarded, in a sense, as being imbedded in the set S.

3.6.2 Vector Space

A vector space V over a field F (the real numbers) is a set of elements, called vectors, such that any two elements x and y of V determine a unique vector $x + y$, and that any vector x from V and any scalar c from F determine a scalar product cx in V, with the following properties.

1. V is an Abelian group under addition.
2. (a) $c(x + y) = cx + cy$,
   (b) $(c + c')x = cx + c'x$.
3. (a) $(c \cdot c')x = c(c'x)$,
   (b) $1 \cdot x = x$.

The set S and the field R of real numbers satisfies all conditions (see Section 3.6.3) above, and therefore constitute a vector space.

3.6.3 Abelian Group

A group G is a system of elements that is closed under a single valued binary operation which is associative, and relative to which G contains an element satisfying the identity law, and with each element another element (inverse) satisfying the inverse law.

Thus the set S amounts to a two-dimensional vector space. The same set is also an additive and a multiplicative Abelian group.

3.6.4 Limit of Convergence

As the dispersion in each of the set of normal random variables converges to zero as a limit, the set S converges to the (decimal) real numbers.

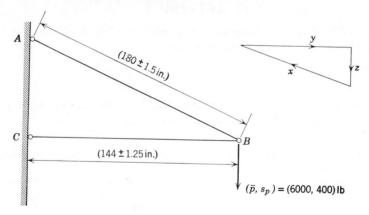

Example. 1 Application of the algebra of normal functions. Ignoring deflection and elongation and assuming that all random variables are normal, compute the cross-section area variates of the wooden member BC and the steel rod AB of the structure ABC, loaded as shown.

The allowable stress for wood is

$$(\bar{f}_w, s_{f_w}) = (160, 12) \text{ psi}$$

and the allowable stress for steel is

$$(\bar{f}_s, s_{f_s}) = (10{,}000; 600) \text{ psi}.$$

Express areas, also, in terms of mean values and tolerances, where tolerance $\Delta = 3\sigma$.

Solution. z must be determined, $z^2 = x^2 - y^2$. X and Y are statistically independent. The right triangle relationships are maintained. Thus it is assumed that

$$\frac{1}{\sin \theta} = \frac{x}{z} = \frac{15}{9} = \frac{5}{3},$$

$$\frac{1}{\tan \theta} = \frac{y}{z} = \frac{12}{9} = \frac{4}{3},$$

$$P_{\text{steel}} = 1.66(6000; 400) = (10{,}000; 667) \text{ lb},$$

$$P_{\text{wood}} = 1.33(6000; 400) = (8000; 533) \text{ lb},$$

$$\bar{A} = \frac{\text{applied load}}{\text{allowable stress}} = \frac{\bar{P} \text{ lb}}{\bar{f} \text{ psi}}.$$

The mean area for steel member $\bar{A}_s$ (Eq. 3.13) is

$$\bar{A}_s = \frac{10{,}000}{10{,}000} = 1.00 \text{ in.}^2,$$

and, applying Eq. 3.14,

$$s_{A_s} = \frac{1}{10^4} \left[\frac{(10^4)^2(600)^2 + (10^4)^2(667)^2}{(10^4)^2 + (600)^2} \right]^{1/2}$$

$$= \frac{1}{10^4} \sqrt{80.5 \cdot 10^4} = 8.96 \cdot 10^{-2}.$$

Thus

$$(\bar{A}_s, s_{A_s}) = (1.00, 0.0896) \text{ in.}^2 = 1.00 \pm 0.269 \text{ in.}^2$$

The mean area of the wooden element

$$\bar{A}_w = 8000/160 = 50.00 \text{ in.}^2$$

The standard deviation s_{A_w} is

$$s_{A_w} = \frac{1}{160} \left[\frac{(8000)^2(12)^2 + (160)^2(533)^2}{(160)^2 + (12)^2} \right]^{1/2}$$

$$= \frac{1}{160} \sqrt{6.40 \cdot 10^5} = 5.0 \text{ in.}^2$$

Thus

$$(\bar{A}_w, s_{A_w}) = (50.00, 5.0) \text{ in.}^2 = 50 \pm 15 \text{ in.}^2$$

PROBLEMS

1. The standard deviation estimates of an attach fitting reamed hole and the outside diameter of the joint pin are both 0.0015 in. The mean center misalignment is $L = 0.0028$ in. Determine the probability of interference if each dimension of the assembled components is normally distributed.
2. Two heat exchanges (H_1 and H_2) are mounted in a duct through which air for space heating is passed. If the mean output capacity of each heat exchanger is $\bar{H}_1 = \bar{H}_2 = 5500$ Btu/hr and the standard deviations are $S_{H_1} = 500$ Btu/hr and $s_{H_2} = 700$ Btu/hr, what is the probability that the combined output will exceed 12,750 Btu/hr?
3. In Problem 2, assuming equal standard deviations, what must this standard deviation be to assure that the probability of a combined heat output exceeding 12,000 Btu/hr will be 0.0075?
4. An airplane is flying in the direction from A to C. Its shadow is observed to pass point A, then reach point C 0.9 ± 0.07 sec. later. The distance A to C is not directly measurable; however, BC is determined to be 350 ± 20 feet and AB as 275 ± 20 ft. Assuming the tolerances on distance and time equal to three standard deviations, determine

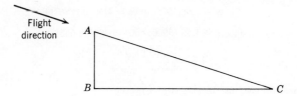

the mean velocity and its standard deviation. What value of velocity will be exceeded only 0.15 percent of the time?
5. Consider a diode with mean useful life $\bar{T} = 245$ hr. Specifications on a certain contract require useful life of from 210 hr to 280 hr with probability 0.99. What is the maximum standard deviation allowable in the manufacturing process?
6. In a circular disc of radius 4.00 ± 0.125 in. and thickness 0.7500 ± 0.0625 in., two holes are drilled. The first hole has radius 1.250 ± 0.025 and that of the second hole 2.250 ± 0.035. The density of the material of the disc is 0.305 lb-in.^{3} What is the mean weight and standard deviation of the discs? Assume all tolerances are three sigma values.
7. Complete the derivation of Eqs. 3.6 and 3.7.
8. Given normally distributed random variables X and Y, with $X = Y$, what is the error in the standard deviation if the assumption of independence is made when in reality the variates are correlated with $r = +1$ (a) for the sum and (b) for the difference?
9. Complete the derivations of Eq. 3.9 and 3.10.
10. Given the same conditions as in Problem 2, what is the error in the standard deviation of the product XY if statistical independence is assumed, whereas X and Y are correlated with $r = +1$?
11. In the example of Section 3.3.3, what would the error have been if, in computing the answer, the numerator and denominator in the expression $R_T = R_1 R_2 / R_1 + R_2$ had been treated as statistically independent?

Algebra of Normal Functions   143

12. Compute the mean and standard deviation of R_T if

$$(\bar{R}_1, S_{R_1}) = (10{,}000;\ 300)\ \text{ohms},$$
$$(\bar{R}_2, S_{R_2}) = (20{,}000;\ 600)\ \text{ohms},$$
$$(\bar{R}_3, S_{R_3}) = (15{,}000;\ 450)\ \text{ohms}.$$

Also derive the expression for S_{R_T}.

13. Why does the axiom of "uniqueness" hold in the algebra of normal functions? Demonstrate this.
14. Why does the "cancellation" hold in the algebra of normal functions? Demonstrate this.
15. Derive the expression for the standard deviation of the random variable $Z = X^3$, assuming that X is normally distributed.
16. Derive the algebraic expression for the correlation coefficient of $Z = XY/X$ where X and Y are statistically independent normally distributed random variables.
17. Ignoring, for this problem, the variation on lengths, determine the necessary cross-section area of the steel tension rod BC, if the allowable working stress $(\bar{f}_t, s_{f_t}) = 15{,}000;\ 900)$ lb-in.² The uniformly distributed load is $(\bar{q}, s_q) = (1000, 70)$ lb-ft.

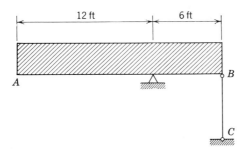

18. Axial elongation of a rod is discussed in Section 8.3. Experiments show that axial elongation is accompanied by lateral contraction of a rod or other section. Further, the ratio is (conventionally) considered constant for a given material, within the elastic limit (see Ref. 19). Thus

$$\frac{\text{unit lateral contraction}}{\text{unit axial elongation}} = \mu$$

μ is called Poisson's ratio. Given that unit elongation ϵ_x is given by

$$\epsilon_x = \frac{f_x}{E} - \mu_0 \left(\frac{f_y}{E}\right),$$

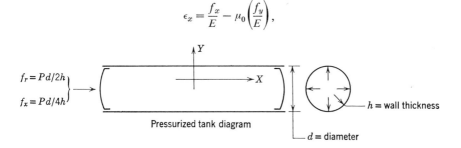

Pressurized tank diagram

determine Poisson's ratio variate, given that

$$(f_y, s_{f_y}) = (600, 500) \text{ psi},$$
$$(f_x, s_{f_x}) = (3000, 250) \text{ psi},$$

$\epsilon_x = 17 \cdot 10^{-5}$ in. with coefficient of variation $= \dfrac{s_{\epsilon_x}}{\bar{\epsilon}_x} = 5.5\%$.

19. **Coefficient of thermal expansion.** The rails of a railway are welded together at $(50.0°, 2.5°)$F. When heated to $(100°, 3°)$F, the stress induced by the temperature change is

$$(f_c, s_{f_c}) = (10{,}500;\ 650) \text{ psi},$$
$$(E, S_E) = (30 \cdot 10^6,\ 1.05 \cdot 10^6) \text{ psi}.$$

Given the relation

$$\delta(t - t_0) = \frac{f_c}{E},$$

where δ is the coefficient of thermal expansion in in. per in. per degree F, determine $(\bar{\delta}, s_\delta)$. See Problem 1, Section 8. *Answer:* $(7.0 \cdot 10^{-6},\ 0.70 \cdot 10^{-6})$.

CHAPTER 4

Determination of Reliability

Many random variables encountered in the physical sciences appear to be normally distributed and the normal distribution gives an adequate approximation to the distribution of many other measurable random variables. "Thus, if a complete theory of statistical inference is developed based on the normal distribution alone, then one has in reality a system which may be employed quite generally, because other distributions can be transformed to or approximated by the normal form" [18], pp. 142–143.

In developing formulas for design problem relationships, we first consider applied stress and allowable stress† of unspecified distribution.

4.1 GENERALLY DISTRIBUTED ALLOWABLE AND APPLIED STRESSES† [2], [3], [11], [18], [19]

The adequacy (reliability) of a mechanical, electrical, thermal, or structural component (or system) is determined from the basic concept that a no-failure probability exists when *allowable stress* (S) is not exceeded by *applied stress* (s), or allowable stress > applied stress. The probability of a stress of value s_1 is equal to the area of the element ds or to A_1 (Fig. 4.1), or where

$$\text{allowable stress} = f(S) = f(X_1, X_2, \ldots, X_n),$$
$$\text{applied stress} = f(s) = g(Y_1, Y_2, \ldots, Y_r),$$
$$p\left(s_1 - \frac{ds}{2} \leq s \leq s_1 + \frac{ds}{2}\right) = f(s_1)\,ds = A_1.$$

Since these are density functions, the probability that $s > s_1$ equals to the shaded area under the allowable stress density curve A_2,

$$p(S > s_1) = \int_{s_1}^{\infty} f(S)\,dS = A_2.$$

† Sometimes referred to as strength and load functions.

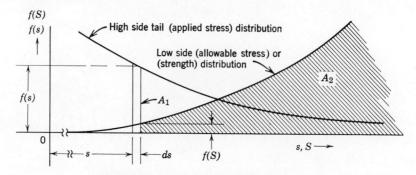

Figure 4.1 Allowable versus applied stress [11].

The reliability R (i.e., the probability of no failure at s_1) is the product of these two probabilities (Eq. 2.14):

$$p\left(s_1 - \frac{ds}{2} \leq s \leq s_1 + \frac{ds}{2}\right) \cdot P(S > s_1)$$

and

$$dR = f(s_1)\, ds \cdot \int_{s_1}^{\infty} f(S)\, dS.$$

Reliability of the component or assembly or device is the probability of strength or allowable stress being greater than the possible values (over the range) of load or applied stress. Thus the basic equation

$$R = \int dR = \int_{-\infty}^{\infty} f(s) \left[\int_{s}^{\infty} f(S)\, dS\right] ds. \qquad (4.1)$$

As already noted, $f(S)$ and $f(s)$ are density functions (Section 2.3.1), thus

$$\int_{-\infty}^{\infty} f(S)\, dS = 1,$$

$$\int_{-\infty}^{\infty} f(s)\, ds = 1.$$

Alternatively, an expression for reliability R may be obtained by considering that a no-failure probability exists when applied stress remains less than some given value of allowable stress. Following the same reasoning as above, the probability of an allowable stress S_1 in an interval dS_1 is

$$p\left(S_1 - \frac{dS_1}{2} \leq S \leq S_1 + \frac{dS_1}{2}\right) = f(S_1)\, dS_1 = A_1'.$$

Determination of Reliability 147

The probability of applied stress less than S_1 is

$$p(s < S_1) = \int_{-\infty}^{S} f(s)\, ds = A'_2.$$

Thus the no-failure probability is the product (Eq. 2.14)

$$(A'_1)(A'_2) = p\left(S_1 - \frac{dS_1}{2} \leq S \leq S_1 + \frac{dS_1}{2}\right) p(s < S_1)$$

and

$$dR = f(S_1)\, dS_1 \int_{-\infty}^{S} f(s)\, ds.$$

The component (or other) reliability is the probability of no-failure for all possible values of allowable stress (see Fig. 4.2),

$$R = \int dR = \int_{-\infty}^{\infty} f(S) \left[\int_{-\infty}^{S} f(s)\, ds \right] dS. \tag{4.2}$$

Probability of failure or of component unreliability which corresponds to the reliabilities given above may be obtained by repeating the previous probability analysis, employing the fact that a failure probability exists when strength is less than a given stress or when stress exceeds a given strength. Thus

$$Q = \int_{-\infty}^{\infty} f(s) \left[\int_{-\infty}^{s} f(S)\, dS \right] ds,$$

or

$$Q = \int_{-\infty}^{\infty} f(S) \left[\int_{S}^{\infty} f(s)\, ds \right] dS, \tag{4.3}$$

$R + Q = 1$ (see Eq. 2.4).

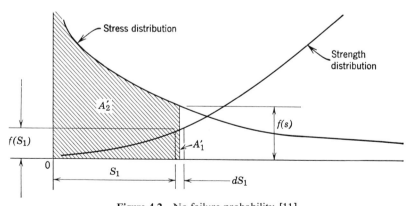

Figure 4.2 No failure probability [11].

4.2 DETERMINATION OF RELIABILITY WHEN STRENGTH AND STRESS DISTRIBUTIONS ARE NORMAL [2], [3], [8], [11]

Exact analytic solutions of Eqs. 4.1 and 4.2 for R for distributions other than normal are not known. The exact solution of these equations, for normally distributed allowable and applied stress, is given in Sections 4.2.1 and 4.2.2. Approximate solutions for combinations of any two distributions are developed in later sections.

4.2.1 Derivation by Difference Function [2], [11]

If s and S are normal random variables, then the density functions are (Eq. 2.28)

$$f(s) = \frac{1}{\sigma_s \sqrt{2\pi}} \exp\left[-\frac{1}{2}\left(\frac{s - \bar{s}}{\sigma_s}\right)^2\right] \quad (-\infty, \infty),\dagger$$

and

$$f(S) = \frac{1}{\sigma_S \sqrt{2\pi}} \exp\left[-\frac{1}{2}\left(\frac{S - \bar{S}}{\sigma_S}\right)^2\right] \quad (-\infty, \infty).\dagger$$

Reliability is the probability that strength exceeds stress, or

$$S - s > 0,$$

$$\zeta > 0,$$

if

$$\zeta = S - s.$$

$f(\zeta)$ is defined as the difference density (Section 3.1.2) of $f(s)$ and $f(S)$. Since $f(s)$ and $f(S)$ are normally distributed, $f(\zeta)$ is normally distributed‡ with density function

$$f(\zeta) = \frac{1}{\sigma_\zeta \sqrt{2\pi}} \exp\left[-\frac{1}{2}\left(\frac{\zeta - \bar{\zeta}}{\sigma_\zeta}\right)^2\right] \quad (-\infty, \infty). \quad (4.4)$$

By Eq. 3.6,

$$\bar{\zeta} = \bar{S} - \bar{s},$$

and by Eq. 3.7

$$\sigma_\zeta = \sqrt{\sigma_S^2 + \sigma_s^2}.\S$$

† See example, Section 2.3.6.
‡ See Section 4.2.2.
§ σ_S and σ_s are used here although these are actually standard derivation estimators s_S and s_s.

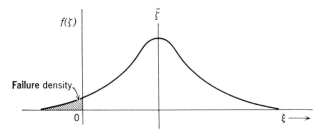

Figure 4.3 Difference distribution: strength minus stress.

As shown (Fig. 4.3), reliability is given by the positive density of ζ. Thus

$$R = p(\zeta > 0) = \int_0^\infty f(\zeta)\, d\zeta$$

$$= \frac{1}{\sigma_\xi \sqrt{2\pi_0}} \int_0^\infty \exp\left[-\frac{1}{2}\left(\frac{\zeta - \bar{\zeta}}{\sigma}\right)^2\right] d\zeta \tag{4.5}$$

The relationship between the distribution of ζ and the standardized normal distribution, tabulated in standard tables of normal functions,† is utilized to evaluate the integral (above). The transformation which relates ζ and the the standard-normalized variable z is

$$z = \frac{\zeta - \bar{\zeta}}{\sigma_\zeta}$$

The new limits of the integrand above, with the change of variable from ζ to z, are as follows: when $\zeta = \leq$,

$$z = \frac{\infty - \bar{\zeta}}{\sigma_\zeta} = \infty;$$

when $\zeta = 0$,

$$z = \frac{0 - \bar{\zeta}}{\sigma_\zeta} = -\left(\frac{\bar{\zeta}}{\sigma_\zeta}\right).$$

Since

$$z = \frac{\zeta - \bar{\zeta}}{\sigma_\zeta}, \quad d\zeta = \sigma_\zeta\, dz.$$

Substituting in Eq. 4.5

$$R = \frac{1}{\sqrt{2\pi}} \int_{-\bar{\zeta}/\sigma_\zeta}^\infty e^{-(z^2/2)}\, dz.$$

† See Table 2.4 and [21].

Thus the lower limit on the integral amounts to

$$z = -\frac{\bar{\zeta}}{\sigma_\zeta} = -\frac{\bar{S} - \bar{s}}{\sqrt{\sigma_s^2 + \sigma_S^2}}$$

or

$$z = \frac{|\bar{S} - \bar{s}|}{\sqrt{s_s^2 + s_S^2}}. \tag{4.6}$$

Equation 4.6 is called a "coupling equation" since it probabilistically relates applied and allowable stresses. For values of z see Table 2.4.

As an example in the application of the results above, consider the normal allowable stress random variable

$$(\bar{S}, s_S) = (27{,}000;\ 4000)\text{psi},$$

and the normal applied stress random variable

$$(\bar{s}, s_s) = (13{,}000;\ 3000)\ \text{psi},$$

$$z = \frac{|27{,}000 - 13{,}000|}{\sqrt{(3000)^2 + (4000)^2}} = \frac{14{,}000}{5000} = 2.8,$$

and

$$R = \int_{\zeta/s}^{\infty} e^{-(z^2/2)}\ dz = \int_{-2.8}^{\infty} e^{-(z^2/2)}\ dz.$$

From Table 2.4, $R = 0.9974$. The failure density or unreliability Q may be obtained by considering the density of $\zeta < 0$, which is

$$Q = \int_{-\infty}^{0} f(\zeta)\ d\zeta,$$

where

$$Q = 1 - R = 1 - \int_{0}^{\infty} f(\zeta)\ d\zeta = 1 - \int_{-\bar{\zeta}/s_\zeta}^{\infty} e^{-(z^2/2)}\ dz.$$

Since both methods are equivalent, they yield the following expression:

$$Q = \frac{1}{\sqrt{2\pi}} \int_{-\infty}^{-\bar{\zeta}/s_\zeta} e^{-(z^2/2)}\ dz.$$

Example. An examination of Table 1.1 shows the result of changes in the various parameters.

PROBLEM. From the difference function $Z = X - Y$ of two normally distributed random variables (statistically independent), the "coupling formula" may be written as follows.

$$z = \frac{\mu_x - \mu_y}{\sqrt{\sigma_x^2 + \sigma_y^2}}.$$

Determination of Reliability   151

Given a load random variable (μ_y, σ_y) or $(\bar{y}, s_y)$ resulting from a dead load component (900 lb) and $(\bar{y}_a, s_a) = (2700, 320)$ lb acting downward and $(\bar{y}_b, s_b) = (1750, 140)$ lb acting upward. The mean strength of the resisting component is $\bar{x} = 3000$ lb. With specified $0.99 = R$, what is the permissible standard deviation s_x? What is the $\pm$ tolerance on strength if no more than 5 percent of the population of components is to fall outside tolerance?

4.2.2   Derivation by Convolutions

Because of its later importance and its example value, the "coupling formula" is derived by the method of convolutions.†

Let X be the applied stress random variable with $x = f(x_1, x_2, \ldots, x_n)$, with density $f(x)$ described by the couple (μ_y, σ_x).

Let Y be the allowable stress random variable with $y = f(y_1, y_2, \ldots, y_r)$, with density $f(y)$ described by the couple (μ_y, σ_y).

$$Z = X - Y.$$

Failure occurs when
$$z > 0,$$

or when applied stress exceeds allowable stress.

The probability density function of z, $g(z)$ is computed. With $g(z)$, the probability of $z > 0$ is computed

$$\int_0^\infty g(z)\, dz.$$

The densities $f(x)$ and $f(y)$ are pictured in Fig. 4.4.

In this derivation, allowable stress is assumed to be statistically independent of applied stress, thus the joint density function of X and Y is (by Eq. 2.31):

$$f(x)f(y)\, dx\, dy.$$

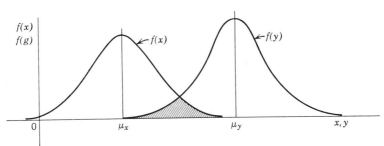

Figure 4.4   Applied stress versus allowable stress; the Warner diagram (after W. K. Warner, a pioneer in the discipline of reliability engineering).

† See Section 2.8.4.

Since interest is in $Z = X - Y$, the joint density function is transformed:

$$f(x)f(x - z) \, dx \, dz.\dagger$$

In this form, the convolution is recognized. The density of Z denoted by $g(z)$ is (in differential form):

$$g(z) \, dz = \int_a^b f(x)f(x - z) \, dx \, dz$$

where a and b give the smallest and largest stresses.

The densities of X and Y are

$$f(x) = \frac{1}{\sigma_x\sqrt{2\pi}} \exp\left[-\frac{1}{2}\left(\frac{x - \mu_x}{\sigma_x}\right)^2\right] \quad (-\infty, \infty),$$

$$f(y) = \frac{1}{\sigma_y\sqrt{2\pi}} \exp\left[-\frac{1}{2}\left(\frac{y - \mu_y}{\sigma_y}\right)^2\right] \quad (-\infty, \infty)$$

where X, Y and Z have finite moments. Thus

$$g(z) = \int_{-\infty}^{\infty} \frac{1}{2\pi\sigma_x\sigma_z} \exp\left[-\frac{(x - \mu_x)^2}{2\sigma_x^2} - \frac{(x - z - \mu_y)^2}{2\sigma_y^2}\right] dx.$$

Now consider the exponent in brackets. Multiplying numerator and denominator by $\sigma_x^2 + \sigma_z^2$

$$= \frac{\sigma_x^2 + \sigma_y^2}{2\sigma_x^2 2\sigma_y^2}\left[\frac{x^2 - 2x(\mu_x\sigma_z^2 + z\sigma_x^2 + \mu_y\sigma_x^2)}{\sigma_x^2 + \sigma_y^2}\right.$$

$$\left. + \mu_x^2\sigma_y^2 + \frac{(z^2 + \mu_y^2 + 2z\mu_z)\cdot\sigma_x^2}{\sigma_x^2 + \sigma_y^2}\right].$$

Multiplying the numerator and denominator of the last term by $\sigma_x^2 + \sigma_y^2$ yields

$$-\frac{\sigma_x^2 + \sigma_y^2}{2\sigma_x^2\sigma_y^2}\left\{\frac{x^2 - 2x[\mu_x\sigma_y^2 + (z + \mu_y)\sigma_x^2]}{\sigma_x^2 + \sigma_y^2}\right.$$

$$\left. + \frac{\mu_x^2\sigma_x^2\sigma_y^2 + (z^2 + 2z\mu_y + \mu_y^2)\sigma_x^4 + \mu_x^2\sigma_y^4 + (z^2 + 2z\mu_y + \mu_x^2)\sigma_x^2\sigma_y^2}{(\sigma_x^2 + \sigma_y^2)^2}\right\}.$$

Separating the two middle terms of the last fraction from the other two terms and adding and subtracting

$$\frac{2\mu_x(z + \mu_y)\sigma_x^2\sigma_y^2}{(\sigma_x^2 + \sigma_y^2)^2}$$

† See [13].

Determination of Reliability

gives

$$-\frac{\sigma_x^2 + \sigma_y^2}{2\sigma_x^2\sigma_y^2}\left[x^2 - 2x\frac{\mu_x\sigma_y^2 + (z+\mu_y)\sigma_x^2}{\sigma_x^2 + \sigma_y^2}\right.$$

$$+ \frac{\mu_x^2\sigma_y^4 + 2\mu_x(z+\mu_y)\sigma_x^2\sigma_y^2 + (z+\mu_y)^2\sigma_x^4}{(\sigma_x^2 + \sigma_y^2)^2}$$

$$\left.+ \frac{\mu_x^2 - 2\mu_x(z+\mu_y) + (z+\mu_y)^2}{(\sigma_x^2 + \sigma_y^2)^2}\sigma_x^2\sigma_y^2\right].$$

Separating the last term from the first three and multiplying by the factor

$$-\frac{\sigma_x^2 + \sigma_y^2}{2\sigma_x^2\sigma_y^2}$$

yields

$$-\frac{\sigma_x^2 + \sigma_y^2}{2\sigma_x^2\sigma_y^2}\left(x - \frac{\mu_x\sigma_y^2 + z + \mu_y)\sigma_x^2}{\sigma_x^2 + \sigma_y^2}\right)^2 - \frac{(z+\mu_y - \mu_x)^2}{2(\sigma_x^2 + \sigma_y^2)}.$$

Substituting into the expression for $g(z)$,

$$g(z) = \frac{1}{\sqrt{2\pi(\sigma_x^2 + \sigma_y^2)}}\exp\left\{-\frac{\sigma_x^2 + \sigma_y^2}{2\sigma_x^2 y^2}\left[x - \frac{\mu_x\sigma_y^2 + (z+\mu_y)\sigma_x^2}{\sigma_x^2 + \sigma_y^2}\right]\right.$$

$$\left.- \frac{(z+\mu_y - \mu_x)^2}{2(\sigma_x^2 + \sigma_y^2)}\right\}dx,$$

$$g(z) = \frac{1}{\sqrt{2\pi(\sigma_x^2 + \sigma_y^2)}}\exp\left[-\frac{(z+\mu_y - \mu_x)^2}{2(\sigma_x^2 + \sigma_y^2)}\right]K,$$

with K set equal to

$$K = \int_{-\infty}^{\infty} \frac{\sqrt{\sigma_x^2 + \sigma_y^2}}{\sqrt{2\pi\sigma_x\sigma_z}} \exp\left\{-\frac{\sigma_x^2 + \sigma_y^2}{2\sigma_x^2\sigma_z^2}\left[x - \frac{\mu_x\sigma_y^2 + (z+\mu_y)\sigma_x^2}{\sigma_x^2 + \sigma_y^2}\right]^2\right\}dx.$$

From the following substitution, $K = 1$,

$$t = \frac{\sqrt{\sigma_x^2 + \sigma_y^2}}{\sigma_x\sigma_y}\left[x - \frac{\mu_x\sigma_y^2 + (z+\mu_y)\sigma_x^2}{\sigma_x^2 + \sigma_y^2}\right],$$

$$dt = \frac{\sqrt{\sigma_x^2 + \sigma_y^2}}{\sigma_x\sigma_y}dx,$$

and

$$K = \int_{-\infty}^{\infty} \frac{1}{2\pi} e^{-(t^2/2)}\, dt = 1.$$

Thus

$$g(z) = \frac{1}{\sqrt{2\pi(\sigma_x^2 + \sigma_y^2)}} \exp\left[-\frac{1}{2}\frac{(z + \mu_y - \mu_x)^2}{(\sigma_x^2 + \sigma_y^2)}\right].$$

Thus Z is a normally distributed random variable, with moments

$$\mu_z = \mu_x - \mu_y,$$
$$\sigma_z = \sqrt{\sigma_x^2 + \sigma_y^2}.$$

This result demonstrates that the difference of two normally distributed random variables is closed (i.e., $Z = X - Y$ is a normally distributed random variable).

The probability that load exceeds strength, Q is given by

$$\int_0^\infty g(z)\, dz.$$

Now let

$$t = \frac{z + \mu_y - \mu_x}{\sqrt{\sigma_x^2 + \sigma_y^2}};$$

then

$$dt = \frac{dz}{\sqrt{\sigma_x^2 + \sigma_y^2}}$$

and the integral becomes

$$Q = \int_0^\infty g(z)\, dz = \int_{\frac{\mu_y - \mu_x}{\sqrt{\sigma_x^2 + \sigma_y^2}}}^\infty \frac{1}{2\pi} e^{-(t^2/2)}\, dt = 1 - R,$$

the same result as Eq. 4.5.

4.3 NONNORMAL DISTRIBUTIONS: TRANSFORM METHOD FOR DETERMINING RELIABILITY [11], [18], [19]

A technique is now presented which is a transform method for solving the general reliability expression. The reliability expression, Eq. 4.1, is

$$R = \int_{-\infty}^\infty \left[\int_s^\infty f(S)\, dS\right] f(s)\, ds.$$

This expression for R may be rewritten by setting

$$G = \int_s^\infty f(S)\, dS,$$

and

$$F = \int_s^\infty f(s)\, ds.$$

Then
$$dF = f(s)\, ds$$
and
$$R = \int_0^1 G\, dF.$$

The limits change since the variables (of integration) have changed. Where the lower and upper limits of s are $-\infty$ and ∞ and of S are s and ∞, respectively, the maximum range for either G or F is 0 to 1, by definition of the new variables.

Consideration of the expression
$$R = \int_0^1 G\, dF,$$
reveals that the reliability is the area under the curve $G = f(F)$, as shown in the upper graph of Fig. 4.5. This area may be planimetered, and its ratio to the total area by the axes and $G = 1$ and $F = 1$ computed, thus obtaining the reliability. Probability of failure in this case is obtained from
$$Q = \int_{-\infty}^{\infty}\left[\int_S^{\infty} f(s)\, ds\right] f(S)\, dS,$$
by putting
$$G(S) = \int_S^{\infty} f(S)\, dS$$
and
$$F(s) = \int_S^{\infty} f(s)\, ds;$$
then
$$Q = \int_0^1 F\, dG$$

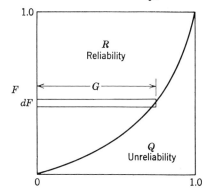
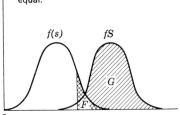

For a given value of S, G, F, except at O and 1, where they are equal.

Figure 4.5 F versus G.

where

$$dG = f(S)\, dS \text{ (see Fig. 4.6).}$$

This method permits the computation of reliability or unreliability for any distribution of stress or strength, and for any combination of two different distributions of stress and strength, provided the partial areas F and G under these densities can be found.

The accuracy of determining the reliability by this transform method depends on the accuracy of evaluating the partial areas of F and G, of plotting and measuring the area for R or Q. The desired accuracy may be obtained by evaluating the areas of F, G, and R or Q by digital computer.

A method employing numerical methods for evaluating R and Q for any distribution (using Simpson's rule) is presented in Section 5.2.1.

4.4. NORMAL CORRELATED ALLOWABLE AND APPLIED STRESS RANDOM VARIABLES

Recalling Eq. 4.6, the mean value estimate was

$$\bar{\zeta} = \bar{S} - \bar{s},$$

and the standard deviation estimation of ζ,

$$s_\zeta = \sqrt{s_S^2 + s_s^2}.$$

With S and s correlated, the standard deviation estimate of ζ is modified as follows:

$$\zeta = S - s;$$

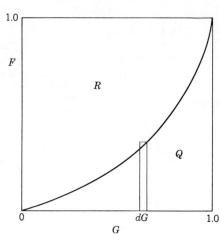

Figure 4.6 F versus G.

for example, let $S = f(x_1, x_2)$ and $s = g(x_1, x_3)$. Then, since the allowable and applied stress expressions have the random variable x_1 in common, applying Eq. 2.61

$$\sigma_3 = \left[\left(\frac{\partial \zeta}{\partial x_1}\right)^2 s_{x_1}{}^2 + \left(\frac{\partial \zeta}{\partial x_2}\right)^2 s_{x_2}{}^2 + \left(\frac{\partial \zeta}{\partial x_3}\right)^2 s_{x_3}{}^2 \right]^{1/2}.$$

Substituting values in Eq. 4.6,

$$z = \frac{\bar{S} - \bar{s}}{[(\partial \zeta/\partial x_1)^2 s_{x_1}{}^2 (\partial \zeta/\partial x_2)^2 s_{x_2}{}^2 (\partial \zeta/\partial x_3)^2 s_{x_3}{}^2]^{1/2}}. \tag{4.7}$$

CHAPTER 5

Numerical Methods[†]

In the first four sections are adequate analytic methods for solving many problems which involve the normally distributed random variable or combinations of normally distributed random variables. Furthermore, many random variables, not normally distributed, may be adequately represented by suitable normal approximations. There are, however, classes of random variables for which transformations and approximations are not adequate. It is with these random variables that this section is concerned.

Two statements [26 and 27] provide justification for approximating given functions by polynomials or by trigonometric series. These statements are the following.

1. Any function that is continuous in an interval (a, b) can be approximated in the interval, to a specified degree of accuracy, by a polynomial. It is possible to write a polynomial $H(x)$ satisfying the condition

$$|f(x) - H(x)_i| < \epsilon,$$

for each value of x in the interval (a, b). ϵ is any preassigned positive number, however small (Fig. 5.1).

2. Any continuous function (with period 2π) can be approximated by a finite trigometric series of the form

$$h(x) = a_0 + a_1 \sin x + a_2 \sin 2x + \cdots + a_n \sin nx$$
$$+ \cdots + b_1 \cos x + b_2 \cos 2x + \cdots b_n \cos nx,[‡]$$

satisfying the condition $|f(x) - h(x)| < \delta$, for all values of x in the interval specified. δ denotes any preassigned positive number, however small.

Geometrically interpreted, these statements imply that, having constructed the curves of $y = f(x)$, $y = f(x) + \epsilon$, and $y = f(x) - \epsilon$ (see Fig. 5.1), we

[†] See [29], Fourier series and Integrals.
[‡] Portions adapted from Numerical Mathematical Analysis by James B. Scarborough, the Johns Hopkins Press, Baltimore; by permission.

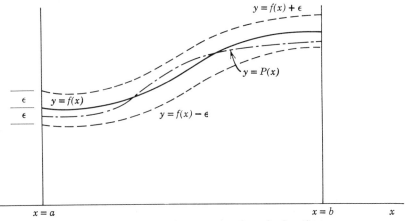

Figure 5.1 Polynomial approximation of a function.

may write a polynomial (or finite trigometric series) whose curve is everywhere bounded by $y = y(x) + \epsilon$ and $y = f(x) - \epsilon$ for values of x between a and b, whatever the value of ϵ. These statements imply that a specified function may be approximated by a polynomial (or finite trigometric series) to a specified degree of accuracy.

The problem is to write suitable polynomials for approximating any specified function over a specified interval. Newton's formula for forward interpolation [26 and 27] provides a method for finding such polynomials. Let $y = f(x)$ denote a function with values $y_0, y_1, y_2, \ldots, y_n$ corresponding to (equidistant) values) $x_0, x_1, x_2, \ldots, x_n$ of the independent variable x.†
Let $\gamma(x)$ be a polynomial of the nth degree.‡ Such a polynomial $\gamma(x)$ is written as

$$\gamma(x) = a_0 + a_1(x - x_0) + a_2(x - x_0)(x - x_1) + \cdots$$
$$\cdots + a_3(x - x_0)(x - x_1)(x - x_2) + \cdots$$
$$\cdots + a_4(x - x_0)(x - x_1)(x - x_2)(x - x_3) + \cdots + \cdots$$
$$\cdots + a_n(x - x_0)(x - x_1)(x - x_2) \times \cdots \times (x - x_{n-1}). \quad (a)$$

The coefficients $a_0, a_1, a_2, \ldots, a_n$ are determined to satisfy the conditions

$$\gamma(x_0) = y_0,$$
$$\gamma(x_1) = y_1,$$
$$\gamma(x_2) = y_2,$$
$$\gamma(x_n) = y_n.$$

† $x_{(i+1)} - x_i = \lambda$, $x_{i+2} - x_i = 2\lambda, \ldots$, etc.
‡ A source of error can arise when a sample is supposed to come from a continuous infinite distribution, see [28], Section 5.

Substituting into Eq. *a*, for $\gamma(x)$, the successive values $x_0, x_1, x_2, \ldots, x_n$ and at the same time put $\gamma(x_0) = y_0$, $\gamma(x_1) = y_1, \ldots$, etc. Next, recall that $x_1 - x_0 = \lambda$, $x_2 - x_1 = \lambda, \ldots$, etc. The coefficient values are then determined by substituting values into Eq. *a*:

$$a_0 = (y_0),$$

$$y_1 = a_0 + a_1(x_1 - x_0) = y_0 + a_1\lambda.$$

$$a_1 = \frac{y_1 - y_0}{\lambda} = \left(\frac{\Delta y_0\dagger}{\lambda}\right),$$

$$y_2 = a_0 + a_1(x_2 - x_0) + a_2(x_2 - x_0)(x_2 - x_1)$$

$$= y_0 + \frac{y_1 - y_0}{\lambda}(2\lambda) + a_2(2\lambda)\lambda.$$

$$a_2 = \frac{y_2 - 2y_1 - y_0}{2\lambda^2} = \left(\frac{\Delta^2 y_0}{2\lambda^2}\right),$$

$$y_3 = a_0 + a_1(x_3 - x_0) + a_2(x_3 - x_0)(x_3 - x_1)$$
$$+ a_3(x_3 - x_0)(x_3 - x_1)(x_3 - x_2).$$

$$= y_0 + \frac{y_1 - y_0}{\lambda}(3h) + \frac{y_2 - 2y_1 + y_0}{2\lambda^2}(3h)(2h)$$
$$+ a_3(3\lambda)(2\lambda)(\lambda).$$

$$a_3 = \frac{y_3 - 3y_2 + 3y_1 - y_0}{6\lambda^3} = \left(\frac{\Delta^3 y_0}{6\lambda^3}\right),$$

$$y_4 = a_0 + a_1(x_4 - x_0) + a_2(x_4 - x_0)(x_4 - x_1) + \cdots$$
$$+ a_3(x_4 - x_0)(x_4 - x_1)(x_4 - x_2) + \cdots$$
$$+ a_4(x_4 - x_0)(x_4 - x_1)(x_4 - x_2)(x_4 - x_3)$$

$$= y_0 + \frac{y_1 - y_0}{\lambda}(4\lambda) + \frac{y_2 - 2y_1 + y_0}{2\lambda^2}(4\lambda)(3\lambda) + \cdots$$
$$+ \frac{y_3 - 3y_2 + 3y_1 - y_0}{6\lambda^2}(4\lambda)(3\lambda)(2\lambda) + \cdots$$
$$+ a_4(4\lambda)(3\lambda)(2\lambda)(\lambda).$$

$$a_4 = \frac{y_4 - 3y_3 + 6y_2 - 4y_1 + y_0}{24\lambda^4} - \left(\frac{\Delta^4 y_0}{24\lambda^4}\right)$$

.
.
.

etc.

† Let $y_{(j+1)} - y_j = \Delta y_j$.

Substitution of these values (into Eq. a) of $a_0, a_1, a_2, \ldots, a_n$ yields

$$y(x) = y_0 + \frac{\Delta y_0}{\lambda}(x - x_0) + \frac{\Delta^2 y_0}{2\lambda^2}(x - x_0)(x - x_1)$$

$$+ \frac{\Delta^3 y_0}{6\lambda^3}(x - x_0)(x - x_1)(x - x_2) + \cdots$$

$$+ \frac{\Delta^4 y_0}{24\lambda^4}(x - x_0)(x - x_1)(x - x_2)(x - x_3) + \cdots$$

$$+ \frac{\Delta^n y_0}{n!\,\lambda^n}(x - x_0)(x - x_1)\cdots(x - x_{n-1}). \tag{b}$$

Equation b is called Newton's formula for forward interpolation (in terms of the variable x) and is simplified by a change of variable to

$$\frac{x - x_0}{\lambda} = u \quad \text{and} \quad x = x_0 + \lambda u.$$

Substitution of values in Eq. b yields [26]

$$y(x) = y(x_0 + \lambda u) = y_0 + u\,\Delta y_0 + \cdots$$

$$+ \frac{u(u-1)}{2}\Delta^2 y_0 + \frac{u(u-1)(u-2)}{6}\Delta^3 y_0 + \cdots$$

$$+ \frac{u(u-1)(u-2)(u-3)}{24}\Delta^4 y_0 + \cdots$$

$$+ \frac{u(u-1)(u-2)\cdots(u-n+1)}{n!}\Delta^n y_0 = k(u). \tag{c}$$

5.1 NUMERICAL INTEGRATION

Numerical integration is an operation for estimating the value of a definite integral from a set of numerical values of the integrand. The estimation (numerical integration) is affected by approximating the integrand by an interpolation formula and then summing between the desired limits. Thus, to estimate the value of a definite integral,

$$\int_A^B y\,dx,$$

$f(x) = y$ is replaced by an interpolation formula, such as one of those presented in Chapter 5, and then integrated between the limits A and B. Similarly, quadrature formulas may be written for the (approximate) integration of any function for which numerical values may be computed.

5.1.1 Simple Integration Formula (for Equidistant Ordinates)

In Newton's interpolation formula, Eq. c, a variable relationship $x = x_0 + \lambda u$ was utilized. From this $dx = \lambda\, du$.

Let Newton's interpolation formula (Eq. c) for $y(x) = y(x_0 + \lambda u) = k(u)$ be integrated over n equidistant intervals of width $\lambda = \Delta x$. The limits of the interval are x_0 and $x_0 + n\lambda$. It is easily shown that the limits of u are 0 and n. This yields

$$\int_{x_0}^{(x_0+n\lambda)} y\, dx = \lambda \int_0^n \left[y_0 + u\,\Delta y_0 + \frac{u(u-1)}{2}\Delta^2 y_0 + \frac{u(u-1)(u-2)}{6}\Delta^3 y_0 \right.$$

$$+ \frac{u(u-1)(u-2)(u-3)}{24}\Delta^4 y_0$$

$$+ \frac{u(u-1)(u-2)(u-3)(u-4)(u-5)}{120}\Delta^5 y_0$$

$$\left. + \frac{u(u-1)(u-2)(u-3)(u-4)(u-5)}{720}\Delta^6 y_0 + \cdots \right] du,$$

or, after integrating and evaluating at the limits

$$\int_{x_0}^{(x_0+n\lambda)} y\, dx = (\lambda)\left[n y_0 + \frac{n^2}{2}\Delta y_0 + \left(\frac{n^3}{3} - \frac{n^2}{2}\right)\left(\frac{\Delta^2 y_0}{2}\right) \right.$$

$$+ \left(\frac{n^4}{4} - n^3 + n^2\right)\left(\frac{\Delta^3 y_0}{6}\right)$$

$$+ \left(\frac{n^5}{5} - \frac{3n^4}{2} + \frac{11n^3}{3} - 3n^2\right)\left(\frac{\Delta^4 y_0}{24}\right)$$

$$+ \left(\frac{n^6}{6} - 2n^5 + \frac{35n^4}{4} - \frac{50n^3}{3} + 12n^2\right)\left(\frac{\Delta^5 y_0}{120}\right)$$

$$\left. + \left(\frac{n^7}{7} - \frac{15n^6}{6} + 17n^5 - \frac{225n^4}{4} + \frac{274n^3}{3} - 60n^2\right)\left(\frac{\Delta^6 y_0}{720}\right) \right]$$

(d)

Utilizing Eq. d, many quadrature formulas may be written by letting $n = 1, 2, 3, \ldots$, etc. See [27, p. 123] where the expressions are given for $n = 1$ through $n = 8$. Two very useful equations are derived by putting $n = 2$ and $n = 6$.

5.1.2 Simpson's Rule†

If the value $n = 2$ is substituted into Eq. d and differences above the second order ignored,‡

$$\int_{x_0}^{(x_0+2\lambda)} y\, dx = (\lambda)\left[2y_0 + 2\Delta y_0 + (\tfrac{8}{3} - 2)\frac{\Delta^2 y_0}{2}\right]$$

$$= (\lambda)[2y_0 + 2y_1 - 2y_0 + \tfrac{1}{3}(y_2 - 2y_1 + y_0)]$$

$$= \left(\frac{\lambda}{3}\right)(y_0 + 4y_1 + y_2).$$

Intervals from x_2 to $x_2 + 2\lambda$ yield

$$\int_{x_2}^{(x_2+2\lambda)} y\, dx = \left(\frac{\lambda}{3}\right)[y_2 + 4y_3 + y_4].$$

For the third pair of intervals

$$\int_{x_4}^{(x_4+2\lambda)} y\, dx = \frac{\lambda}{3}[y_4 + 4y_5 + y_6]$$

.
.
.

etc.

Summing all expressions from x_0 to x_n (with n an even number) yields

$$\int_{x_0}^{(x_0+n\lambda)} y\, dx = \left(\frac{\lambda}{3}\right)[y_0 + 4y_1 + y_2 + y_2 + 4y_3 + y_4 + y_4 + \cdots$$
$$+ 4y_5 + y_6 + y_6 + 4y_7 + \cdots].$$

Simpson's rule's

$$\int_{x_0}^{(x_0+n\lambda)} y\, dx = \left(\frac{\lambda}{3}\right)[y_0 + 4y_1 + 2y_2 + 4y_3 + 2y + \cdots$$
$$+ 2y_{n-2} + 4y_{n-1} + y_n]$$

$$= \left(\frac{\lambda}{3}\right)[y_0 + 4(y_1 + y_3 + \cdots + y_{n-1}) + \cdots$$
$$+ 2(y_2 + y_4 + \cdots + y_{n-2}) + y_n].\S \quad (e)$$

† See [27], Article 33, Eq. (1), also [26].
‡ *Note.* The interval is x_0 to $x_0 + 2\lambda$. In this specific interval are the functional values y_0, y_1, and y_2. With but three values difference above the second do not exist.
§ See [27], p. 121, and [26].

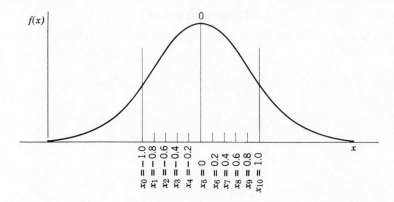

Simpson's rule is now applied in three examples. The examples are selected deliberately to provide comparisons of exact answers and approximations computed by Simpson's rule.

Example 1. Integrate the standard normal function over the interval $\pm 1\sigma$: 1. by Simpson's rule and 2. by utilizing the values found in Table 2.4.

Table 5.1 Application of Simpson's Rule, Normal Distribution

x		$y^a = f(x)$	
x_0	-1.0	y_0	0.2420
x_1	-0.8	y_1	0.2897
x_2	-0.6	y_2	0.3332
x_3	-0.4	y_3	0.3683
x_4	-0.2	y_4	0.3910
x_5	0.0	y_5	0.3989
x_6	0.2	y_6	0.3910
x_7	0.4	y_7	0.3683
x_8	0.6	y_8	0.3332
x_9	0.8	y_9	0.2897
x_{10}	1.0	y_{10}	0.2420

$$f(x) = \frac{1}{\sqrt{2\pi}} \cdot e^{-(x^2/2)},$$
$$\lambda = 0.2,$$
$$n = 10.$$

[a] See Table 2.4 and set $x = z$ and $y = f(z)$.

by Simpson's rule (Eq. e)

$$\int_{-1}^{-1} f(x)\,dx = \frac{\lambda}{3}[y_0 + 4(y_1 + y_3 + y_5 + y_7 + y_9) + \cdots + 2(y_2 + y_4 + y_6 + y_8) + y_{10}]$$

$$= \frac{0.2}{3}[0.2420 + 4(1.7139) + 2(1.4484) + \cdots + 0.2420] = 0.6824.$$

Answer from Table 2.4: $0.6827 = 1.0000 - 2R(z)$ (with $z = 1.0$); error $= 0.0002$.

Example 2. Integrate the exponential function (Eq. 2.24) analytically and by Simpson's rule.

$$f(x) = \frac{1}{\beta} e^{-(x/\beta)} \quad (\beta > 0; 0 \leq x < \infty).$$

First, by conventional methods of integration, integrate Eq. 2.24 between 0 and ∞.

$$\int_0^\infty f(x)\, dx = \int_0^\infty \frac{1}{\beta} e^{-(x/\beta)}\, dx = \frac{1}{\beta} \int_0^\infty e^{-(1/\beta)x}\, dx$$

$$= \frac{1}{\beta} \cdot \left[\frac{e^{-(x/\beta)}}{(-1/\beta)}\right]_0^\infty = \left[-e^{-(x/\beta)}\right]_0^\infty$$

$$= [-e^{-(\infty/\beta)} - (-e^{-(0/\beta)})] = 0 + e^0 = 1.$$

Next compare results by exact and numerical methods. To be definite let

$$f(x) = \tfrac{1}{2} e^{-(x/2)} (0, \infty)$$

and let the range of interest to 0 to 10. Thus by integration

$$\int_0^{10} f(x)\, dx = \int_0^{10} \tfrac{1}{2} e^{-(x/2)}\, dx = \left[-e^{-(x/2)}\right]_0^{10}$$

$$= [-e^{-(10/2)} + 1] = -0.006738 + 1 = 0.993262.$$

Now Table 5.2 is prepared by computing values of y for abscissa values $x = 0, 1, 2, \ldots, 10$

Table 5.2 Exponential Function Integration (by Simpson's Rule)

x		y		
x_0	0	y_0	0.500000	
x_1	1	y_1	0.303265	
x_2	2	y_2	0.183934	
x_3	3	y_3	0.111565	
x_4	4	y_4	0.067668	
x_5	5	y_5	0.041042	$y = f(x) = \tfrac{1}{2} e^{-(x/2)}$
x_6	6	y_6	0.024893	
x_7	7	y_7	0.015098	
x_8	8	y_8	0.009158	
x_9	9	y_9	0.005054	$n = 10$
x_{10}	10	y_{10}	0.003369	$\lambda = 1$

Writing Eq. e for $n = 10$,

$$\int_0^{10} f(x)\, dx = \frac{\lambda}{3} [y_0 + 4(y_1 + y_2 + y_3 + y_5 + y_7 + y_9 + \cdots$$

$$+ 2(y_2 + y_4 + y_6 + y_8) + y_{10}].$$

Substituting values from Table 5.2, and with $\lambda = 1$,

$$\int_0^{10} f(x)\, dx = \tfrac{1}{3}[0.500000 + 4(0.476024) + 2(0.285658) + 0.003369]$$
$$= 0.992927.$$

Error
$$e = 0.000335.$$

Example 3. Compare the results of integrating the gamma function exactly and by Simpson's rule. The gamma density function (Eq. 2.23) is

$$f(x) = \frac{1}{\beta^{\alpha+1}\Gamma(\alpha + 1)} \cdot x^\alpha e^{-(x/\beta)} \qquad (\alpha > -1; \beta > 0; 0 \le x < \infty).$$

From Eq. 2.23, the distribution function is

$$F(x) = \int_0^\infty \frac{1}{\beta^{\alpha+1}\Gamma(\alpha + 1)} x^\alpha e^{-(x/\beta)}\, dx \qquad (0 \le x < \infty),$$

The density function of the gamma distribution can be integrated exactly (see Section 2.3.3) if α is an integer. For this example let $\alpha = 2$ and $\beta = 0.1$. With $\alpha = 2$ and $\beta = 0.1$, the following constants are computed:

$$\beta^{(\alpha+1)} = (0.1)^3 = 0.001$$

and

$$\Gamma(\alpha + 1) = \Gamma(3) = (3 - 1)! = 2.\dagger$$

With

$$\frac{1}{\beta^{(\alpha+1)}\Gamma(\alpha + 1)} = \frac{1}{(0.001)2} = 500,$$

$f(x)$ is first shown to be a density function.

$$F(x) = 500 \int_0^\infty x^2 e^{-(1/0.1)x}\, dx = 500 \int_0^\infty x^2 e^{-10x}\, dx$$
$$= 500 \left[\frac{x^2 e^{-10x}}{(-10)}\right]_0^\infty - \frac{(2)(500)}{(-10)} \int_0^\infty x e^{-10x}\, dx$$
$$= 500 \left[\frac{x^2 e^{-10x}}{(-10)}\right]_0^\infty + 100 \left[\frac{e^{-10x}}{(-10)^2}(-10x - 1)\right]_0^\infty$$
$$= \left[-50x^2 e^{-10x} - e^{-10x}(10x + 1)\right]_0^\infty = 1.$$

Integrating between limits of 0 and 1, first by conventional methods, then applying Simpson's rule,

$$500 \int_0^1 x^2 e^{-10x}\, dx \approx 1.\ddagger$$

† See footnote, Section 2.6.3.
‡ *Note.* Evaluation of the derivative indicates that the maximum point on the distribution occurs at approximately $x = 0.2$.

Evaluated by Simpson's rule,

	x		$y = f(x)$	
	x_0	0	y_0	0
	x_1	0.1	y_1	1.8394
	x_2	0.2	y_2	2.7068
	x_3	0.3	y_3	2.2404
$n = 10$	x_4	0.4	y_4	1.4653
$\lambda = 0.1$	x_5	0.5	y_5	0.8422
	x_6	0.6	y_6	0.4462
	x_7	0.7	y_7	0.2234
	x_8	0.8	y_8	0.1074
	x_9	0.9	y_9	0.0499
	x_{10}	1.0	y_{10}	0.0227

$$\int_0^1 \frac{1}{\beta^{\alpha+1}\Gamma(\alpha+1)} x^2 e^{-(x/\beta)} \, dx = \frac{\lambda}{3} [y_0 + 4(y_1 + y_3 + y_5 + y_7 + y_9)$$
$$+ 2(y_2 + y_4 + y_6 + y_8) + y_{10}]$$

$$500 \int_0^1 x^2 e^{-10x} = 0.0333[0 + 4(5.2907) + 2(4.7490) + 0.0227] = 0.999405.$$

Error $= 1.000000 - 0.99405 = 0.000595$

5.1.3 Errors in Simpson's Rule

Consider first the general formula, Eq. d. Let $f(x)$ be a function finite and continuous in the interval $x = x_0 - \lambda$ to $x = x_0 + \lambda$. Assume further that $f(x)$ possesses continuous derivatives of all orders through the fourth (in the specified interval). Let $F(x)$ be the integral of $f(x)$. Thus

$$F(x) = \int_c^x f(x) \, dx; \quad F'(x) = f(x); \quad f''(x) = f'(x).\dagger$$

Then, denoting the exact value of the integral by I,

$$I = \int_{(x_0-\lambda)}^{(x_0+\lambda)} f(x) \, dx = F(x_0 + \lambda) - F(x_0 - \lambda).$$

By Simpson's rule, the approximate value of the same integral I_s is

$$I_s = \frac{\lambda}{3} [f(x_0 - \lambda) + 4f(x_0) + f(x_0 + \lambda)].$$

The difference, I_s, is the error E_s introduced by applying Simpson's rule.

$$E_s = (I - I_s) = F(x_0 + \lambda) - F(x_0 - \lambda)$$
$$- \frac{\lambda}{3} [f(x_0 - \lambda) + 4f(x_0) + f(x_0 + \lambda)]. \quad (f)$$

† See Section 2.3.1, Density Functions.

168 Foundation

Equation f is of little value, as shown above. It requires changing to a more usable form. Note that Eq. f, for E_s is a function of λ, since by this equation

$$\lim_{\lambda \to 0} E_s = 0.$$

By the procedure of Vallee-Poussin Eq. f may be written in the following way to reflect the preceding observation:

$$\gamma(\lambda) = F(x_0 + \lambda) - F(x_0 - \lambda) - \frac{\lambda}{3}[f(x_0 - \lambda) + 4f(x_0) + f(x_0 + \lambda)].$$

Differentiating with respect to λ yields

$$\gamma'(\lambda) = f(x_0 + \lambda) - f(x_0 - \lambda) - \frac{\lambda}{3}[f'(x_0 + \lambda) - f'(x_0 - \lambda)]$$
$$- \tfrac{1}{3}[f(x_0 + \lambda) + 4f(x_0) + f(x_0 - \lambda)],$$

$$\gamma''(\lambda) = f'(x_0 + \lambda) - f'(x_0 - \lambda) - \frac{\lambda}{3}[f''(x_0 + \lambda) - f''(x_0 - \lambda)]$$
$$- \tfrac{2}{3}[f'(x_0 + \lambda) - f'(x_0 - \lambda)],$$

$$\gamma'''(\lambda) = \frac{\lambda}{3}[f'''(x_0 + \lambda) - f'''(x_0 - \lambda)].$$

The theorem of the Mean.† If $f(x)$ is continuous on the closed interval $[A, B]$† and if $f(x)$ is differentiable in the open interval (A, B), then there is some point ξ of the open interval (A, B) such that

$$f'(\xi) = \left[\frac{f(B) - f(A)}{(B - A)}\right].$$

Thus the expression

$$|f'''(x_0 + \lambda) - f'''(x_0 - \lambda)|$$

for small values of λ and where

$$x_0 - \lambda < \xi < x_0 + \lambda$$

is approximately equal to $2\lambda f^{iv}(\xi)$. Consequently

$$\gamma'''(\lambda) = -\frac{2\lambda^2}{3} f^{iv}(\xi).‡$$

Setting $\lambda = 0$ is these equations [for $\gamma(\lambda), \gamma'(\lambda), \gamma''(\lambda)$],

$$\gamma(\lambda) = \gamma'(\lambda) = \gamma''(\lambda) = 0.$$

The expression for $\gamma'''(\lambda)$ is now integrated three times with respect to λ, at the same time making use of the fact that $\gamma(\lambda) = 0$, $\gamma'(\lambda) = 0$ and $\gamma''(\lambda) = 0$ when $\lambda = 0$. Although $f^{iv}(\xi)$ is influenced by the size of the interval $(0, \lambda)$, it

† See [17] and also the discussion in Section 2.3.
‡ See [26].

may be approximated by its mean value (in the interval) and moved outside the integral sign. The results are

$$\gamma'''(\lambda) = -\frac{2\lambda^2}{3}f^{\text{iv}}(\xi),$$

$$\gamma''(\lambda) = -\frac{2\lambda^3}{9}f^{\text{iv}}(\xi),$$

$$\gamma'(\lambda) = -\frac{\lambda^4}{18}f^{\text{iv}}(\xi),$$

$$\gamma(\lambda) = -\frac{\lambda^5}{90}f^{\text{iv}}(\xi)$$

$\gamma(\lambda)$ being the error (inherent) for an interval of width 2λ, the error for an interval of width λ is $\gamma(\lambda)/2$, or

$$\left[-\frac{\lambda^5}{180}f^{\text{iv}}(\xi)\right].$$

Now, since $(B - A) = n \cdot \lambda$, the error (inherent) over the entire interval $(B = A)$ is

$$E_s = n\left[-\frac{\lambda^5}{180}f^{\text{iv}}(\xi)\right] = -\frac{(B-A)}{180}\lambda^4 f^{\text{iv}}(\xi),$$

where ξ now lies between A and B, or $A < \xi < B$. An error formula may be expressed in terms of the ordinates. This is stated without the algebraic development.

For the number of subintervals $n \geq 6$

$$E_s = \frac{\lambda}{90}[y_{-1} + y_{n+1} - 4(y_0 + y_n) + 7(y_1 + y_{n-1})$$
$$- 8(y_2 + y_4 + \cdots + y_{n-2}) + 8(y_3 + y_5 + \cdots + y_{n-3})].$$

It is illuminating to examine an error formula stated in terms of two computed results.

$$\frac{e_1}{e_2} = \frac{\lambda_1^4}{\lambda_2^4} \quad \text{or} \quad e_1 = \frac{\lambda_1^4}{\lambda_2^4}E_2$$

Thus in the hypothetical case in which

$$\lambda_2 = \frac{\lambda_1}{2},$$

$$e_1 = 16e_2;$$

for example, let I be the correct value of a specified integral. For the two computations the results are

$$I = A_1 + e_1 = A_1 + 16e_2,$$
$$I = A_2 + e_2.$$

170   Foundation

Taking the difference of the two equations above and solving for e_2 yields

$$e_2 = \frac{A_2 - A_1}{15},$$

where A_1 and A_2 are the computed answers, λ_1 and λ_2 are the respective intervals, and e_1 and e_2, the errors. This result implies that, if the value of a definite integral is computed by using a specified value for λ and the computation is repeated by employing twice the original number of subdivisions, *the error in the second answer will be approximately 0.066 of the difference in the answers.*

Example 1. In Example 1, Section 5.1.2, the integration of a standard normal distribution over the interval $\pm 1\sigma$, with equal intervals of $\lambda = 0.2$ (standard deviations) and $n = 10$, the error in density due to Simpson's rule was approximately 0.0002.

By increasing n to 20 and reducing λ to 0.1 over the same interval, the error $E_s = I - I_s$ would be reduced to approximately

$$E_s = 0.0002(0.066) = 0.000013.$$

5.2   DOUBLE NUMERICAL INTEGRATION

Now we discuss a method for approximating the numerical value of an integral of a function of two independent variables. The method amounts to an application which may be considered an extension of Simpson's rule to functions of two variables.

To derive a method for double integration, start first with Eq. g. This expression is integrated over two intervals in the y-direction and two intervals in the x-direction. First, from the formula, all terms involving the differences Δ^{3+0}, Δ^{0+3}, Δ^{4+0}, Δ^{3+1}, Δ^{1+3}, Δ^{0+4} are eliminated, since these differences involve values of the function outside the rectangle over which the integration is being made.

The formula for double interpolation† is

$$\begin{aligned}w = f(x, y) = f(x_0 + \lambda u, y_0 + \rho v) &= w_{00} + u\,\Delta^{1+0}w_{00} + v\,\Delta^{0+1}w_{00}‡\\ &+ \tfrac{1}{2}[u(u-1)\,\Delta^{2+0}w_{00} + 2uv\,\Delta^{1+1}w_{00} + v(v-1)\,\Delta^{0+2}w_{00}]\\ &+ \tfrac{1}{6}[u(u-1)(u-2)\,\Delta^{3+0}w_{00} + 3u(u-1)v\,\Delta^{2+1}w_{00}\\ &+ 3uv(v-1)\,\Delta^{1+2}w_{00} + v(v-1)(v-2)\,\Delta^{0+3}w_{00}]\\ &+ \tfrac{1}{24}[u(u-1)(u-2)(u-3)\,\Delta^{4+0}w_{00}\\ &+ 4u(u-1)(u-2)v\,\Delta^{3+1}w_{00} + 6u(u-1)v(v-1)\,\Delta^{2+2}w_{00}\\ &+ 4uv(v-1)(v-2)\,\Delta^{1+3}w_{00} + v(v-1)(v-2)(v-3)\,\Delta^{0+4}w_{00}]\\ &+ R_n(x_0, y_0).\end{aligned} \qquad (g)$$

† See [26] for complete development.
‡ $\rho = y_{(j+1)} - y_j$.

The remainder term R is

$$R_n(x_0, y_0) = \frac{1}{(n+1)!} [u(u-1)(u-2)\cdots(u-n)\Delta^{n+1}w_{00} + \cdots$$

$$+ (n+1)u(u-1)(u-2)\cdots(u-(n-1))v\,\Delta^{n+1}w_{00}$$

$$+ \frac{(n+1)n}{2!} u(u-1)(u-2)\cdots(u-(n-2))v(v-1)\,\Delta^{(n+1)+2}w_{00}$$

$$+ \cdots + v(v-1)(v-2)\cdots(v-n)\,\Delta^{0+(n+1)}w_{00}]$$

Equation g corresponds to Newton's interpolation formula and reduces to it when either $u = 0$ or $v = 0$ (see Chapter 5).

The changes of variable employed in Eq. g are as follows.

$$u = \frac{x - x_0}{\lambda} \quad \text{or} \quad x = x_0 + \lambda u,$$

$$\frac{x - x_1}{\lambda} = \frac{x - (x_0 + \lambda)}{\lambda} = u - 1,$$

$$\frac{x - x_2}{\lambda} = \frac{x - (x_0 + 2\lambda)}{\lambda} = u - 2,$$

.
.
.

$$v = \frac{y - y_0}{\rho} \quad \text{or} \quad y = y_0 + \rho v,$$

$$\frac{y - y_1}{\rho} = \frac{y - (y_0 + \rho)}{\rho} = v - 1,$$

$$\frac{y - y_2}{\rho} = \frac{y - (y_0 + 2)}{\rho} = v - 2.$$

.
.
.

5.2.1 Two Way Differences

Let $w = f(x, y)$ denote any function of two independent variables x and y. Further (see Table 5.3), let

$$w_{rs} = f(x_r, y_s).$$

Table 5.3 Function Table, Double Numerical Integration

	x_0	x_1	x_2	x_3	x_4	x_m
y_0	w_{00}	w_{10}	w_{20}	w_{30}	w_{40}	w_{m0}
y_1	w_{01}	w_{11}	w_{21}	w_{31}	w_{41}	w_{m1}
y_2	w_{02}	w_{12}	w_{22}	w_{32}	w_{42}	w_{m2}
y_3	w_{03}	w_{13}	w_{23}	w_{33}	w_{43}	w_{m3}
y_4	w_{04}	w_{14}	w_{24}	w_{34}	w_{44}	w_{m4}
⋮	⋮	⋮	⋮	⋮	⋮	⋮
y_n	w_{0n}	w_{1n}	w_{2n}	w_{3n}	w_{4n}	w_{mn}

Explanation of the notation employed in Eq. g is as follows

$$\Delta^{1+0} w_{00} = \Delta_x w_{00} = (w_{10} - w_{00}),$$
$$\Delta^{1+0} w_{01} = \Delta_x w_{01} = (w_{11} - w_{01}),$$
$$\Delta^{1+0} w_{02} = \Delta_x w_{02} = (w_{12} - w_{02}),$$
$$\Delta^{0+1} w_{00} = \Delta_y w_{00} = (w_{01} - w_{00}),$$
$$\Delta^{0+1} w_{10} = \Delta_y w_{10} = (w_{11} - w_{10}),$$
$$\Delta^{0+1} w_{20} = \Delta_y w_{20} = (w_{21} - w_{20}).$$

Generally,

$$\Delta^{1+0} w_{rs} = \Delta_x w_{rs} = [w_{(r+1,s)} - w_{(r,s)}],$$
$$\Delta^{0+1} w_{rs} = \Delta_y w_{rs} = [w_{(r,s+1)} - w_{(r,s)}].$$

Returning now to derivation of a double integration formula, since $dx = \lambda \, du$ and $dy = \rho \, dv$, the expression after eliminating the terms previously mentioned is

$$I = \int_{x_0}^{(x_0+2\lambda)} \int_{y_0}^{(y_0+2\rho)} w \, dy \, dx = \lambda \rho \int_0^2 \int_0^2 \{w_{00} + u \, \Delta^{1+0} w_{00} + v \, \Delta^{0+1} w_{00}$$
$$+ \tfrac{1}{2}[u(u-1) \, \Delta^{2+0} w_{00} + 2uv \, \Delta^{1+1} w_{00} + v(v-1) \, \Delta^{0+2} w_{00}]$$
$$+ \tfrac{1}{6}[3u(u-1)v \, \Delta^{2+1} w_{00} + 3uv(v-1) \, \Delta^{1+2} w_{00}]$$
$$+ \tfrac{1}{24}[6u(u-1)v(v-1) \, \Delta^{2+2} w_{00}]\} \, dv \, du.$$

Performing the indicated integrations and replacing the double differences by their values as already shown in this section,

$$I = \frac{\lambda \rho}{9} [w_{00} + w_{02} + w_{22} + w_{20} + 4(w_{01} + w_{12} + w_{21} + w_{10})$$
$$+ \cdots + 16 w_{11}]. \tag{h}$$

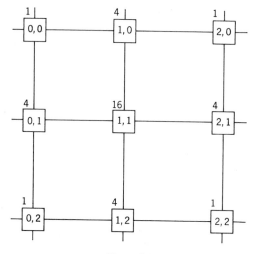

Figure 5.2

Equation *h*, for *I*, corresponds to Simpson's rule for a function of one variable. It is shown diagrammatically in Fig. 5.2. By adding any number of unit blocks of the kind shown above, a formula may be obtained for double integration. Such construction corresponds to Simpson's rule for *n* intervals in single integration. Equation *h*, for *I*, may be rewritten as follows:

$$I = \frac{\lambda}{3}\left[\frac{\rho}{3}(w_{00} + 4w_{01} + w_{02}) + 4\frac{\rho}{3}(w_{10} + 4w_{11} + w_{12})\right.$$
$$\left. + \cdots + \frac{\rho}{3}(w_{02} + 4w_{12} + w_{22})\right], \quad (i)$$

or, an alternate expression,

$$I = \frac{\lambda}{3}\left[\frac{\rho}{3}(w_{00} + 4w_{10} + w_{20}) + 4\frac{\rho}{3}(w_{01} + 4w_{11} + 4w_{12})\right.$$
$$\left. + \cdots + \frac{\rho}{3}(w_{02} + 4w_{12} + w_{12})\right]. \quad (j)$$

Example 1. Approximate by numerical integration the function $w = 1/xy$ with limits $4.0 \le x \le 5.2$ and $2 \le y \le 3.2$ from which the integral I is

$$I = \int_{4}^{5.2}\int_{2}^{2.3}\left(\frac{1}{xy}\right)dy\,dx.$$

In this example let $\lambda = 0.20$ and $\rho = 0.30$ and compute Table 5.4 of values $w = 1/xy$.

Table 5.4 Double Integration—Table of Values

y \ x	4.0	4.2	4.4	4.6	4.8	5.0	5.2	
2.0	0.125000	0.119048	0.113636	0.108696	0.104167	0.100000	0.096154	I_0
2.3	0.108696	0.103620	0.098814	0.094518	0.090580	0.086956	0.083612	I_1
2.6	0.096154	0.091575	0.087413	0.083612	0.080128	0.076923	0.073964	I_2
2.9	0.086207	0.082102	0.078370	0.074962	0.071839	0.068965	0.066313	I_3
3.2	0.078125	0.074405	0.071023	0.067935	0.065104	0.062500	0.060096	I_4

Weddles' rule may be derived (for $n = 6$) from Eq. d. It is somewhat more accurate than Simpson's rule [26].

$$I = \int_{x_0}^{x_0+n\lambda} y \, dx = \frac{3\lambda}{10} [y_0 + 5y_1 + y_2 + 6y_3 + y_4 + 5y_5$$
$$+ 2y_6 + 5y_7 + y_8 + 6y_9 + y_{10} + 5y_{11} + 2y_{12} + \cdots$$
$$+ 2y_{n-6} + 5y_{-5} + y_{n-4} + 6y_{n-3} + y_{n-2} + 5y_{n-1} + y_n] \quad (k)$$

Applying Eq. k to each horizontal row of Table 5.4,

$$I_0 = 0.131182,$$
$$I_1 = 0.114072,$$
$$I_2 = 0.100909,$$
$$I_3 = 0.090470,$$
$$I_4 = 0.081989.$$

For the second integration Simpson's rule (Eq. e) is applied to the I's, with $w_0 = I_0$, $w_1 = I_1, \ldots$, etc.

$$\int_x^{x_0+n\lambda} y \, dx = \frac{\lambda}{3} [w_0 + 4(w_1 + w_3 + \cdots + w_{n-1})$$
$$+ 2(w_2 + w_4 + \cdots + w_{n-2}) + w_n]. \quad (l)$$

Substituting values into Eq. l,

$$I_s = \frac{0.30}{3} [0.131182 + 4(0.114072) + 2(0.100909)$$
$$+ 4(0.090470) + 0.081989],$$
$$I_s = 0.123316$$

By conventional integration

$$\int_4^{5.2} \int_2^{3.2} \left(\frac{1}{xy}\right) dy \, dx = (\ln 1.3)(\ln 1.6) = 0.123321.$$

Error $= I - I_s = 0.00005$.

Numerical Methods 175

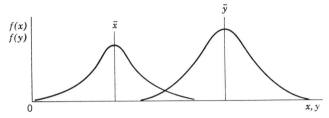

Figure 5.3 Warner diagram (stress-strength).

Example 2. Consider the following hypothetical example. Strength is described by the normally distributed random variable Y, $(\bar{y}, s_y) = (10{,}000;\ 1000)$ psi (see Fig. 5.3). The normally distributed stress function X is described by, $(\bar{x}, s_x) = (7000;\ 700)$ psi. Compute the probability that strength will exceed stress. Applying Eq. 4.6 and solving for z,

$$z = \frac{10{,}000 - 7000}{\sqrt{(1000)^2 + (700)^2}} = \frac{3000}{1220} = 2.458.$$

From Table 2.4, at the value $z = 2.458$,

$$F(z) = R = 0.993014.$$

Next, R is computed by numerical methods. By Eq. 4.1,

$$R = \int_{-\infty}^{\infty} \int_{s}^{\infty} f(s)\, f(S)\, dS\, ds.$$

Making a change of variable,

$$R = \int_{-\infty}^{\infty} \int_{x}^{\infty} f(x)\, f(y)\, dy\, dx.$$

Strength density function is

$$f(y) = \frac{1}{s_y \sqrt{2\pi}} \exp\left[-\frac{1}{2}\left(\frac{y - \bar{y}}{s_y}\right)^2\right]$$

$$= \frac{1}{\sqrt{2\pi} \cdot 10^3} \exp\left[-\frac{1}{2}\left(\frac{y - 10{,}000}{1000}\right)^2\right].$$

Stress density function is

$$f(x) = \frac{1}{s_x \sqrt{2\pi}} \exp\left[-\frac{1}{2}\left(\frac{x - \bar{x}}{s_x}\right)^2\right]$$

$$f(x) = \frac{1}{700\sqrt{2\pi}} \exp\left[-\left(\frac{x - 7000}{700}\right)^2\right].$$

Thus for $\lambda = 1000$ and $\rho = 700$ the limits on Eq. 4.1 are

$$R = \int_{4900}^{9100} \int_{7000}^{13{,}000} f(x)\, f(y)\, dy\, dx.$$

Compute Table 5.5 using Eq. j.

Table 5.5 Double Integration, Normal Variates

		x_0	x_1	x_2	x_3	x_4	x_5	x_6
	x	4900	5600	6300	7000	7700	8400	9100
y		0.004432 σ_x	0.053991 σ_x	0.241971 σ_x	0.398942 σ_x	0.241971 σ_x	0.053991 σ_x	0.004432 σ_x
y_0	7000 0.004432 σ_y	00001964 $\sigma_x\sigma_y$	0002393 $\sigma_x\sigma_y$	0010724 $\sigma_x\sigma_y$	00176811 $\sigma_x\sigma_y$	0010724 $\sigma_x\sigma_y$	0002393 $\sigma_x\sigma_y$	00001964 $\sigma_x\sigma_y$
y_1	8000 0.053991 σ_y	0002393 $\sigma_x\sigma_y$	0028506 $\sigma_x\sigma_y$	0129091 $\sigma_x\sigma_y$	0215392 $\sigma_x\sigma_y$	0129091 $\sigma_x\sigma_y$	0028506 $\sigma_y\sigma_y$	00002393 $\sigma_x\sigma_y$
y_2	9000 0.241971 σ_y	0010724 $\sigma_x\sigma_y$	0129091 $\sigma_x\sigma_y$	0585499 $\sigma_x\sigma_y$	0965324 $\sigma_x\sigma_y$	0585499 $\sigma_x\sigma_y$	0129091 $\sigma_x\sigma_y$	0001724 $\sigma_4\sigma_y$
y_3	10,000 0.398942 σ_y	00176811 $\sigma_x\sigma_y$	0215392 $\sigma_x\sigma_y$	0965324 $\sigma_x\sigma_x$	1591547 $\sigma_x\sigma_y$	0965324 $\sigma_x\sigma_y$	0215392 $\sigma_x\sigma_y$	00176811 $\sigma_x\sigma_y$
y_4	11,000 0.241971 σ_y	0010724 $\sigma_x\sigma_y$	0129091 $\sigma_x\sigma_y$	0585499 $\sigma_x\sigma_y$	0965324 $\sigma_x\sigma_y$	0585499 $\sigma_x\sigma_y$	0129091 $\sigma_x\sigma_y$	0010724 $\sigma_x\sigma_y$
y_5	12,000 0.053991 σ_y	0002393 $\sigma_x\sigma_y$	0028506 $\sigma_x\sigma_y$	0129091 $\sigma_x\sigma_y$	0215392 $\sigma_x\sigma_y$	0129091 $\sigma_x\sigma_y$	0028506 $\sigma_x\sigma_y$	0002393 $\sigma_x\sigma_y$
y_6	13,000 0.004432 σ_y	00001964 $\sigma_x\sigma_y$	0002393 $\sigma_x\sigma_y$	0010724 $\sigma_x\sigma_y$	00176811 $\sigma_x\sigma_y$	0010724 $\sigma_x\sigma_y$	0002393 $\sigma_x\sigma_y$	00001964 $\sigma_x\sigma_y$

Applying Simpson's rule, Eq. *e*,

$$n = 6 \quad \text{and} \quad \lambda = \sigma_x:$$

$$I = \frac{\lambda}{3}[y_0 + 4(y_1 + y_3 + y_5) + 2(y_2 + y_4) + y_6],$$

$$I_0 = \frac{\sigma_x}{3}\left[\frac{0.000019}{\sigma_x\sigma_y} + 4\left(\frac{0.000238}{\sigma_x\sigma_y} + \frac{0.001756}{\sigma_x\sigma_y} + \frac{0.000238}{\sigma_x\sigma_y}\right)\right.$$
$$\left. + 2\left(\frac{0.001065}{\sigma_x\sigma_y} + \frac{0.001065}{\sigma_x\sigma_y}\right) + \frac{0.000019}{\sigma_x\sigma_y}\right],$$

$$I_0 = \frac{0.0044386}{\sigma_y},$$

$$I_1 = \frac{0.0536922}{\sigma_y},$$

$$I_2 = \frac{0.2319156}{\sigma_y},$$

$$I_3 = \frac{0.3995324}{\sigma_y},$$

$$I_4 = \frac{0.2319156}{\sigma_y},$$

$$I_5 = \frac{0.0536922}{\sigma_y},$$

$$I_6 = \frac{0.0044386}{\sigma_y},$$

$$R_s = I_s = \frac{k}{3}\left[\frac{0044386}{\sigma_y} + 4\left(\frac{0.5069168}{\sigma_y}\right)\right.$$

$$\left. + 2\left(\frac{0.4638312}{\sigma_y}\right) + \frac{0044386}{\sigma_y}\right] = 0.98806.$$

Error $= 0.0049$. Since the interval considered was $\pm 3\sigma$, a correction can be made by adding the density from 3σ to infinity (0.00699), which reduces the error to $E_s = 0.00204$. This amount of error could be greatly reduced as shown in Section 5.1.3 by increasing n, thereby reducing λ and ρ.

CHAPTER 6

Monte Carlo Methods†

Monte Carlo methods comprise that part of experimental mathematics concerned with experiments on random numbers. The problems discussed here are probabilistic, the outcome of random processes.

With probabilistic problems, the simplest Monte Carlo approach is to observe random numbers, selected in such a way that they directly simulate the physical random processes of the problem at hand, and to deduce the required solution from the behavior of these numbers.

A basic power of mathematics is its preoccupation with abstraction and generality. Utilizing mathematical symbols, equations may be written which withdraw the substance and show the basic structure of a problem. With this power is associated an intrinsic weakness—usually, with generality and formal language, theory loses immediate ability to provide quantitative answers to specific problems. The idea behind the Monte Carlo approach to troublesome problems is to utilize this strength of abstract mathematics and simultaneously avoid its intrinsic weakness—by replacing theory by experiment whenever theory wavers. For instance, assume a problem which can be set down in a system of equations but cannot be solved by known procedures. The problem (being deterministic) has no direct association with random processes; nonetheless, with the basic problem structure exposed by theory, it frequently happens that the model developed also describes a known apparently unrelated random process, hence is solvable numerically by Monte Carlo simulation.

The process of deducing general laws based on specific observations associated with them results in uncertain conclusions, since specific observations are only a somewhat representative sample from the population of possible observations. Sound experimental practice attempts to assure that the sample is representative. Good presentation of conclusions clearly shows the probability of being in error and by how much. Monte Carlo solutions

† Portions adapted from J. M. Hammersly and D. C. Handscomb, *Monte Carlo Methods*, Wiley, New York, 1964.

involve uncertainty since they arise from raw observational data consisting of random numbers. They can be useful depending on assurance that uncertainty is very small, that is, that error is negligible.

One way of reducing error in solutions is to increase the base to greater numbers of observations. This, however, is rarely a cheap course of action. Roughly, there is a square-law relationship between error in an answer and the required number of observations. To reduce error by a factor of two requires a four-fold increase in the observations. The basic procedure of the Monte Carlo method is the manipulation of random numbers. These should be employed with care. Each random number is a possible source of added uncertainty in the final answer. It usually pays to study each part of the Monte Carlo experiment with the view of replacing any possible parts with exact theoretical analysis that contributes no error. As progressing work provides added insight into the problem nature and suggests appropriate theory, good Monte Carlo practice, to this extent, may be self-liquidating.

Before starting the practical solution of a design problem, it is necessary to consider the number of samples necessary to assure that the answers will be of the minimum required accuracy. In practical design application, the tails of the frequency distributions of the random variable parameters are of particular interest. The tail areas are the areas at the ends of the range containing perhaps one percent or less of the area under the curves. The probability that any sample will give a value of the statistic that lies within one of these extreme areas is small; for example, a sample of 200 (values of a specified random variable parameter) provides no information about the 1 percent points and very little about the 5 percent points. A sample of 1000 values provides little information about the 5 percent points and very little about the 1 percent points. A sample of 2000 values provides little information about the 1 percent points and a reasonable estimate of the 5 percent points [28].†

A necessary feature, common to all Monte Carlo computations, is that at some point a set of actual values must be substituted for a random variable. This set of values must possess the statistical properties of the random variable.

6.1 RANDOM NUMBERS

The values substituted for the random variables are numbers—on the argument that such numbers could be produced by chance by suitable random processes. The question of whether these numbers are correctly

---

† G. B. Hey, A New Method of Experimental Sampling Illustrated on Certain Nonnormal Populations, *Biometrica*, **30**, pp. 68–80.

distributed is answered by suitable statistical tests performed on the numbers themselves. In Monte Carlo work, we are satisfied if the final results of the tests are reasonably good, if the answers come out approximately correct.

The term "random number," when used without qualification, is understood to refer to the standardized rectangular distribution,

$$F(y) = \begin{cases} 0, y < 0, \\ y, 0 \leq y \leq 1, \\ 1, y > 1. \end{cases}$$

The symbol ξ (with or without prefixes) is reserved for such a number. The same symbol is used for a substitute number (e.g., pseudorandom number) which in practice often assumes the role of a standardized rectangular random number. It is sometimes convenient to think of numbers as continuously distributed when actually they take on values that are multiples of some very small number, such as 10^{-5} or 10^{-7}, the result of rounding off in numerical computation.

For the Monte Carlo work that they may be adapted to there are published tables of random numbers (The Rand Corporation or the C.R.C. Standard Mathematical Tables) which are generated by physical processes that are random in the strictest sense (as far as one can tell). These lists of numbers are also subjected to statistical tests.

Essentially, there are two methods for generating random numbers, referred to as the physical process and the arithmetical process.

For work that requires the use of digital computers, it is many times convenient to compute a sequence of numbers as needed (one at a time), by a specified rule (mathematical formula). Such rules are devised so that reasonable statistical tests do not detect any significant deviation from randomness. Such a sequence of numbers is called pseudorandom. One advantage of a specified rule is the reproducibility of the sequence for checking purposes.

Ordinarily, a pseudorandom sequence ξ is computed from a sequence of positive integers x as follows:

$$\xi_i = \frac{x_i}{n},$$

where n is a suitable positive integer.

One early method for producing a pseudorandom sequence was the *Midsquare Method*. According to this method, each number is generated by squaring its predecessor and utilizing the middle digits of the result. Thus X_{i+1} in the equation above consists of the middle digits of X_i^2. The pseudorandom number generator utilizes a fixed length in the digital computer. This method has largely been abandoned as unsatisfactory and has been replaced by congruential methods, to be discussed later.

A pseudorandom sequence that can be generated by a recurrence relation (due to Lehmer) is

$$X_{i+1} = aX_i \pmod{n},$$

where the X_{i+1} is the remainder when X_i is divided by m. The expression above has been generalized to

$$X_{i+1} = (aX_i + c)(\mod n);$$

n is a large integer whose magnitude is determined by the limitations of the digital computer available. n is usually some large power of 10 or of 2, and a, c, and X_{i+1} are integers between zero and $n-1$. The numbers $X_{i/n}$ are employed as pseudorandom numbers. The two formulas above for X_{i+1} are called congruential methods for the generation of pseudorandom numbers.

The expression

$$X_{i+1} = aX_i \pmod{n}$$

is termed the multiplicative congruential method. Sequences as generated by congruential methods are repetitive after (at most) n steps; that is, they are periodic. For example, the expression

$$X_{i+1} = (3X_i + 1)(\mod 16),$$

with $x_0 = 2$, generates the sequence

2, 7, 6, 3, 10, 15, 14, 11, 2, 7, 6, 3, 10, ...

Thus the period of this particular pseudorandom sequence is 8. Clearly, the period must be as long as longer than the number of random numbers needed in a specific problem.

If the equation

$$X_{i+1} = (aX_i + c)(\mod n)$$

is employed as the generator, the total period of n may be realized, if the following conditions exist:

1. c and n have no common divisor.
2. $a \equiv 1 \pmod{p}$ for every prime factor p of n.
3. $a \equiv 1 \pmod{4}$ if n is a multiple of 4.

A case of frequent occurrence is $n = 2^\alpha$, because the computations are easily performed on a digital computer working in the binary scale. The number 8 is frequently in the value range 30 to 40. In such case, the period of $X_{i+1} = aX_i \pmod{n}$ is $n/4$ provided a differs by 3 from the nearest multiple of 8 and x_0 is an odd number; furthermore $X_{i+1} = aX_i + c \pmod{n}$ displays a period of n if c is an odd number and a is one larger than a multiple of 4.

Another source of semirandom numbers is the sequence of digits in transcendental numbers [28]. The numbers of decimal digits published are, for example, the following:

1. 2556 digits of e, National Bureau of Standards (1951b).
2. 2000 digits of π, Reitwiesner (1950).
3. 1312 digits of $\sqrt{3}$, Uhler (1951b).

6.2 NONRECTANGULAR DISTRIBUTION SAMPLING

The discussion to this point has been concerned with rectangularly distributed random numbers. In design by reliability when Monte Carlo methods are indicated, sampling is often required from distributions other than rectangular distributions. The procedure is usually to begin with rectangularly distributed random numbers and make appropriate transformations.

As before, ξ denotes a rectangularly distributed number. Now, let n denote a number with density function $f(y)$ and cumulative distribution function $F(y)$. We need an expression of n as an explicit function of the numbers (Fig. 6.1).

When the function F possesses a known inverse function F^{-1}, the problem is straightforward. The procedure is to take

$$n = F^{-1}(\xi),$$

as shown in Fig. 6.1, for example, the distribution of the exponential function

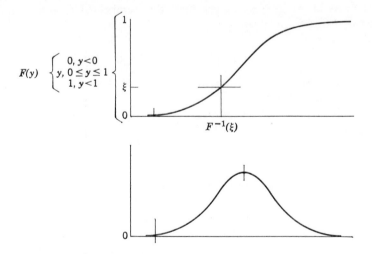

Figure 6.1 Rectangular distribution transformation.

is $\xi = e^{-\lambda n}$.

$$|\log_e \xi| = |-\lambda n \log_e e| = |-\lambda n|.$$

and

$$n = \left(\frac{\log_e \xi}{\lambda}\right).$$

6.2.1 Direct Simulation

Direct simulation of a probabilistic problem is possibly the simplest form of the Monte Carlo method. Nonetheless, it remains one of the principal forms of Monte Carlo practice, for it naturally appears often in practical engineering problems.

The majority of direct simulation problems cannot be discussed adequately without entering into many small practical details. Selected as an example is the estimation of reliability of a structural skin panel by direct Monte Carlo simulation.†

Example 1. In this skin-panel analysis the procedure is as follows.

STEP 1. Select one random value of each random variable in each relevant equation using the distribution, mean value, and standard deviation. Compute the allowable stress, F_i, and the actual stress, f_i. F_i may be allowable local crippling stress and f_i the actual compressive stress, or F_i may be allowable shear stress and f_i the actual shear stress, etc.

Figure 6.2 shows the detail skin-panel design. The skin panel performs several functions: that of an aerodynamic surface and a protective cover; it also transfers shear stress, transfers external pressure to stringers and frames, and carries axial compression loads.

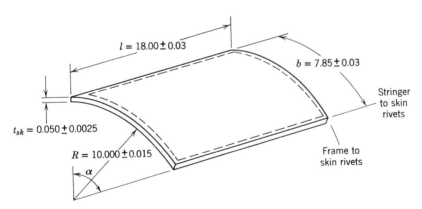

Figure 6.2 Skin panel geometry.

† "Structural Reliability," C. Dicks and S. Wilson, 11th National Symposium on Reliability and Quality Control, January 1965.

184 Foundation

Shear flows, q_i (in the skin panel), are given by

$$q_i = \frac{V}{\pi R} \sin \alpha_i, \qquad (a)$$

where α_i = the included angle,
 R = outer radius,
 V = transverse shear load.

The shear stress f_s is

$$f_{s_i} = \frac{q_i}{t_{skin}}. \qquad (b)$$

A combination of Eqs. *a* and *b* gives the expression for actual shear stress:

$$f_{s_i} = \frac{V/\pi R \sin \alpha_i}{t_{skin}} = \frac{V \sin \alpha_i}{\pi R t_{skin}}. \qquad (c)$$

Allowable shear stress is given by Eq. *d*, the expression for a semimonocoque cylinder panel (simply supported with clamped edges).

$$F_s = 0.1 E \frac{t}{R} + 6.25 E \left(\frac{t}{b}\right)^2, \qquad (d)$$

where E = modulus of elasticity,
 t = skin thickness,
 R = outside skin radius,
 b = stringer spacing.

Summary of Step 1. The computer data from an input data table (for Eqs. *c* and *d*) is as follows.

Loads			Statistical Properties
V	14,275 lb, $+1\%$, -5%		Negative lognomal, mode at 14,275
			-3σ at -5%
			$+3\sigma$ at $+1\%$
Geometry			
t	0.050 in., ± 0.0025		Normal, mean -0.050, $3\sigma = 0.0025$
R	10.000 in., ± 0.015		Negative lognomal, mode = 10,000
			$+3\sigma = 10.015$
			$-2.327\sigma = 9.985$
Materials			
E	10.5×10^6 lb/in.^{2}, -0%, $+6\%$		Normal where -2.327σ is located at 10.5×10^6

Random-number generators are used to provide one value from each random variable. The corresponding applied stress and allowable stress are computed, one applied stress value and one allowable stress value. These values constitute a trial and are the values used in Step 2.

STEP 2. Allowable and applied stresses are compared for success, $F_i > f_i$, or for failure, $F_i < f_i$.

Data from Step 1 are used directly. They are utilized in the structural equations and logic.

Equations c and d are expressed as stress ratios:

$$R_p = \frac{P}{P'} = \frac{\text{applied external pressure}}{\text{allowable external pressure}} \quad (e)$$

and

$$R_s^2 = \frac{f_s^2}{F_s^2} = \frac{(\text{applied shear stress})^2}{(\text{allowable shear stress})^2} \quad (f)$$

These stress ratios (additively) define failure modes and conditions for structural adequacy. The structural interaction equation combines external pressure and the transverse shear loads as follows.

$$R_p + R_s^2 = 1. \quad (g)$$

The skin panel is adequate in a trial when

$$R_p < 1 \quad (h)$$

$$R_s < 1 \quad (j)$$

$$R_p + R_s^2 - 1 \leq 0. \quad (k)$$

Thus individual pressure stresses and shear stresses are used to compute individual stress ratios, and individual combined stress ratios as defined in Eqs. g and h. The program contains structural engineering logic to record an individual trial as adequate or not. Individual stress ratios are stored and used if no failures occur.

For instance, the results of an individual trial are $p_i = 2.015, p'_i = 7.050, f_{s_i} = 2,305.000$, and $F_{s_i} = 8,050.000$. Thus $R_{p_i} < 0$, $R_{s_i}^2 < 0$ and $R_{p_i} + R_{s_i}^2 - 1 < 0$, a successful trial,

STEP 3. Record the comparison and repeat Steps 1, 2, and 3 for the number n of trials needed ($n = 10,000$ in this problem). The combined stresses interaction distribution obtained from Eq. k is shown in Fig. 6.3.

STEP 4. Utilizing the data from 10,000 trials, compute the relative frequency† (called reliability). If failures have occurred in the n trials, Eq. l is used.

$$R_{\text{structures}} = 1 - \frac{\text{number of failures}}{\text{number of failures} + \text{number of successes}} = 1 - \left[\frac{\sum_{\lambda=1}^{n}(F_i - f_i)}{n}\right] \quad (l)‡$$

If no failures occurred in the n trials, the distributions of F and f are used with Eq. 4.6 to compute reliability:

$$F = \Pr[(F_i - f_i) \geq 0].$$

The adequacies are calculated by Eq. m, and difference distribution and Eq. 4.6 (illustrated in Fig. 6.3). The lowest 1000 values in the tail are placed in a cumulative distribution from, $F_{(x)}$, and are used to determine the constants of Eq. m. Then $F_{(0)} = ke^a$ is used to compute the unreliability.

$$F_{(x)} = Ke^{a+bx+cx}. \quad (m)$$

† (Section 2.1.2).
‡ Equation 2.4.

186 Foundation

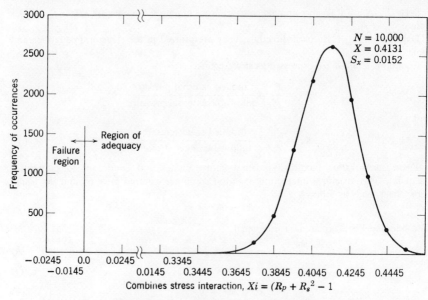

Figure 6.3 Distribution of stress interaction.

STEP 5. Use adequacy relative frequency in the overall reliability system model.

Using the reliability block diagram members and logic to calculate the overall system's reliability, and using (Eq. 2.14) (see Figs. 6.4 and 6.5):

$$R_{\text{overall}} = R_{\text{electronics}}xR_{\text{propulsion}}xR_{\text{structure}}x, \ldots$$

Example 2. [23] Monte Carlo simulation is employed in this example to determine variates $Z = XY$, $Z = X/Y$, and $Z = X^2$ in selected combinations of normal independent random variables X and Y. Mean values and standard deviations of each Z are computed by formulas given in Chapter 3. The computed mean values and standard deviations are compared (Tables 6.1, 6.2, and 6.3) with those resulting from the Monte Carlo simulations. For each variate pair studied, a histogram of the synthesized population of each Z is plotted under the idealized normal curve of Z. Finally, each simulated random variable,

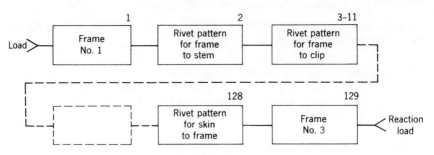

Figure 6.4 Structural reliability block diagram (simplified).

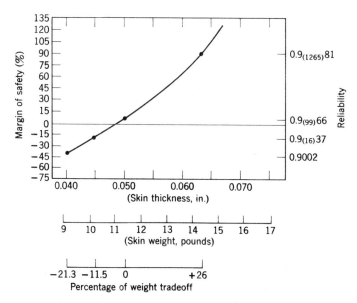

Figure 6.5 Quantitative relationships of structural reliability, weight and margin of safety when safety factor = 1.4.

characterized by the synthesized set $z_1, z_2, \ldots, z_r$, is chi-square tested for normality at the one percent significance level. The particular Monte Carlo program used does the following.

1. It randomly selects a value x_i and a value y_i from sample populations $x_1, x_2, \ldots, x_n$ and $y_1, y_2, \ldots, y_n$, of X and Y ($n = 32$). One thousand twenty-four random pairs (x_i, y_i) are formed for each combination of X and Y studied.

2. The operation of multiplication or division is carried out on each random pair (x_i, y_i).

$$z_i = x_i y_i$$

or

$$z_i = \frac{x_i}{y_i}.$$

The consequence of this operation is a sample population

$$z_i, z_2, \ldots, z_r,$$

(where $r = 1024$).

3. The sample mean value and sample standard deviation of Z are computed (called Monte Carlo moments in the tabulated results).

4. The chi-square test for normality is performed on the simulated population. The purpose is to locate, as closely as possible, the magnitude of standard deviation versus mean value for which the function Z of X and Y remains essentially normal.

Tables 6.1, 6.2, and 6.3 and Figs. 6.6, 6.7, and 6.8 are exhibits of the simulation study described.

Table 6.1 Products $X = YZ$

	Variates		Coefficient of Variation		Pass X^3 Test	Computed Moments		Monte Carlo Moments	
	X	Y	x	y		μ_z	σ_z	$\bar{z}$	s_z
1	(100, 2.5)	(50, 5)	2.5	10.0	No	5000.0	515.5	4950.0	485.9
2	(100, 5.0)	(50, 5)	5.0	10.0	No	5000.0	599.6	4956.6	521.9
3	(100, 7.5)	(50, 5)	7.5	10.0	[a]	5000.0	616.1	4963.3	576.5
4	(100, 10)	(50, 5)	10.0	10.0	Yes	5000.0	708.9	4969.9	645.1
5	(100, 15)	(50, 5)	15.0	10.0	Yes	5000.0	904.5	4983.2	809.4
6	(100, 20)	(50, 5)	20.0	10.0	Yes	5000.0	1122.5	4996.5	994.5
7	(100, 5)	(50, 1.25)	5.0	2.5	Yes	5000.0	279.6	4999.2	249.0
8	(100, 5)	(50, 2.5)	5.0	5.0	Yes	5000.0	353.8	4985.0	322.4
	(100, 5)	(50, 5)	5.0	10.0	No	5000.0	559.6	4956.6	521.9
9	(100, 5)	(50, 7.5)	5.0	15.0	No	5000.0	791.5	4928.2	743.9
10	(100, 5)	(50, 10)	5.0	20.0	No	5000.0	1032.0	4899.8	973.2

[a] Marginal.

Table 6.2 Quotients $Z = X/Y$

	Variates		Coefficient of Variation		Pass X^2 Test	Computed Moments			Monte Carlo Moments	
	X	Y	x	y		μ_z	σ_z		$\bar{z}$	s_z
11	(1.0, 0.025)	(5, 0.10)	2.5	2	Yes	0.20	0.0064		0.20079	0.0058
12	(1.0, 0.025)	(5, 0.25)	2.5	5	Yes	0.20	0.0064		0.20079	0.0058
13	(1.0, 0.025)	(5, 0.50)	2.5	10	No	0.20	0.0205		0.2044	0.0197
14	(1.0, 0.025)	(10, 0.20)	2.5	2	Yes	0.10	0.0032		0.10039	0.0029
15	(1.0, 0.025)	(10, 0.50)	2.5	5	Yes	0.10	0.0056		0.10092	0.0052
16	(1.0, 0.025)	(10, 1.0)	2.5	10	No[a]	0.10	0.0103		0.10219	0.0098

[a] Marginal.

Table 6.3 Squares $Z = X^2$

	Variates	Coefficients of Variation		Pass X^3 Test	Computed Moments		Monte Carlo Moments	
	X Y	x	y		μ_z	σ_z	$\bar{z}$	s_z
38	(10, 0.20)	2.0		Yes	100.04	4.0006	100.03	3.9171
39	(10, 0.50)	5.0		Yes	100.25	10.0094	100.24	9.7960
40	(10, 0.75)	7.5		Yes	100.56	15.0316	100.54	14.702
41	(10, 1.00)	10.0		Yes	101.00	20.0749	100.96	19.621
42	(10, 1.50)	15.0		[a]	102.25	30.252	102.16	29.511
43	(100, 2)	2.0		Yes	10,004.0	400.06	10,003.0	391.71
44	(100, 5)	5.0		Yes	10,025.0	1000.9	10,024.0	979.60
45	(100, 7.5)	7.5		Yes	10,056.0	1503.2	10,054.0	1470.2
46	(100, 10)	10.0		Yes	10,100.0	2007.5	10,960.0	1962.1
47	(100, 15)	15.0		[a]	10,225.0	3025.2	10,216.0	2951.1

σ/μ = coefficient of variation
$\mu_z = \mu_x^2 + \sigma_x^2$
$\sigma_z = \sqrt{4\mu_x^2\sigma_x^2 + 2\sigma_x^4}$
[a] Marginal.

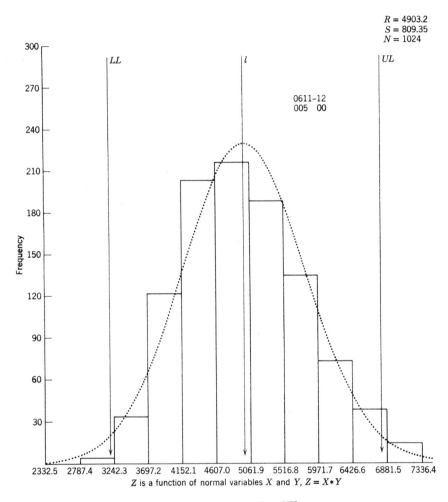

Figure 6.6 Product $Z = XY$.

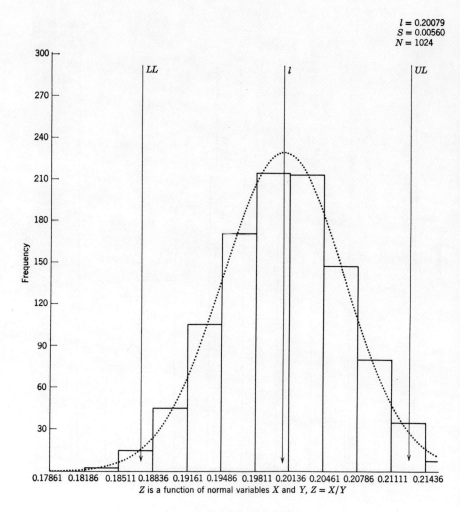

Figure 6.7 Quotient $Z = X/Y$.

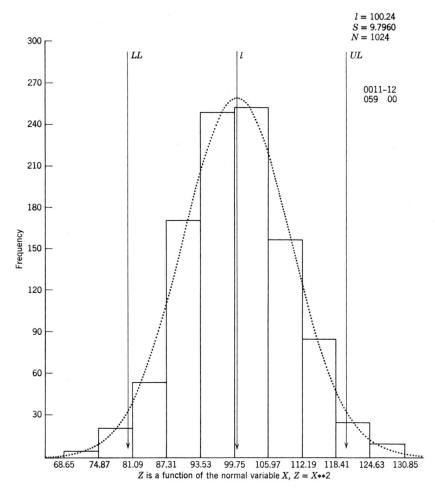

Figure 6.8 Variate squared $Z = X^2$.

6.3 MONTE CARLO GENERAL PRINCIPLES

Any Monte Carlo computation that yields quantitative results may be considered as estimating the value of a multiple integral. If no computation requires more than $N = 10^8$ random numbers, for example, then the results will be a vector valued function

$$R(\xi_1, \xi_2, \ldots, \xi_N)$$

of the sequence of random numbers $\xi_1, \xi_2, \ldots, \xi_N$. This is an unbiased estimation of

$$\int_0^1 \cdots \int_0^1 R(x_1, x_2, \ldots, x_N)\, dx_1 \cdot dx_2 \cdots dx_N.$$

The problem of evaluating integrals provides a means of showing various Monte Carlo techniques of more general application. For simplicity consider as a standard example the one-dimensional integral

$$= \int_0^1 f(x)\, dx.$$

Assume that $f \in L^2(0, 1)$ or

$$\int_0^1 [f(x)]^2\, dx$$

exists, hence θ exists. *Relative efficiency* may be defined by the following expression:

$$\frac{n_1 \sigma_1^2}{m_2 \sigma_2^2}$$

as that of Method 2 relative to Method 1; σ_1^2 and σ_2^2 are the respective variances of the two methods and n_1 and n_2 are the respective numbers of times that $f(\cdot)$ is evaluated in each method.

6.3.1 Crude Monte Carlo

If $\xi_1, \xi_2, \ldots, \xi_n$ are independent rectangularly distributed random numbers with range $(0, 1)$, then the quantities

$$f_i = f(\xi_i)$$

are independent random variates with expectation θ. Thus

$$\bar{f} = \frac{1}{n} \sum_{i=1}^{m} f_i$$

is an unbiased estimation of $\bar{f}$ and its variance is

$$\frac{1}{n}\int_0^1 [f(x) - \theta]^2 \, dx = \sigma_n^2$$

(called the variance in the mean).
The standard deviation in the mean is then

$$\sigma_{\bar{f}} = \frac{\sigma}{\sqrt{n}}.$$

$\bar{f}$ may be called a crude Monte Carlo estimator of σ.

6.3.2 Hit or Miss Monte Carlo

Assume $0 \leq f(x) \leq 1$ when $0 \leq x \leq 1$.
The curve $y = f(x)$ may be drawn in the unit square (as shown in Fig. 6.9).
Thus

$$0 \leq x$$
$$y \leq 1.$$

The $n \int_0^1 f(x) \, dx$ is the proportion of area under the curve, or, more formally,

$$f(x) = \int_0^1 g(x, y) \, dy,$$

where

$$g(x, y) = 0 \quad \text{if} \quad f(x) < y$$
$$= 1 \quad \text{if} \quad f(x) \geq y.$$

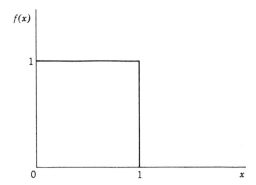

Figure 6.9 Hit or miss Monte Carlo.

θ may then be evaluated as follows:

$$\theta = \int_0^1 f(x)\,dx = \int\int_0^1 g(x,y)\,dx\,dy$$

by the estimator

$$\bar{g} = \frac{1}{n}\sum_{i=1}^{n} t(\xi_{2-i1}, \xi_{2i}) = \frac{n'}{n},$$

where n' is the number of occasions on which $f(\xi_{2i-1}) \geq \xi_{2i}$.

In words, the procedure amounts to taking n points at random in the unit square and counting the number of them which lie below the curve $z = f(x)$. Such procedure amounts to sampling from the binomial distribution where $p = \theta$ and the standard deviation is

$$\left[\frac{\theta(1-\theta)}{n}\right]^{1/2}.$$

"Statisticians found that distribution sampling is a technique that is better than nothing but less desirable than analytical techniques. The same conclusion has been reached in attempts to apply general techniques of sampling distributions to numerical computations, e.g., matrix inversion, solution of multiple integrals, etc. In these sampling has been found to be generally inferior to direct numerical methods. Therefore, as in statistical problems, the technique should be used only as a last resort." [28]

PART TWO

APPLICATION

CHAPTER 7

Mechanical Elements

7.1 ELEMENTS OF FORCE SYSTEMS

The equations of static equilibrium are used in solving for unknown forces or relations or unknown stresses. In addition, consideration must be given to the variable nature of important variables.

To define a force completely, its magnitude, direction, and point of application must be known. To be known is to be described statistically, since this amounts to the most realistic description now conceivable. These facts are generally referred to as the characteristics of the force.

A force (variate) acting in space is completely defined if its component variates in three directions and its moment variates about three axes are known; for example, F_x, F_y, and F_z and M_x, M_y, and M_z. For the equilibrium of a force system (as conventionally stated), there can be no residual force and thus the equilibriums are obtained by equating the force and moment equations to zero. As in preceding sections, this requires not only that the forces be identical but also that they be absolutely correlated,† a rather unlikely possibility in real situations.

Thus the system of Eqs. *a*

$$\sum F_x = 0 \qquad \sum M_1 = 0$$
$$\sum F_y = 0 \quad \text{or} \quad \sum M_2 = 0 \qquad (a)$$
$$\sum F_z = 0 \qquad \sum M_3 = 0$$

represents an idealized statement.

Since magnitude, direction, and point of application are each random variables (characterized by unique distributions), the system of Eqs. *a* needs modification to reflect actual circumstances. Thus a partial answer to realistic

† See footnote, Section 3.1.1.

description of a force system is of the following kind:

$$\sum \bar{F}_x = 0; \quad \sum \bar{M}_1 = 0,$$
$$\sum \bar{F}_y = 0; \quad \sum \bar{M}_2 = 0, \qquad (b)$$
$$\sum \bar{F}_z = 0; \quad \sum \bar{M}_3 = 0.$$

Although the statements above relate to physical forces, they may be extended and generalized to include stresses of other kinds, such as thermal, electrical, and environmental stresses.

Inevitably, attached to each of the equilibrium equations in (b) is a probability that equilibrium will not exist or that P(applied stress > allowable stress). For the case where normal variates model the parameters reasonably well, related to the system of Eqs. b is a set of statements of the following kind:

$$\begin{array}{cc} \textit{Forces} & \textit{Moments} \\ \sum s_x \geq 0 & \sum s'_1 \geq 0, \\ \sum s_y \geq 0 & \sum s'_2 \geq 0, \\ \sum s_z \geq 0 & \sum s'_3 \geq 0. \end{array} \qquad (c)$$

Systems of Eqs. b and c are a statistical analogue of Eq. a.

7.2 CENTROIDS

The force caused by gravity acting on a body may be considered as the resultant of many parallel forces acting on each element of the body. The magnitude variate of the resultant of a family of parallel force variates equals to the sum variate of such a family (see Section 3.1.1, linear combinations). With the position variate of the resultant is associated a moment variate, about any axis, which equals to the sum variate of the moments of the component force variates. Thus the resultant gravity force variate on a body is its weight variate ($\bar{w}, s_w$), which must equal to the sum of the weight variates ($\bar{w}_i, s_{wi}$) of the total of elements of the body. If the forces of gravity act parallel to the z axis in Fig. 7.1, the moment variates of all force variates about the x and y axis must equal to the moment variate of the resultant:

$$(\bar{M}_1, s_{M_1}) = (\bar{W}, s_w)(\bar{x}, s_x) = \sum(\bar{x}_i, s_{x_i})(\bar{W}_i, s_W),$$
$$(\bar{M}_2, s_{M_2}) = (\bar{W}, s_w)(\bar{y}, s_y) = \sum(\bar{y}_i, s_{y_i})(\bar{W}_i, s_W).$$

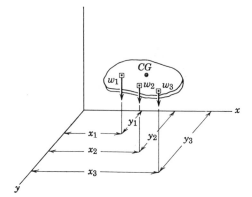

Figure 7.1 Centroids.

7.3 MOMENT OF INERTIA

Moment of inertia of a plane area with respect to an axis in its plane is described by an integral of the type [33]

$$I_z = \int_A y^2 \, dA.$$

Each element dA is multiplied by the square of its distance from the z axis and integration is carried out over the entire cross-section area A (see Fig. 7.2).

For many simple geometric shapes moments of inertia are readily computed analytically. As an example, consider the rectangular section shown in

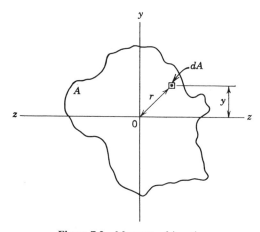

Figure 7.2 Moment of inertia.

202 Application

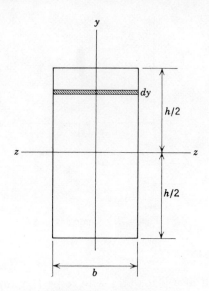

Figure 7.3 Moment of inertia: rectangular section.

Fig. 7.3. The moment of inertia of a rectangle is computed with respect to the axis of symmetry $z-z$, by dividing it into infinitesimal elements, as shown shaded in Fig. 7.3, and summing. Now for an idealized figure [33]

$$I_z = 2\int_0^{h/2} y^2\, dA = 2\int_0^{h/2} by^2\, dy = \frac{bh^3}{12}.$$

Since the defining dimensions of the rectangle are random variables, a differential element of area is

$$dA = (\bar{b}, s_b)\, dy,\dagger$$

and the limits of integration are zero to (the random variable) $h/2$,

$$\tfrac{1}{2}(\bar{h}, s_h).$$

The expression for moment of inertia I_z is thus

$$(\bar{I}_z, s_{I_z}) = 2\int_0^{\frac{1}{2}(\bar{h}, s_h)} y^2(\bar{b}, s_b)\, dy = \tfrac{1}{12}(\bar{b}, s_b)(\bar{h}, s_h)^3.$$

† *Note.* $\lim_{\Delta y \to dy} \sigma_z = 0$ and $\lim_{\Delta A \to dA} \sigma_A = 0.$

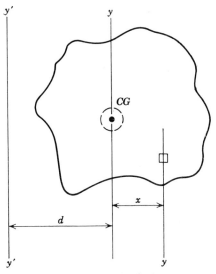

Figure 7.4 Parallel axis theorem.

Parallel axis theorem (see Fig. 7.4).

$$I'_y = \int (d + x)^2 \cdot dA = d^2 \int dA + 2d \int x\, dA + \int x^2\, dA$$

$$x\, dA = 0,$$

since $z - z$ is the centroid of the body, and

$$I'_2 = I + Ad^2$$

$$(I'_y, s_{I_{y'}}) = (\bar{I}, s_I) + (\bar{A}, s_A)(\bar{d}, s_d)^2.$$

Example 1 (see Fig. 7.5).

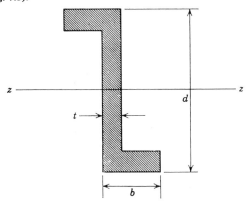

Figure 7.5 Zee section: moment of inertia.

First, it is noted that, although B, D, and T are independent random variables, the right side of the expression above is not termwise independent (see Section 3.1.1, footnote on correlation).

$$I_z = \tfrac{1}{12}(6b\,d^2t - 12b\,dt^2 + 8bt^3 + d^3t - 6d^2t^2 + 12\,dt^3 - 8t^4)$$

The methods of Chapter 3 are applied to obtain the standard deviation.

7.4 RADIUS OF GYRATION

The radius of gyration ρ of a solid body is

$$I = \rho^2 A \quad \text{or} \quad \rho = \sqrt{\frac{I}{A}}.$$

Care must be exercised in computing the random variable ρ from random variables I and A for I and A are not statistically independent; for example, in the case of a rectangular section

$$A = bh$$

and

$$I_z = \tfrac{1}{12}bh^3.$$

Hence

$$\rho = \left(\frac{I}{A}\right)^{1/2} = \left(\frac{bh^3}{12bh}\right)^{1/2} = \left(\frac{h^2}{12}\right)^{1/2} = \left(\frac{h}{2\sqrt{3}}\right)$$

and

$$(\bar{\rho}, s_\rho) = \frac{1}{2\sqrt{3}}(\bar{h}, s_h).$$

When the variate I/A must be determined as a quotient, Eqs. 3.22 and 3.23 are employed to determine $(\bar{\rho}, s_\rho)$.

7.5 METHODS OF ANALYSIS

As in any design method, certain assumptions must be made and the design held valid within the boundaries of these assumptions. For the probabilistic designs that follow, these assumptions are made:

1. The load is independent of the strength.
2. The external load, material strength, and geometric variation are normally distributed random variables or may be adequately approximated by normal random variables.
3. Variations in cross-sectional geometry are symmetrical with respect to the neutral axis.
4. The level of (probability) adequacy is specified.

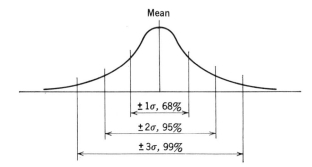

Figure 7.6 Intervals in the normal distribution.

7.5.1 Estimating Variance

Associated with mechanical or structural geometry are tolerances. Since, in many instances, statistical descriptions of geometrical variables do not exist, estimates must be made. The specified dimension may be taken as the mean value and the tolerance interpreted in terms of variance. For example, if samples fall within the tolerance range $\pm \Delta$ 95 percent of the time (see Fig. 7.6), these are interpreted as a $\pm 2\sigma$ interval

$$\Delta x = 2s_x; \quad s_x = \frac{\Delta x}{2}.$$

Each measurable parameter is treated as a random variable. The assumption is that design parameters can be adequately represented by normal random variables. The reasons for this assumption are (a) that many parameters are known to be normal variates, (b) the comparative best fits of data to model distributions such as given in [34], and (c) that the analytic method at hand permits the illustration of probabilistic design methods. When variables appear that do not fit the assumptions of normality, other methods are available to resolve such problems.

PROBLEMS

1. Determine the area, moment of inertia, and radius of gyration variates for the ring shown. Assume that r and t are normally distributed and statistically independent random variables. Use the approximate expression for moment of inertia

$$I = \pi r^3 \cdot t \text{ in.}^4$$

Given

$(\bar{t}, s_t) = (0.125, 0004)$ in.

$(\bar{r}, s_r) = (1.150, 0.030)$ in.

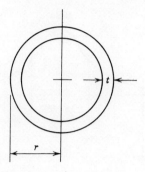

2. Repeat the exercise of Problem 1, except that the section is a circle, with

$$(\bar{r}, s_r) = (1.150, 0.030) \text{ in.}$$

and

$$I = \pi r^4/4 \text{ in.}^4$$

3. Determine the area, moment of inertia, and radius of gyration of the rectangular tube shown. Determine I_{z-z}. Do not neglect the effect of correlation.

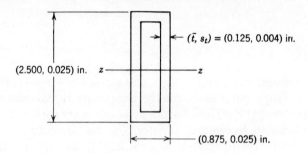

4. Determine the area, moment of inertia, and radius of gyration of the angle shown.

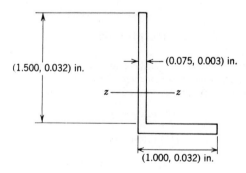

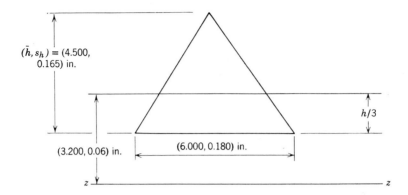

5. Determine the moment of inertia of the figure shown with respect to the $z-z$ axis.

CHAPTER 8

Elements in Tension [35]

In this chapter the design and analysis of tensions elements is discussed. When the adequacy (reliability) is specified and statistical descriptions of external loads and material strength properties are known, the required cross-section area can be computed. Also, the variability of the cross-section area (stated as tolerance or standard deviation) can be computed. If the variability of the cross-section area (for example, machining tolerance) is known, the mean cross-section area may be computed. The analysis of an existing or proposed design is accomplished by computing the probability of adequacy. Since, in some applications, the length of a tension element is of importance, length effects are examined.

8.1 DESIGN OF A TENSION ELEMENT [36]

With a tension element, the design task is relatively simple. If the loading to be supported is described by the normal random variable $(\bar{P}, s_p)$ and the cross-section area by $(\bar{A}, s_A)$, the applied stress random variable is

$$(\bar{f}_t, s_{f_t}) = \frac{(\bar{P}, s_P)}{(\bar{A}, s_A)} \text{ psi.}$$

By Eq. 3.13, the mean value estimate of a quotient is

$$\bar{f}_t = \bar{P}/\bar{A} \text{ psi}$$

and, by Eq. 3.14, the standard deviation estimate of a quotient is

$$s_{f_t} = \frac{1}{\bar{A}} \left(\frac{\bar{P}^2 s_A^2 + \bar{A}^2 s_P^2}{\bar{A}^2 + s_A^2} \right)^{\frac{1}{2}} \text{ psi.}$$

If ultimate strength F_{tu} is taken as the failure criterion, the allowable stress random variable is simply the ultimate strength random variable (of the material of construction). Thus

$$(\bar{F}_{tu}, s_{F_{tu}}) \text{psi.}$$

The ultimate strength random variable is assumed to be normally distributed.

Elements in Tension

The design solution is completed by substituting values into Eq. 4.6,

$$z = \frac{\bar{P}/\bar{A} - \bar{F}_{tu}}{\sqrt{s_{f_t}^2 + s_{F_{tu}}^2}}.$$

Example 1. Design a tensile element, given the following (see Fig. 8.1):

Load variate = $(\bar{P}, s_p) = (6000, 90)$ lb
Tensile ultimate (4130 steel, 160 K Ht. Trt.) strength variate

$$(\bar{F}_{tn}, s_{F_{tu}}) = (156{,}000;\ 4300)\text{ psi.}$$

Probability of adequacy: failure rate $P_f = \frac{1}{1000}$ (maximum), reliability $R = 0.999$ (from Table 2.4 the equivalent is $z \approx 3$). The element is a rod of circular cross section.

Solution. The cross-sectional area is

$$A = \pi r^2,$$

and, applying Eq. 2.61, the standard deviation estimate s_A,

$$s_A = 2\pi \bar{r} s_r,\quad \bar{A} = \pi \bar{r}^2.$$

The applied stress is

$$(\hat{f}_t, s_{f_t}) = \frac{(\bar{P}, s_P)}{(\bar{A}, S_A)} = \frac{(6000, 90)}{(\pi \bar{r}^2, 2\pi \bar{r} s_r)}\text{ psi.}$$

From consideration of manufacturing tolerances, the standard deviation is estimated.

$$s_r = \frac{0.015\bar{r}}{3}\ (99\%\text{ interval}),$$

$$s_r = 0.005\bar{r}.$$

Thus by Eq. 3.13

$$\hat{f}_t = \frac{\bar{P}}{\pi \bar{r}^2} = \frac{6000}{\pi \bar{r}^2}\text{ psi.}$$

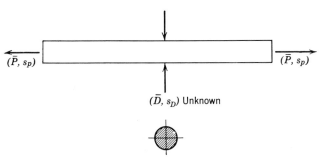

Figure 8.1 Tension element.

210 Application

Noting that $\bar{A}^2 \approx \bar{A}^2 + s_A{}^2$, the following simplification is possible. Substituting into Eq. 3.14,

$$s_{f_t}{}^2 \approx \frac{1}{\pi^2 \bar{r}^4} \left[\frac{0.0081\pi^2 \bar{r}^4 + 36[2\pi r(0.005r)]^2}{\pi^2 \bar{r}^4} \right] (1000 \text{ lb})^2$$

$$s_{f_t}{}^2 \approx \frac{0.0117}{\pi^2 \bar{r}^4} (1000 \text{ lb})^2.$$

The allowable stress is

$$(\bar{F}_t, s_{F_t}) = (156{,}000; 4300) \text{ psi.}$$

Substituting the values determined above into the "coupling formula," Eq. 4.6,

$$3 = \frac{(6/\pi \bar{r}^2) - 156}{[0.0117/\pi^2 \bar{r}^4 + (4.3)^2]^{\frac{1}{2}}}.$$

Solving the quadratic in $\bar{r}^2$,

$$\bar{r}^2 = 0.0135 \text{ in.}^2; \qquad \bar{r} = 0.116 \text{ in.},$$

$$\text{diameter } D = 0.232 \pm 0.002 \text{ in.}$$

8.2 ANALYSIS OF TENSION ELEMENTS

Analysis is after the fact of cross-sectional area selection. Once z has been determined, $p(\text{strength} > \text{stress})$ is found by consulting Table 2.4 or some standard tables of normal functions. In Examples 2a and 2b the only operation that precedes substitution of values into Eq. 4.6 is that of describing the applied stress random variables.

The examples of analysis (2a and 2b) illustrate (in addition to the calculation of reliability) the impact of a change of variate on reliability. Recalling that, in Example 1, the specified reliability was 0.999 (equivalent to $z = 3$), in Example 2a the standard deviation (s_p) on load is increased from $s_p = 90$ lb to $s_p = 400$ lb. All other values are as in Example 1. In Example 2b the mean allowable stress of $\bar{F}_{tu}$ is changed from $\bar{F}_{tu} = 156{,}000$ psi. to 150,000 psi. Increase in standard deviation on load (Example 2a) results in degradation of reliability from $R = 0.999$ to $R = 0.871$. Decreasing the allowable mean strength (Example 2b) results in degradation of reliability from $R = 0.999$ to $R = 0.894$.

Example 2a. Analysis of a Tension Element. Given the following:

$$\text{load } (\bar{P}, s_p) = (6000; 400) \text{ lb.}$$

Rod diameter $d = 0.230 \pm 0.003$ in.^{2} (see design example).
Strength, tensile ultimate $(\bar{F}_{t_u}, s_{F_{t_u}}) = (156{,}000; 4300)$ psi.

Solution.

$$\bar{A} = \frac{\pi \bar{d}^2}{4} = \frac{3.1416}{4}(0.230)^2 = 0.0416 \text{ in.}^2,$$

$$s_A = \frac{\pi}{4} 2\bar{d}s_d = \frac{3.1416}{4}(2)(0.230)(0.001) = 0.000362 \text{ in.}^2$$

With load $(\bar{P}, s_p)$ and area $(\bar{A}, s_A)$, the applied stress random variable is

$$(\bar{f}_t, s_{f_t}) = \frac{(\bar{P}, s_p)}{(\bar{A}, s_A)} = \frac{(6000; 400) \text{ lb}}{(0.0416; 0.000362) \text{ in.}^2}$$

and mean applied stress is (by Eq. 3.13)

$$\bar{f}_t = \frac{\bar{P}}{\bar{A}} = \frac{6000 \text{ lb}}{0.0416 \text{ in.}^2} = 144{,}200 \text{ psi.}$$

Table 8.1 Change of Reliability (R) with Standard Deviation of Load (s_p)

s_p	R	Mean Load $\bar{P}$
90 lb	0.999	6000 lb
200 lb	0.981	6000 lb
400 lb	0.871	6000 lb

The standard deviation of applied stress (Eq. 3.14) is

$$s_{f_t} = \frac{1}{\bar{A}}\left(\frac{\bar{A}^2 s_p^2 + \bar{P}^2 s_A^2}{\bar{A}^2 + s_A^2}\right)^{1/2} = \frac{1}{0.0416}\cdot\left[\frac{(0.0416)^2(400)^2 + (6000)^2(0.000326)^2}{(0.0416)^2 + (0.000362)^2}\right]^{1/2}$$

$$s_{f_t} = 9530 \text{ psi}$$

Thus applied stress random variable

$$(\bar{f}_t, s_{f_t}) = (144{,}200; 9530) \text{ psi}$$

and allowable stress random variable

$$(\bar{F}_{tu}, s_{F_{tu}}) = (156{,}000; 4300) \text{ psi.}$$

Substituting values into Eq. 4.6

$$z = \frac{[\bar{f}_t - \bar{F}_{tu}]}{[s_{F_{tu}}^2 + s_{f_t}^2]^{1/2}} = \frac{|144{,}200 - 156{,}000|}{[(9.53)^2 + (4.3)^2]^{1/2}(1000)} = 1.13.$$

From Table 2.4, the equivalent of $z = 1.13$ is $R = 0.871$ (see Table 8.1).

Example 2b. Given the following,

$$\text{load } (\bar{P}, s_p) = (6000; 90) \text{ lb.}$$

$$\text{rod area } (\bar{A}, s_A) = (0.0416, 0.000362) \text{ in.}^2,$$

$$\text{strength, tensile ult. } (\bar{F}_{tu}, s_{F_{tu}}) = (150{,}000; 4{,}300) \text{ psi.}$$

Compute reliability with the reduced $\bar{F}_{tu} = 150{,}000$ psi.

Solution. Having load $(\bar{P}, s_p)$ and area $(\bar{A}, s_A)$, the applied stress random variable is calculated

$$(\bar{f}_t, s_{f_t}) = \frac{(\bar{P}, s_p)}{(\bar{A}, s_A)} = \frac{(6000; 90)}{(0.0416; 0.000362)} \text{ psi,}$$

$$\bar{f}_t = \frac{\bar{P}}{\bar{A}} = \frac{6000 \text{ lb}}{0.0416 \text{ in.}^2} = 144{,}200 \text{ lb/in.}^2 \text{ (by Eq. 3.13).}$$

Table 8.2 Change of Reliability (R) with Mean Allowable Stress($\bar{F}_{tu}$)[2]

$\bar{F}_{tu}$	R	Standard Deviation Allowable Stress
156,000 psi	0.999	4300 psi
153,000 psi	0.991	4300 psi
150,000 psi	0.894	4300 psi
147,000 psi	0.863	4300 psi

By Eq. 3.14

$$s_{f_t} = \frac{1}{\bar{A}} \cdot \left(\frac{\bar{A}^2 s_p^2 + \bar{P}^2 s_A^2}{\bar{A}^2 + s_A^2}\right)^{1/2} = \frac{1}{0.0416}\left[\frac{(0.0416)^2(90)^2 + (0.00362)^2(6000)^2}{(0.0416)^2 + (0.000362)^2}\right]^{1/2},$$

$s_{f_t} = 1760$ psi,

$$(\bar{f}_t, s_{f_t}) = (144{,}200; 1760) \text{ psi,}$$

$$(\bar{F}_{tu}, s_{F_{tu}}) = (150{,}000; 4300) \text{ psi.}$$

Substituting values into Eq. 4.6 and solving for z,

$$z = \frac{|\bar{f}_t - \bar{F}_{tu}|}{[s_{f_t}^2 + s_{F_{tu}}^2]^{1/2}} = \frac{|144{,}200 - 150{,}000|}{[(1.76)^2 + (4.3)^2]^{1/2}(1000)} = 1.2480.$$

From standard tables of normal functions or from Table 2.4, $z = 1.248$ is equivalent to $R = 0.894$. (See Table 8.2.)

8.3 CONSIDERATION OF LENGTH

Since, in applications such as truss design, the length of tension elements is of importance, a statistical application of Hooke's law is given. The equation for Hooke's law is

$$\text{elongation} = \delta = \frac{Pl}{AE},$$

and

$$\text{load random variable} = (\bar{P}, s_p) \text{ lb.}$$
$$\text{length random variable} = (\bar{l}, s_l) \text{ in.}$$
$$\text{cross-sectional area random variable} = (\bar{A}, s_A) \text{ in.}^2$$
$$\text{modulus of elasticity† random variable} = (\bar{E}, s_E) \text{ psi.}$$
$$\text{elongation random variable} = (\bar{\delta}, s_\delta) \text{ in.}$$

Thus the elongation random variable $(\bar{\delta}, s_\delta)$ of a metal bar under tensile loading is

$$(\bar{\delta}, s_\delta) = \frac{(\bar{P}, s_p)(\bar{l}, s_l)}{(\bar{A}, s_A)(\bar{E}, s_E)} \text{ in.}$$

Example 3. Elongation of a Tension Element. Given the following,

$$\text{load } (\bar{P}, s_p) = (15{,}000; 750) \text{ lb,}$$
$$\text{rod cross-sectional area } (\bar{A}, s_A) = (1, 0.05) \text{ in.}^2,$$
$$\text{length } (\bar{l}, s_l) = (250, 2.5) \text{ in.,}$$
$$\text{modulus of elasticity } (\bar{E}, s_E) = (30 \cdot 10^6; 4.50 \cdot 10^5).$$

Solution.

$$(\bar{\delta}, s_\delta) = \frac{(15{,}000; 750)(250{,}750)}{(1, 0.05)(30 \cdot 10^6, 4.5 \cdot 10^5)} \text{ in.} = \frac{(\bar{u}, s_u)}{(\bar{v}, s_v)}.$$

Mean elongation is

$$\bar{\delta} = \frac{\bar{P}\bar{l}}{\bar{A}\bar{E}} = \frac{15{,}000 \cdot 250}{1.30 \cdot 10^6} = 0.125 \text{ in.}$$

† The standard deviation of E is estimated as $1\frac{1}{2}$ percent $E = s_E$.

214 Application

Before computing s_δ, s_u and s_v must be computed by applying Eq. 3.10. Thus

$$s_u = \sqrt{\bar{P}^2 s_1^2 + \bar{1}^2 s_p^2 + s_p^2 s_1^2},$$

$$s_u = \sqrt{(1.5 \cdot 10^4)2(2.5)2 + (2.5 \cdot 10^2)2(7.5 \cdot 10)^2 + (750)^2(2.5)^2} = 19.12 \cdot 10^4,$$

$$s_v = \sqrt{\bar{A}^2 s_E^2 + \bar{E}^2 s_A^2 + s_A^2 s_E^2},$$

$$s_v = \sqrt{(4.5 \cdot 10^5)^2 + (30 \cdot 10^6)^2 (0.05)^2 + (4.5 \cdot 10^5)^2 (0.05)^2} = 15.66 \cdot 10^5,$$

$$(\bar{\delta}, s_\delta) = \frac{(\bar{P}, s_p)(\bar{1}, s_1)}{(\bar{A}, s_A)(\bar{E}, s_E)} = \frac{(3.75 \cdot 10^6; 1.912 \cdot 10^5)}{(30 \cdot 10^6; 15.66 \cdot 10^5)} \text{ in.}$$

$$= \frac{(3.75; 1.912)}{(300; 15.66)} \text{ in.}$$

$$s_\delta = \frac{1}{300} \cdot \left[\frac{9 \cdot 10^4 (1.912)^2 + (37.5)^2 (15.66)^2}{(300)^2 + (15.66)^2} \right]^{\frac{1}{2}}.$$

Since $(300)^2 \approx (300)^2 + (15.66)^2$, a simplification is possible.

$$s_\delta = \frac{1}{(300)^2} \cdot [32.9 \cdot 10^4 + 344.75 \cdot 10^3]^{\frac{1}{2}} = \frac{125.1}{(300)^2} = 0.00139 \text{ in.}$$

The elongation random variable is

$$(\bar{\delta}, s_\delta) = (0.125, 0.00139) \text{ in.}$$

or mean with tolerance (where $\Delta = 3\sigma_\delta$)

$$\text{elongation} = 0.125 \pm 0.0042 \text{ in.}$$

8.4 GENERALIZATION OF SECTION

The class of cross sections not definable by single random variables requires a technique different than that used in Example 1. The equations are solved for mean cross-sectional area ($\bar{A}$). The standard deviation on area (s_A), representing a second variable, is given a specified relationship to mean area ($\bar{A}$) by means of the coefficient of variation (V_A),†

$$V_A = s_A / \bar{A}.$$

Example 4. Compute the tensile element area random variable. Given,

$$\text{load } (\bar{P}, s_p) = (6000, 90) \text{ lb},$$

$$\text{material strength } (F_{tu}, s_{F_{tu}}) = (156, 4) \text{ ksi},$$

$$\text{required reliability equivalent to } z = 5.0.$$

† See Section 3.4.10.

Solution For first calculation of cross-sectional area let
$$v_A = s_A/\bar{A} = 0.01.$$
Substitute into Eq. 4.6
$$z = \frac{\bar{f}_t - \bar{F}_{tu}}{(s_{f_t}^2 + s_{F_{tu}}^2)^{1/2}} = \frac{\bar{P}/\bar{A} - 156}{[s_{f_t}^2 + (4)^2]^{1/2}} = 5.0. \quad (a)$$

The standard deviation s_{f_t} is needed. Thus by Eq. 3.14
$$s_{f_t} = \frac{1}{\bar{A}} \cdot \left[\frac{\bar{A}^2(0.09)^2 + (6)^2(0.01\bar{A})^2}{\bar{A}^2 + (0.01\bar{A})^2}\right]^{1/2}$$

$\bar{A}^2 + (0.001\bar{A})^2 \approx \bar{A}^2$.

$$s_{f_t} = \frac{0.108}{\bar{A}}.$$

Substitute into Eq. a
$$5 = \frac{6/\bar{A} - 156}{(0.0117/\bar{A}^2 + 16)^{1/2}}$$

and
$$25 \frac{[0.0117 + 16\bar{A}^2]^{1/2}}{\bar{A}^2} = \frac{[6 - (156)\bar{A}^2]^2}{\bar{A}^2},$$

$$0.2925 + 400\bar{A}^2 = 36 + 24{,}336\bar{A}^4 - 1872\bar{A}^2,$$

$$\bar{A} = 0.141, 0.271 \text{ in.}^2,$$

$$s_A = v_A \bar{A} = (0.01), (0.141),$$

$$s_A = 0.0014 \text{ in.}^2$$

With a circular section
$$\bar{d} = 0.23 \text{ in.},$$

$$s_d = 0.001 \text{ in.}$$

PROBLEMS

1. Force per unit area called applied stress (S) is given by $s = P/A$ psi. The elongation of a rod per unit length (ϵ) is given by $\epsilon = \Delta/l$ inches per inch. Using the equations per stress and elongation, Hooke's law may be represented as, $\epsilon = s/E$, where E is the modulus of elasticity of the material of the rod.

   If the total elongation (Δ) of a steel rod $(\bar{l}, s_l) = (25.0, 0.125)$ in. long and under tensile stress $(\bar{f}_t, s_{f_t}) = (15 \cdot 10^3, 9 \cdot 10^2)$ psi is $\frac{1}{80}$ in., with coefficient of variation $s_\Delta/\bar{\Delta} = 0.065$, determine the modulus of elasticity variate $(\bar{E}, s_E)$ for the steel. [Answer: $(30 \cdot 10^6, 2.76 \cdot 10^6)$]

2. The formula for the membrane stress in a spherical pressure vessel is given [19] as
$$\text{applied stress} = f = \frac{Pr}{2t},$$

   $P = $ pressure, psi,

   $r = $ mean radius, in.,

   $t = $ wall thickness, in.

216 Application

The material of the vessel is an aluminum alloy for which the tensile yield strength is

$$(\bar{F}_{ty}, s_{F_{ty}}) = (52{,}000, 4000) \text{ psi.}$$

Given

$$(\bar{r}, s_r) = (20.00, 0.125) \text{ in.,}$$

$$\text{nominal wall} = \bar{t} = 0.156 \text{ in.,}$$

$$\text{nominal pressure} = \bar{P} = 2600 \text{ psi (at } T = 75°\text{F)}.$$

Pressure varies (random) due to temperature change, the 3σ value is $\Delta \bar{T} = \pm 35°\text{F}$

Determine the tolerance range, where tolerance $\Delta = \pm 3s_t$.

CHAPTER 9

Simple Beams: Concentrated Loading [37]

In this chapter (and in Chapters 10 to 12), the design and analysis of simple beams, under assumptions of determinacy, are discussed. In the section on cantilever beams, the indeterminate propped cantilever beam is discussed. In the section on uniformly loaded beams, the case of a beam with built-in ends is treated.

9.1 SIMPLE BEAM: SINGLE LOAD

As shown in Fig. 9.1, the model beam is free to rotate at A and B, with rollers under the support at B. The weight of the beam is assumed to be negligible. The known random variables are

$$\text{load} = (\bar{p}, s_p),$$
$$\text{beam length} = (\bar{l}, s_l),$$
$$\text{load location} = (\bar{a}, s_a).$$

In determining maximum stress at the point of maximum moment, the first step requires a determination of the reactions at A and B in response

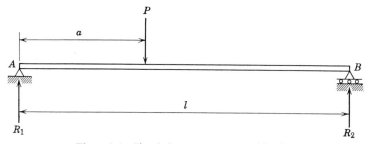

Figure 9.1 Simple beam: concentrated load.

218 Application

to the load (P, σ_P):

$$\text{reaction at } A = (\bar{R}_1, s_{R_1}),$$
$$\text{reaction at } B = (\bar{R}_2, s_{R_2}).$$

Taking moments about the support at A,

$$(\bar{R}_2, s_{R_2})(\bar{l}, s_l) - (\bar{P}, s_P)(\bar{a}, s_a) = 0,$$

$$(\bar{R}_2, s_{R_2}) = \frac{(\bar{P}, S_P)(\bar{a}, S_a)}{(\bar{l}, s_l)} \text{ lb},$$

$$(\bar{R}_1, s_{R_1}) = (\bar{P}, s_P) - (\bar{R}_2, s_{R_2})$$

$$= (\bar{P}, s_P) - \frac{(\bar{P}, s_P)(\bar{a}, s_a)}{(\bar{l}, s_l)},$$

$$(\bar{R}_1, s_{R_1}) = \left[(1, 0) - \frac{(\bar{a}, s_a)}{(\bar{l}, s_l)}\right] \cdot (\bar{P}, s_P) \text{ lb}.$$

Example 1. Reactions. The right reaction $(\bar{R}_R, s_{R_R})$ of Example 3 is computed. Given

$$= (\bar{P}, s_P) = (6000; 300) \text{ lb}$$
$$= (\bar{l}, s_l) = (10.0, 0.0033) \text{ ft}$$
$$= (\bar{a}, s_a) = (4.0, 0.0033) \text{ ft},$$

the right reaction random variable is

$$(\bar{R}_R, s_{R_R}) = \frac{(\bar{P}, s_P)(\bar{a}, s_a)}{(\bar{l}, s_l)} \text{ lb}$$

$$= \frac{(6000; 300)(4.0, 0.0033)}{(10.0, 0.0033)} \text{ lb}$$

$$= \frac{(6.0, 0.3)(4., 0.0033)}{(10.0, 0.0033)} 1000 \text{ lb}.$$

The product RV† in the numerator is computed (Eqs. 3.9 and 3.10).

$$(6, 0.3)(4, 0.0033)1000 = (24, 1.12)1000 \text{ ft-lb}.$$

The quotient RV is computed (Eqs. 3.13 and 3.14):

$$(\bar{R}_R, s_{R_R}) = \frac{(24, 1.20)1000}{(10., 0.0033)} = (2.4, 0.12)1000 \text{ lb}$$

where

$$s_{R_R} = \frac{1}{10}\left[\frac{10^2(1.2)^2 + 24^2(0.0033)^2}{10^2 + (0.0033)^2}\right]^{\frac{1}{2}} 1000\text{lb} = 0.12(1000) \text{ lb}.$$

† RV equals random variable.

9.2 STRESSES IN BEAMS

Next we examine the stresses produced in the beam during bending (Fig. 9.2). (*Note.* The dotted lines in Fig. 9.2 indicate variation.)

Not only must external load $(\bar{P}, s_P)$ and reaction $(\bar{R}_1, s_{R_1})$ be considered, but also the internal force variates distributed over $m - n$. These represent the variable actions of the right portion of the beam on the left portion. The internal force variates must "equal" external force variates $(\bar{P}, s_P)$ and $(\bar{R}_1, s_{R_1})$. The magnitude of the vertical force variate is

$$(\bar{V}, s_V) = (\bar{R}_1, s_{R_1}) - (\bar{P}, s_P).$$

The resisting couple is

$$(\bar{M}, s_M) = (\bar{R}_1, s_{R_1})(\bar{x}, 0).$$

At the x-section (Fig. 9.2) directly under the load $(\bar{P}, s_P)$

$$\sum Y = 0,$$

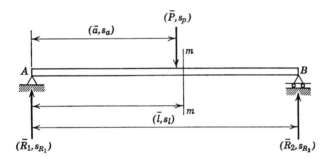

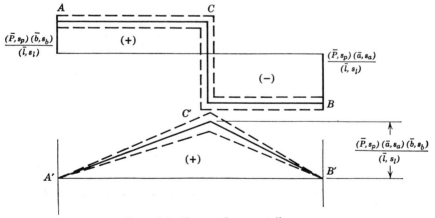

Figure 9.2 Shear and moment diagram.

and the formulas for vertical shear variate $(\bar{V}, s_V)$ and moment variate $(\bar{M}, s_M)$ are [19]:

$$(\bar{V}, s_V) = \frac{(\bar{P}, s_P)(\bar{b}, s_b)}{(\bar{l}, s_l)}$$

and

$$(\bar{M}, s_M) = \frac{(\bar{P}, s_P)(\bar{a}, s_a)(\bar{b}, s_b)}{(\bar{l}, s_l)}.$$

With the above formulas, it is possible to solve a problem of beam design. Example 2 illustrates a beam whose cross section can be expressed as a function of one variable. The section is specified as tubular, and the radius variate $(\bar{r}, s_r)$ is computed.

Example 2. Design of Beam: Single Load. The load vector is assumed perpendicular to the axis of the beam (see Fig. 9.3).

The load random variable is

$$(\bar{P}, s_P) = (6070; 200) \text{ lb},$$

and the dimensions are

$$l = 120 \pm \tfrac{1}{8} \text{ in.}, \quad \Delta l = \tfrac{1}{8} \text{ in.},$$

$$a = 72 \pm \tfrac{1}{8} \text{ in.}, \quad \Delta a = \tfrac{1}{8} \text{ in.}$$

Tolerances cover the 99 percent interval (see Section 7.6)

$$3s_l = 3s_a = 0.125 \text{ in.},$$

$$s_l = s_a = \frac{0.125}{3} = 0.0416 \text{ in.}$$

Thus

$$\text{length } (\bar{l}, s_l) = (120, 0.0416) \text{ in.},$$

$$(\bar{a}, s_a) = (72, 0.0416) \text{ in.},$$

$$\bar{b} = \bar{l} - \bar{a} = 48 \text{ in. (by Eq. 3.6)},$$

$$s_b = \sqrt{s_l^2 + s_a^2} = 0.0566 \text{ in. (by Eq. 3.7)},$$

$$(\bar{b}, s_b) = (48, 0.0566) \text{ in.}$$

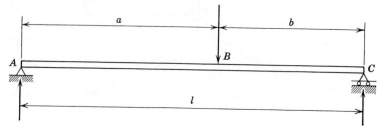

Figure 9.3 Example 3.

From the data given

$$(\bar{M}, s_M) = \frac{(P, s_P)(a, s_a)(b, s_b)}{(l, s_l)}$$

$$= \frac{(6070; 200)(72; 0.0416)(48; 0.0566)}{(120, 0.0416)}.$$

First compute the $RVu = ab/l$:

$$u = \frac{ab}{l} = \frac{a(l-a)}{l} = \frac{al-a^2}{l}.$$

Applying Eq. 2.61,†

$$\frac{\partial u}{\partial a} = \frac{l-2a}{l}; \quad \frac{\partial u}{\partial l} = \frac{al-(al-a^2)}{l^2} = \frac{a^2}{l^2}$$

$$s_u = \left[\left(\frac{l-2\bar{a}}{l}\right)^2 s_a^2 + \left(\frac{\bar{a}^2}{l^2}\right)^2 s_l^2\right]^{\frac{1}{2}} = 0.0356 \text{ in.}$$

$$\bar{u} = \frac{\bar{a}\bar{b}}{\bar{l}} = 28.8 \text{ in.}$$

$(\bar{M}, s_M) = (6070; 200)(28.8, 0.0356)$ in.-lb (Eqs. 3.9 and 3.10)
$= (174{,}816; 5764)$ in.-lb.

The maximum fiber stress occurs in the extreme fibers.

$$s = \frac{Mc}{I}$$

where s = applied (fiber) stress, psi,
 M = external bending moment, in.-lb,
 c = distance, neutral axis to extreme fibers, in.,
 I = moment of inertia of beam cross section about the neutral axis, in.⁴

$$f = \frac{M}{(I/c)}$$

$$(\bar{f}, s_f) = \frac{(\bar{M}, s_M)}{(\bar{I}/c, s_{I/c})}.$$

An expression for applied stress is written in terms of the unknown cross-section geometry. A tubular cross section is specified and the mean radius ($\bar{r}$) computed.

The approximate expression for moment of inertia (I) of a tubular section is

$$I = \pi r^3 t,$$

where r = outside radius,
 t = wall thickness.

Assume, on the basis of known requirements for stability of a section that

$$\frac{\bar{r}}{t} = 50; \quad t = \frac{\bar{r}}{50};$$

$r = c$, where c is the distance from the neutral axis to the extreme fibers. For I/c

$$\frac{I}{c} = \frac{\pi \bar{r}^3 t}{\bar{r}} = \pi \bar{r}^2 \left(\frac{\bar{r}}{50}\right) = 0.0628 \bar{r}^3.$$

† See Section 3.1.1, correlation.

222 Application

Based on known manufacturing tolerances, a reasonable standard deviation estimate is

$$s_r = 0.015\bar{r}.$$

Applying Eq. 2.61, the standard deviation (approximate) of r^3 is

$$s_{(r^3)} = 3\bar{r}^2 s_r = 0.045\bar{r}^3$$

and

$$s_{I/c} = 0.0628 \cdot (0.045\bar{r}^3) = 0.00283\bar{r}^3.$$

The expression for *applied stress* (s) is

$$(\bar{s}, s_s) = \frac{(174{,}820;\ 5764)}{(0.0628\bar{r}^3,\ 00283\bar{r}^3)} \text{ psi},$$

$$(\bar{s}, s_s) = \left(\frac{2.7837}{\bar{r}^3} \cdot 10^6;\ \frac{8.550}{\bar{r}^3} \cdot 10^4\right) \text{psi}.$$

Allowable stress (S) is given by the random variable for material strength. This, for 4130 steel, is

$$(\bar{S}, s_S) = (170{,}100;\ 4760)\text{ psi}.$$

The required reliability is taken as $R = 0.999999$ which corresponds to $z \approx 5$. Substituting into the "coupling formula," Eq. 4.6,

$$z = \frac{\bar{s} - \bar{S}}{\sqrt{s_s^2 + s_S^2}} = 5 = \frac{(2.7837/\bar{r}^3)10^6 - 171{,}000}{[(8.550/\bar{r}^3 \cdot 10^4)^2 + (4760)^2]^{1/2}}.$$

Solving for $\bar{r}$,

$$\bar{r} = 2.72 \text{ in.}, \quad t = 0.0544 \text{ in.}$$

and

$$r = 2.72 \pm 0.041 \text{ in.}, \quad t = 0.0544 \text{ in.}$$

Example 3. Design of Beam: Single Load (section not specified). The section modulus (I/c) with allowable tolerance is computed (see Fig. 9.4).

Known:

Load $(\bar{P}, s_P) = (6{,}000;\ 300)$ lb

Materials strength $(\bar{S}, s_S) = (171, 5)$ ksi

Beam length, $L = 120 \pm 0.125$ in.

Standard deviation estimate of L, $s_L = \dfrac{\Delta L}{3} = \dfrac{0.125}{3} = 0.0416$ in.

$$= 0.0033 \text{ ft}$$

Beam length $(\bar{L}, s_L) = (10, 0.0033)$ ft

Load position a (Fig. 9.4):

$$a = 48 \pm 0.125 \text{ in.}$$

Standard deviation estimate, $s_a = \dfrac{\Delta a}{3} = \dfrac{0.125}{3} = 0.0033$ ft

Load position $(\bar{a}, s_a) = (4.0, 0.0033)$ ft

Load position b:

$$b = 72 \text{ in.} = 6 \text{ ft}$$

$$s_b = \sqrt{s_L^2 + s_a^2} = 0.0047 \text{ ft}$$

Load position $(\bar{b}, s_b) = (6, 0.0047)$ ft

Simple Beams: Concentrated Loading   223

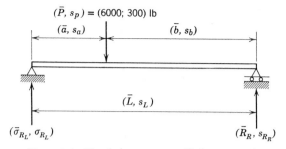

Figure 9.4 Simple beam: unspecified cross section.

Required:
Member section modulus to satisfy $z = 5.0$,
Reactions,

$$\text{Mean left reaction, } \bar{R}_L = \frac{\bar{P}\bar{b}}{L} = \frac{6 \cdot 6}{10} = 3.6 \ (1000 \text{ lb}).$$

Applying Eq. 3.14,

$$\text{standard deviation on } R_L, \ s_{R_L} = \frac{1}{L} \left[\frac{(\bar{P}\bar{b})^2 s_L^2 + L^2 s_{Pb}^2}{L^2 + s_L^2} \right]^{1/2}.$$

Compute the product function $(\bar{P}\bar{b}, s_{Pb})$. Mean of product $= \bar{P}\bar{b}$.
Applying Eq. 3.10,

$$\text{standard deviation } s_{Pb} = [\bar{P}^2 s_b^2 + \bar{b}^2 s_P^2 + s_b^2 s_P^2]^{1/2}.$$

Thus variance of Pb is (approximate)

$$s_{Pb}^2 = \bar{P}^2 s_b^2 + \bar{b}^2 s_P^2.$$

Substituting into the equation for s_{R_L},

$$s_{R_L} = \frac{1}{L} \cdot \left[\frac{(\bar{P}\bar{b})^2 s_L^2 + L^2(\bar{P}^2 s_b^2 + \bar{b}^2 s_P^2)}{L^2 + sL^2} \right]^{1/2}$$

$$= \frac{1}{10} \left[\frac{(6 \cdot 6)^2(0.0033)^2 + 10^2(6^2 \cdot 0.0047^2 + 6^2 \cdot 0.3^2)}{10^2 + (0.0033)^2} \right]^{1/2}.$$

Note. $s_L^2 = (0.0033)^2 = 1.089 \cdot 10^{-5} \ll 10^2$.

$$s_{R_L} = \frac{1}{100} [1.41 \cdot 10^{-2} + 3 \cdot 24 \cdot 10^2]^{1/2} \ 1000 = \frac{18000}{100} = 0.18 \ (1000) \text{ lb}$$

$(\bar{R}_L, s_{R_L}) = (3.6, 0.18)(1000)$ lb.

The coefficient of variation (V_{R_L}) is (see Section 3.4.10):

$$V_{R_L} = s_{R_L}/\bar{R}_L = 0.18/3.6 = 0.05$$

$$\text{Mean right reaction, } \bar{R}_R = \frac{\bar{P}\bar{a}}{L} = \bar{P} - \bar{R}_L = 6 - 3.6 = 2.4 \ (1000) \text{ lb}$$

$$(\bar{R}_R, s_{R_R}) = \frac{(\bar{P}, s_P)(\bar{a}, s_a)}{(L, s_L)}.$$

$(\bar{R}, s_R) = (2.4, 0.12)$ kps

224 Application

The coefficient of variation (V_{R_R}) is

$$V_{R_R} = \frac{s_{R_R}}{\bar{R}_R} = \frac{0.12}{2.4} = 0.05.$$

Maximum moment

$$(\bar{M}, s_M) = (\bar{R}_L, s_{R_L})(\bar{a}, s_a) \text{ ft-kps}$$
$$= (3.6, 0.18)(4, 0.0033) \text{ ft-kps.}$$

Mean moment
$$\bar{M} = 3.6 \cdot 4 = 14.4 \text{ ft-kps.}$$

Standard deviation

$$s_M = (\bar{R}_L^2 \bar{s}_a^2 + \bar{a}^2 s_{R_L}^2)^{\frac{1}{2}}$$
$$= [(3.6)^2(0.003)^2 + (4)^2(0.18)^2]^{\frac{1}{2}} = 0.72 \text{ ft-kps}$$
$$(\bar{M}, s_M) = (14.4, 0.72) \text{ ft-kps.}$$

Applied Stress. At this point we may solve for applied stress F, by utilizing Eq. 6.4.

$$z = \frac{|\bar{F} - \bar{S}|}{\sqrt{s_F^2 + s_S^2}}$$

$$(\bar{S}, s_S) = (171, 5) \text{ ksi.}$$

Applying Eq.4.6

$$5 = \frac{|\bar{F} - 171|}{[(0.03\bar{F})^2 + 5^2]^{\frac{1}{2}}}$$

$$\bar{F}^2 - 342\bar{F} + 28{,}616 = 0$$

Solving the quadratic in $\bar{F}$

$$\bar{F} = \frac{342 \pm 50}{2} = 146; 196 \text{ ksi.}$$

Using $\bar{F} = 146$ ksi,

$$s_F = V F = 0.03 \cdot F = 0.03 \cdot 146 = 4.38 \text{ ksi.}$$

Section modulus:

$$(I/c, \sigma_{I/c}) = \frac{(\bar{M}, s_M)}{(\bar{F}, s_F)}.$$

Mean section modulus:

$$(\bar{I}/\bar{c}) = \frac{\bar{M}}{\bar{F}} = \frac{14.4 \,(12)}{146.} = 1.182 \text{ in.}^3$$

Standard deviation estimate of section modulus,

$$\sigma_{I/c} = \frac{1}{\bar{F}} \cdot \left(\frac{\bar{M}^2 s_F^2 + \bar{F}^2 s_M^2}{\bar{F}^2 + s_F^2} \right)^{\frac{1}{2}}$$

$$\sigma_{I/c} = \frac{1}{146} \left[\frac{173^2 (4 \cdot 38)^2 + 146^2 (0.72 \cdot 12)^2}{146^2 + 4.38^2} \right]^{\frac{1}{2}} = 0.047 \text{ in.}^3$$

9.3 MULTIFORCE BEAM

Next the simple beam acting under the influence of more than one concentrated load (see Fig. 9.5) is discussed.

Imagine the beam $A - B$ cut at $m - n$, a distance x from support A. Not only the external loads variates $(\bar{P}_1, s_{P_1})$, $(\bar{P}_2, s_{P_2})$ and reaction variate $(\bar{R}_1, s_{R_1})$ are considered, but also the internal force variates distributed over $m - n$. These represent the action of the right portion of the beam on the left portion. Internal force variates must be of such magnitude as to equal the external variates. The magnitude of the vertical force variate is given by

$$(\bar{V}, s_V) = (\bar{R}_1, s_{R_1}) - (\bar{P}_1, s_{P_1}) - (\bar{P}_2, s_{P_2}).$$

The magnitude of the resisting couple is

$$M = R_1 x - P_1(x - C_1) - P_2(x - C_2)$$

$$(\bar{M}, s_M) = (\bar{R}_1, s_{R_1})(\bar{x}, 0) - (\bar{P}_1, s_{P_1})[(\bar{x}, 0) - (\bar{C}_1, s_{C_1})]$$
$$- (\bar{P}_2, s_{P_2})[(\bar{x}, 0) - (\bar{C}_2, s_{C_2})]$$

$$(\bar{M}, s_M) = \overline{(xR_1, xs_{R_1})} - (\bar{P}_1, s_{P_1})\overline{(x - C_1, s_{C_1})}$$
$$- (\bar{P}_2, s_{P_2})\overline{(x - C_2, s_{C_2})}.$$

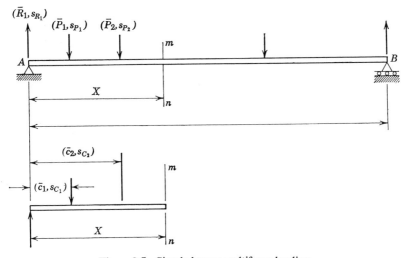

Figure 9.5 Simple beam: multiforce loading.

CHAPTER 10

Simple Beams: Distributed Loads

The concept of uniform loading is an idealization. Because of unavoidable variations, loading over the span of any beam is not uniformly distributed. The magnitude of variation under different design conditions will be large or small, depending on specific circumstances, but never ideally uniform. Consequently, although uniform loading intensity may be stated as q pounds per inch over the span of a beam, it is understood that some measure of tolerance must be added.

It is instructive to present a simple conventional design example prior to discussion of the probabilistic method. This provides the opportunity to view some aspects of the design problem where treatment of its statistical nature adds new insight.

Example 1. Conventional Design. The loading in this example is an ultimate value[†], and the flexural strength (yield) is a minimum guaranteed value. It is noted that each design variable value in the example is a discrete number.

As shown in Fig. 10.1, the beam is free to rotate about points A and B, and the end at B is supported on frictionless rollers. Compute the required cross section, given

length of span 12 ft
loading 300 lb/ft
material: wood (allowable flexural stress $f_t = 1200$ psi)
depth of cross section (d), twice the width (b)
weight of beam: negligible

Solution.
Total weight, $W = 12 \times 300 = 3600$ lb

Maximum positive moment (M) $M = \dfrac{Wl}{8} = \dfrac{(3600)(144)}{8} = 64{,}800$ in.-lb

[†] For discussion, see Chapter 1.

Simple Beams: Distributed Loads 227

Figure 10.1 Simple beam: uniform loading.

Required value of I/c (section modulus), $\left(\dfrac{I}{c}\right) = \dfrac{M}{f_t} = \dfrac{64{,}800}{1200} = 54 \text{ in.}^3$

The moment of inertia of a rectangular section is $I = b\,d^3/12$ and $c = b = d/2$.

$$\left(\dfrac{I}{c}\right) = \dfrac{b\,d^3}{12}\left(\dfrac{2}{d}\right) = \dfrac{b\,d^2}{6} = \dfrac{d^3}{12} \text{ in.}^3,$$

$$\dfrac{I}{c} = \dfrac{d^3}{12} = 54 \text{ in.}^3,$$

$$d^3 = 648; \quad d = 8.65 \text{ in.} \quad \text{and} \quad b = 4.32 \text{ in.}$$

The system discussed is that shown in Fig. 10.2.

Random variable parameters† that influence performance under uniformly distributed loading are

modules of elasticity	$(\bar{E}, s_E)$ psi,
moment of inertia	$(\bar{I}, s_I)$ in.^{4},
beam span	$(\bar{l}, s_l)$ in.,
load	$(\bar{q}, s_q)$ lb/lineal in.,
c and x	$(\bar{c}, s_c)$ and $(\bar{x}, s_x)$ in.

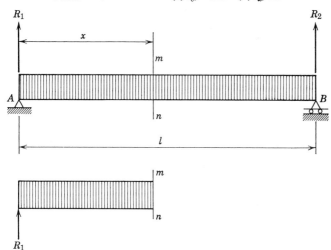

Figure 10.2 Uniform loading system.

† The word parameter is sometimes used in lieu of variable.

10.1 REACTIONS

Reactions at the support points A and B (Fig. 10.2) are

$$(\bar{R}_1, s_{R1}) = \frac{(\bar{q}, s_q)(\bar{l}, s_l)}{(2, 0)} \text{ lb}$$

and

$$(\bar{R}_2, s_{R_2}) = (\bar{q}, s_q)(\bar{l}, s_l) - (\bar{R}_1, s_{R_1})$$

$$= (\bar{q}, s_q)(\bar{l}, s_l) - \frac{(\bar{q}, s_q)(\bar{l}, s_l)}{(2, 0)}$$

$$= \frac{(\bar{q}, s_q)(\bar{l}, S_l)}{(2, 0)}.$$

From the relationship above, and by Eq. 3.9,

$$\bar{R}_1 = \bar{R}_2 = \frac{\bar{q}\bar{l}}{2} \text{ lb.}$$

The standard deviation estimators, by Eq. 3.10, are

$$s_{R_1} = s_{R_2} = \frac{\sqrt{\bar{q}^2 s_l^2 + \bar{l}^2 s_q^2 + s\bar{l}^2 \cdot s_q^2}}{(2, 0)} \text{ lb.}$$

Example 2. Reaction Computation. Given a beam supported as in Fig. 10.2, with loading

$$q = 15 \pm 1.0 \text{ lb/in.},$$

Complete the reaction random variables R_1 and R_2. Assume that tolerance covers a 95 percent interval, thus

$$\Delta q = 2s_q.$$

The estimate of standard deviation s_q is

$$s_q = \frac{\Delta q}{2} = \frac{(1)\text{lb}}{2} = 0.5 \text{ lb/in.}$$

Length of beam is 120 ± 0.75 in. The estimate of length standard deviation s_l is

$$s_l = \frac{\Delta l}{2} = \frac{0.75}{2} = 0.375 \text{ in.}$$

The reaction random variables are

$$(\bar{R}_1, s_{R_1}) = (\bar{R}_2, s_{R_2}) = \tfrac{1}{2}(\bar{q}, s_q)(l, s_l)$$
$$(\bar{R}_1, s_{R_1}) = (\bar{R}_2, s_{R_2}) = \tfrac{1}{2}(15, 0.50)(120, 0.375) \text{ lb.}$$

Applying Eqs. 3.9 and 3.10,

$$(\bar{R}_1, s_{R_1}) = (\bar{R}_2, s_{R_2}) = (900, 32) \text{ lb.}$$

10.2 SHEAR

To investigate the stress variates over a cross section $m - n$ (Fig. 10.2), consider the equilibrium of the left portion of the beam. The algebraic sum of the force variates to the left of $m - n$, called shearing force random variable V, is

$$(\bar{V}, s_v)\dagger = (\bar{q}, s_q)[\tfrac{1}{2}(\bar{l}, s_l) - (\bar{x}, 0)].$$

Applying Eqs. 3.6 and 3.7

$$(\bar{V}, s_v) = (\bar{q}, s_q)\left(\frac{\bar{l}-\bar{x}}{2}, s_{l/2}\right) \cdot \ddagger$$

10.3 MOMENT

The expression for the moment variate M at the cross section $m - n$ is

$$(\bar{M}, s_M) = \left[(\bar{R}_1, s_{R_1})(\bar{x}, 0) - \frac{(\bar{q}, s_q)(\bar{x}, 0)^2}{(2, 0)}\right] \text{in-lb}$$

$$= \frac{(\bar{q}, s_q)(\bar{x}, 0)}{(2, 0)} \cdot [(\bar{l}, s_l) - (\bar{x}, 0)]$$

$$(\bar{M}, s_M) = (\overline{qx}, s_{qx})(\bar{l} - \bar{x}, s_l) \text{ in-lb.}$$

10.3.1 Bending Moment Signs

The sign convention as shown in Fig. 10.3 is used. Moments which produce bending convex downward are considered positive.

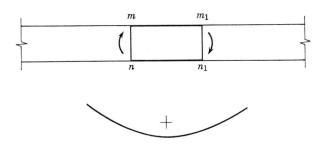

Figure 10.3 Bending moment signs.

† *Note.* x is treated as a constant or degenerate variate $x = (\bar{x}, 0)$.
‡ See Section 7.1.

230 Application

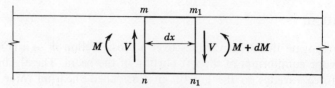

Figure 10.4 Shearing force signs.

10.3.2 Shearing Force Signs

According to theory, bending moments are not equal at adjacent cross sections. From Fig. 10.4

$$dM = V\,dx$$

and

$$\frac{dM}{dx} = V.$$

It was shown that

$$(\bar{V}, s_v) = (\bar{q}, s_q)\left(\frac{\bar{l}}{2} - \bar{x},\, s_{l/2}\right).$$

Shear is a random variable, $(\bar{V}, s_v)$, and thus the derivative dM/dx is also a random variable,

$$\left[\overline{\frac{dM}{dx}},\, s_{dM/dx}\right] = (\bar{V}, s_v).$$

The derivative dM/dx may be visualized as in Fig. 10.5, where

$$\overline{\tan \theta} = \frac{\bar{a}}{\bar{b}}$$

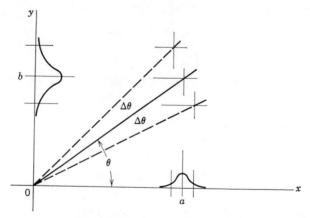

Figure 10.5 Angular dispersion.

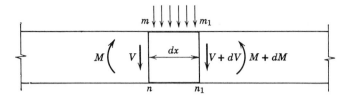

Figure 10.6 Action: load elements.

and

$$\bar{\theta} = \tan^{-1}\left(\frac{\bar{a}}{\bar{b}}\right).$$

Shearing force equals to rate of change of bending moment. Consider a beam with "uniform distributed load" of intensity q acting between the cross sections mn and m_1n_1 (Fig. 10.6). The total load acting on the element dx is $(\bar{q}, s_q)\, dx$. If $(\bar{q}, s_q)$ is positive, it is concluded (from the equilibrium of the element) that the shearing force at a cross section $m_1 - n_1$ is different from that at $m - n$ by the amount

$$dv = -(\bar{q}, s_q)\, dx$$

$$\frac{dv}{dx} = -(\bar{q}, s_q).$$

Thus rate of change of shearing force is equal to the intensity of the load variate (with negative sign). Also as above, we have

$$\left(\overline{\frac{dv}{dx}}, s_{dv/dx}\right) = -(\bar{q}, s_q).$$

Taking the moments of all force variates on the element,

$$dM = (\bar{V}, s_v)\, dx - (\bar{q}, s_q)\, dx\, (dx/2).$$

Dropping the second term on the right as a small second-order quantity

$$\left(\overline{\frac{dM}{dx}}, s_{dM/dx}\right) = (\bar{V}, s_v).$$

In practical applications it is of importance to find the cross section at which the bending moment has its maximum or minimum values. According to theory, the slope of the bending moment diagram at any point is equal to the shearing force. The bending moment diagram indicates a maximum or minimum value at the cross section where the shearing force changes sign. Actually, we cannot with certainty locate the cross section at which moment is maximum. Utilizing the mean values of each parameter, we may compute

232 Application

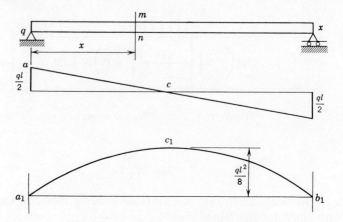

Figure 10.7 Shear and moment diagram.

the most probable location of the maximum moment. Similarly, utilizing the variance, we may compute bounds in the location of maximum moment. As is seen from

$$v = q\left[\left(\frac{l}{2}\right) - x\right]$$

$$M = \frac{qx}{2}(l - x).$$

The curve of the mean bending moment tends to be parabolic, with vertical axis near the ¢ of symmetry of the beam (Fig. 10.7).

The moments at the ends vanish and the maximum occurs (usually) near the middle (where V changes sign). Thus at

$$(\bar{x}, s_x) = (0, 0),$$

$$(\bar{M}, s_M) = \frac{(\bar{q}, s_q)(0, 0)}{(2, 0)}[(\bar{l}, S_l) - (0, 0)] = (0, 0)$$

and

$$(\bar{x}, s_x) = (\bar{l}, s_l),$$

$$(\bar{M}, s_M) = \frac{(\bar{q}, s_q)(\bar{l}, s_l)}{(2, 0)}[(\bar{l}, s_l) - (\bar{l}, s_l)] = (0, 0).$$

Maximum moment is obtained by substituting $x = l/2$ in the expression

$$M = \frac{qx}{2}(l - x) = \frac{ql^2}{2}$$

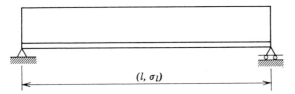

Figure 10.8 Example 3: beam diagram.

and

$$(\bar{M}, s_M) = \frac{(\bar{q}, s_q)(x, 0)}{2}[(\bar{l}, s_l) - (x, 0)]$$

$$= \frac{(\bar{q}, s_q)(\bar{l}, s_l)^2}{(2, 0)}.$$

Example 3. Moment Computation. Because of the usual method of support for beams, the variation in l is often somewhat greater than dimensions would indicate. A knife-edge support at the ends of a beam (see Fig. 10.8) is an idealization.

Given

$$(\bar{l}, s_l) = (120, 6) \text{ in.}$$
$$(\bar{q}, s_q) = (15, 1) \text{ lb/in.},$$

compute the maximum moment — $(\bar{M}, s_M)$.

$$(\bar{M}, s_M) = \frac{(\bar{q}, \bar{s}_q)(\bar{l}, s_l)^2}{(8, 0)} \text{ in-lb}.$$

$$(\bar{M}, s_M) = \frac{(15, 1)(120, 6)^2}{(8, 0)} \text{ in-lb}.$$

First compute the RV† of l^2, employing Eqs. 3.17 and 3.21.

$$(\bar{l}, s_l)^2 = [(\bar{l}^2 + s_l^2), \sqrt{4\bar{l}^2 s_l^2 + 2s_l^4}],$$
$$(\bar{l}, s_l)^2 = (14{,}400; 1440) \text{ in.}^2$$

Next compute the RV of ql^2, employing Eqs. 3.9 and 3.10

$$(\bar{q}, s_q)(\bar{l}, s_l)^2 = (15, 1)(14{,}400; 1445) \text{ in-lb},$$
$$\bar{q}\bar{l}^2 = (15)14{,}400 = 216{,}000 \text{ in-lb},$$
$$s_{(ql^2)} = \sqrt{(\bar{l}^2)s_q^2 + \bar{q}^2 s_{l^2}^2 + s_q^2 s_{l^2}^2} \text{ in-lb}$$
$$= 25{,}960 \text{ in-lb}.$$

Thus

$$(\bar{M}, s_M) = \left(\frac{216{,}000}{8}, \frac{25{,}960}{8}\right) \text{ in-lb}.$$

† Random variable.

10.4 UNIFORM LOAD ON PART OF SPAN

If a uniform load covers part of the span (Fig. 10.9), three sections of the beam, namely a, b, and c, must be considered.

The uniformly distributed load is replaced by its equivalent, qb. Equations are written next for moments with respect to A and B.

The reaction (R_2) is [36]:

$$R_2 = \frac{qb}{l}\left(a + \frac{b}{2}\right).$$

Random variables l, b, and c are independent.

The length a is

$$a = l - b - c.$$

Applying Eqs. 3.6 and 3.7,

$$\bar{a} = \bar{l} - \bar{b} - \bar{c}.$$

$$s_a = \sqrt{s_b^2 + s_c^2 + s_l^2}.$$

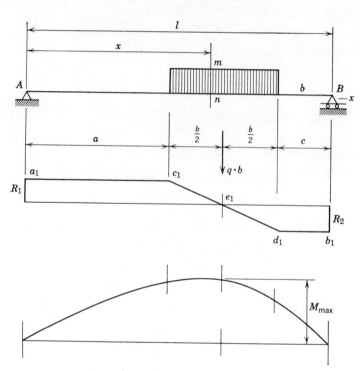

Figure 10.9 Shear and moment diagram.

The reaction R_1 is

$$R_1 = \frac{q}{l}\left(bc + \frac{b^2}{2}\right) \text{ lb.}$$

Consider the term $bc + b^2/2$. By the methods of moment generating functions (Section 2.9.3), the moments of $(xy + x^2/2)$ are

$$\mu_{(xy+x^2/2)} = \left(\mu_x \mu_y + \frac{\mu_x^2 + s_x^2}{2}\right)\dagger$$

$$s_{(xy+x^2/2)} = \sqrt{\mu_y^2 \mu_x^2 + 2\mu_x^2 \mu_y^2 s_x^2 + \mu_x^2 s_x^2 + s_x^2 s_y^2}.\dagger$$

$$(\bar{R}_1, s_{R_1}) = \frac{(\bar{q}, s_q)(\mu_{ac+b^2/2}, s_{bc+b^2/2})}{(\bar{l}, s_l)} \text{ lb.}$$

The reaction at R_2 is

$$R_2 = \frac{q}{l}\left(ba + \frac{b^2}{2}\right) \text{ lb}$$

and $a = l - b - c$.

$$R_2 = \frac{q}{l}\left[b(l - c - b) + \frac{b^2}{2}\right],$$

$$R_2 = \frac{q}{l}\left[b(l - c) - \frac{b^2}{2}\right],$$

where $b < l - c$.

$$z = xy - \frac{x^2}{2}.$$

By the methods of moment generating functions (Section 2.9.3),

$$\mu_z = \left(\mu_x \mu_y - \frac{\mu_x^2 + s_x^2}{2}\right),$$

and

$$s_z = \sqrt{\mu_y^2 s_x^2 - 2\mu_x \mu_y s_x^2 + \mu_x^2 s_x^2 + \mu_x^2 s_y^2}.$$

The expression for R_2 is

$$(\bar{R}_2, s_{R_2}) = \frac{(\bar{q}, s_q)}{(\bar{l}, s_l)}(\mu_z, s_z) \text{ lb.}$$

Shearing force and bending moment variates (for $0 < x < a$) are

$$(\bar{V}, s_V) = (\bar{R}_1, s_{R_1})$$

and

$$(\bar{M}, s_M) = (\bar{R}_1, s_{R_1})(x, 0)$$

† Details of the derivations are left as exercises.

236 Application

Considering a cross section $m - n$, shearing force is the difference
$$V = R_1 - q(x - a),$$
and, since
$$R_1 = \frac{q}{l}\left(bc + \frac{b^2}{2}\right),$$
$$V = \frac{q}{l}\left(bc + \frac{b^2}{2}\right) - q(x - l + b + c),$$
$$V = q\left(\frac{bc}{l} + \frac{b^2}{2l} - x + l - b - c\right),$$
$$E(z) = \mu_z = \frac{\mu_a \mu_c}{l} + \frac{\mu_b^2 + \sigma_b^2}{2l} - \mu_b - \mu_c + c,$$
$$\sigma_z^2 = \left[\frac{\mu_c^2 \sigma_b^2}{l^2} + \frac{2\mu_a \mu_c \sigma_b^2}{l^2} - \frac{\mu_c \sigma_b^2}{l} - \frac{2\mu_b \sigma_b^2}{l} + \frac{\mu_b^2 \sigma_b^2}{l^2}\right.$$
$$\left. + \frac{\mu_a^2 \sigma_c^2}{l^2} - \frac{2\mu_b \sigma_c^2}{l} + \frac{2\sigma_b^2 \sigma_c^2}{l^2} + \sigma_b^2 + \sigma_c^2 + \frac{\sigma_b^4}{l^2}\right].$$

Thus
$$(\bar{V}, s_V) = (\bar{q}, s_q)(\bar{z}, s_z).$$

The bending moment in the cross section $m - n$ is
$$M = q\left[\frac{bx}{l}\left(c + \frac{b}{2}\right) - \frac{(x - a)^2}{2}\right].$$

10.5 DESIGN COMPUTATION

Example 4. Compute the rectangular cross section required for a wooden beam to satisfy the following conditions:

Load 255 ± 45 lb/in-ft (see Example 1). See Fig. 10.10.
Assume $3s_q = 45$ lb/ft.
Load random variable $(\bar{q}, s_q)$:

$$(\bar{q}, s_q) = (255, 15) \text{ lb/ft} = (21.25, 1.25) \text{ lb/in.}$$

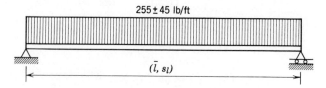

Figure 10.10 Example 4: beam design diagram.

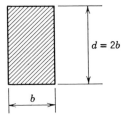

Figure 10.11 Beam cross section.

Beam length RV (Example 1):

$$(\bar{l}, s_l) = (12, 0.5) \text{ ft} = (144, 6) \text{ in.}$$

Material (wood), allowable flexural stress 1350 ± 150 psi $= F_t$.
Assume $3s_{F_t} = 150$ psi,

$$(\bar{F}_t, s_{F_t}) = (1350, 50) \text{ psi}$$

Cross-sectional requirements, width $= b$ and depth $d = 2b$. See Fig. 10.11.
Reliability 0.998 (equivalent to $z \approx 3.0$).

The tolerance (hence, standard deviation) on width and depth are assumed equal. Let tolerance $\Delta = b/12$, the standard deviation estimate is

$$3s_b = \frac{b}{12}; \quad s_b = \frac{b}{36}.$$

From Section 10.3, maximum moment is

$$M = \frac{ql^2}{8} \text{ in-lb,}$$

$$(\bar{M}, s_M) = \frac{(\bar{q}, s_q)(\bar{l}, s_l)^2}{(8, 0)} \text{ in-lb.}$$

By Eqs. 3.17 and 3.21

$$(\bar{l}, s_l)^2 = (\bar{l}^2 + s_l^2, \sqrt{4\bar{l}^2 s_l^2 + 2S_l^4}).$$

Applying Eqs. 3.9 and 3.10,

$$(\bar{M}, s_M) = \frac{(21.25, 1.25)(144, 6)^2}{(8, 0)} = (55{,}040; 5610) \text{ in-lb.}$$

The expression for applied stress $(\bar{f}, s_f)$ [36] is

$$\frac{M}{f} = \frac{I}{c} \quad \text{or} \quad f = \frac{M}{I/c}.$$

The moment of inertia (I) is (Section 7.4) is

$$c = \frac{d}{2} = b \quad \text{and} \quad I = \frac{bd^3}{12} = \frac{b(2b)^3}{12} = \frac{2b^4}{3}$$

$$f = \frac{M}{\left(\dfrac{2b^4}{3/b}\right)} = \frac{3M}{2b^3} \text{ psi.}$$

238 Application

Applying partial derivative methods to obtain s_{b^3} (Eq. 2.61),

$$s_{b^3} = 3b^2 s_b.$$

Thus

$$(\bar{f}, s_f) = \frac{(\bar{M}, s_M)}{(b^3, 3bs_b)} = \frac{3}{2} \frac{(55{,}040;\ 5610)}{(b^3,\ b^3/12)}\ \text{psi}.$$

By Eqs. 3.13 and 3.14

$$(\bar{f}, s_f) = \left(\frac{82{,}500}{b^3},\ \frac{7240}{b^3}\right)\ \text{psi}.$$

The allowable stress distribution $(\bar{F}_t, s_{F_t})$ is

$$(\bar{F}_t, s_{F_t}) = (1350;\ 50)\ \text{psi}.$$

Utilizing Eq. 4.6,

$$z = \frac{|\bar{f} - \bar{F}_t|}{\sqrt{s_f^2 + s_{F_t}^2}}$$

$$3.0 = \frac{82{,}500/b^3 - 1350.}{\sqrt{(7240/b^3)^2 + (50)^2}}.$$

After expanding and simplifying

$$0 = b^6 - 12.5 \cdot 10\, b^3 + 34.78 \cdot 10^2,$$

$$b^3 = \frac{125.0 \pm 41.4}{2} = 166.4,\ 83.6,$$

$$b = 5.46\ \text{in.},$$

$$d = 2b = 10.92\ \text{in.},$$

$$\text{tolerance} = \frac{b}{12} = 0.36\ (\text{max})\ \text{in}.$$

10.6 DEFLECTION

There is usually interest in the amount of deflection that results from the loading. As expected, deflection is a random variable arising from variations of strength properties, geometry and loads. The results are expressed as mean deflection and deflection standard deviation. The specification of maximum deflection is impossible; such a statement will be replaced by a 3σ value.

The differential equation of the beam neutral axis is (Fig. 10.12)

$$\frac{1}{r} = \frac{M}{EI}, \qquad (a)$$

M, E, and I are random variables; therefore $1/r$ is a random variable. From [36]

$$\frac{1}{r} = \frac{d^2 y}{dx^2}. \qquad (b)$$

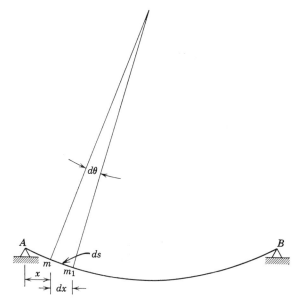

Figure 10.12 Beam deflection diagram.

Substituting Eq. *b* into Eq. *a*,

$$EI \frac{d^2y}{dx^2} = -M. \quad (c)$$

$$(\bar{y}, s_y) = \frac{5}{384} \frac{(\bar{q}, s_q)(\bar{l}, s_l)^4}{(\bar{E}, s_E)(\bar{I}, s_I)}. \quad (d)$$

The slope $(\bar{\theta}, s_\theta)$ at the left end is [36]:

$$\left(\frac{dy}{dx}\right)_{\max} = \frac{ql^3}{24EI}, \quad (e)$$

$$(\bar{\theta}, s_\theta) = \frac{(\bar{q}, s_q)(\bar{l}, s_l)^3}{24(\bar{E}, s_E)(\bar{I}, s_I)}.$$

Example 5. Deflection Computation.

PROBABILISTIC APPROACH. Compute the deflection RV, given

$\bar{E} = 1.5 \cdot 10^6$ psi and s_E(estimate) $= 2\%(\bar{E}) = 3.0 \cdot 10^4$ psi.
$\bar{l} = 16$ ft $= 192$ ft and s_l(estimate) $= 1$ percent $(192$ in.$) = 1.92$ in.
$\bar{q} = 155$ lb/ft $= 12.82$ lb/in. and s_q(estimate) $= \frac{4.5}{3}(\frac{1}{12}) = 1.25$ lb/in.

Solution. From Fig. 10.13, the moment of inertia (I) RV is

$$\bar{I} = \frac{b\bar{h}^3}{12} = \frac{3(14)^3}{12} = 686 \text{ in.}^4$$

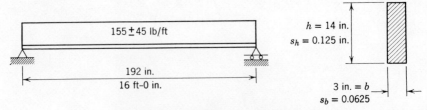

Figure 10.13 Example 5: beam deflection.

Applying Eq. 2.61,
$$(\bar{h}^3, s_{(h^3)}) = (\bar{h}^3, 3\bar{h}^2 s_h) = (2744, 73.5) \text{ in.}^3$$

By Eq. 3.10,
$$s_I = \tfrac{1}{12}\sqrt{b^2 s_{(\bar{h}^3)} + (\bar{h}^3)^2 s_b^2}$$
$$(\bar{I}, s_{\bar{I}}) = (686, 23.3) \text{ in.}^4$$

The deflection RV is (Eq. d):
$$(\bar{y}, s_y) = \frac{(12.82, 1.25)(192, 1.92)^4}{(1.5 \cdot 10^6, 3.0 \cdot 10^4)(686, 23.3)} \text{ in.}$$

$$\bar{y} = \frac{5}{384} \cdot \frac{(12.82)(192)^4}{(1.5 \cdot 10^6)(686)} = \frac{86.93 \cdot 10^9}{3.951 \cdot 10^{11}} = 0.220 \text{ in.}$$

The standard deviation estimate s_y is determined by first computing the numerator standard deviation
$$RV_N = (12.82, 1.25)(192, 1.92)^4.$$
$$(\bar{I}, s_{\bar{I}})^4 = (192, 1.92)^4 = (1.359 \cdot 10^9, 5.436 \cdot 10^7).$$
$$RV_N = (12.82, 1.25)(1.359 \cdot 10^9, 5.436 \cdot 10^7) = (1.742 \cdot 10^{10}, 1.835 \cdot 10^9)$$

The RV of the denominator is
$$(1.5 \cdot 10^6, 3.0 \cdot 10^4)(686, 23.3) = (1.029 \cdot 10^9, 4.056 \cdot 10^7).$$

The standard deviation estimate (s_y) is (Eq. 3.14)
$$s_y = \frac{5}{384} \cdot \frac{1}{1.029 \cdot 10^9} \cdot \left[\frac{(1.742 \cdot 10^{10})^2 (4.056 \cdot 10^7)^2 + (1.029 \cdot 10^9)^2 (1.835 \cdot 10^9)^2}{(1.029 \cdot 10^9)^2 + (4.056 \cdot 10^7)^2} \right]^{1/2}$$
$$= 0.025 \text{ in.}$$

The RV of y is
$$(\bar{y}, s_y) = (0.220, 0.025) \text{ in.}$$

For a 2σ interval (95 percent of the population)
$$y(95\% \text{ interval}) = 0.220 + 2(0.025) = 0.270 \text{ in.}$$

CONVENTIONAL APPROACH. The deflection is computed by conventional methods with the following data.

Wooden beam 3 in. wide × 14 in. deep.
Span = 16 ft.
$E = 1.5 \cdot 10^6$ psi (modulus of elasticity).
Load = 200 lbs/lineal ft (ultimate).

Solution.
$$y_{max} = \frac{5}{384} \cdot \frac{ql^4}{EI} \text{ in.,}$$
$$y_{max} = 0.287 \text{ in.}$$

Note. In this example a factor of safety is not utilized.

10.7 BEAM: BOTH ENDS BUILT-IN

Example 6. Consider the beam with built-in ends, shown in the sketch.

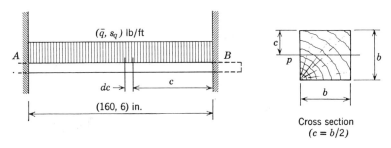

Cross section
$(c = b/2)$

1. Compute the cross-sectional requirements and tolerances.
2. Compute the maximum deflection variate. Given R (specified) $= 0.9999$, $z \approx \sqrt{13.7}$,

$$z^2 = 13.7,$$
$$(\bar{l}, s_l) = (160, 6) \text{ in.,}$$
$$(\bar{q}, s_q) = (150, 35) \text{ lb/ft,}$$
$$(\bar{m}_s, s_s) = (6000, 500) \text{ psi (allowable stress)}$$

Solution. From symmetry (see sketch)

$$R_A = R_B = \frac{ql}{2}$$
$$M_A = M_B,$$
$$\frac{M_B l}{EI} + \frac{ql^3}{6EI} = \frac{R_A l^2}{2EI},$$
$$M_B l = \frac{ql^3}{4} - \frac{ql^3}{6} = \frac{ql^3}{12},$$
$$M_B = \frac{ql^2}{12}.$$

242 Application

Moment at center of span

$$M_{ctr} = -\frac{ql^2}{12} - \frac{ql}{2}\left(\frac{l}{4}\right) + \frac{ql}{2}\left(\frac{l}{2}\right)$$

$$= ql^2\left(-\frac{1}{12} - \frac{1}{8} + \frac{1}{4}\right) = \frac{ql^2}{24}.$$

Maximum applied stress

$$f = \frac{6M}{b^3} = \frac{6ql^2}{12b^3} = \frac{l^2}{2b^3}.$$

Mean applied stress

$$\bar{f} = \frac{\bar{q}\bar{l}^2}{\bar{b}^3} = \frac{16 \cdot 10^4}{\bar{b}^3} \text{ psi}$$

By Eq. 2.61

$$s_f^2 = \frac{15.5 \cdot 10^8}{\bar{b}^3} \text{ psi}$$

Applying Eq. 4.6 and squaring,

$$z^2 = 13.7 = \frac{[(6000\bar{b}^3 - 16 \cdot 10)^4/\bar{b}^3]^2}{25 \cdot 10^4 \bar{b}^6 + 15 \cdot 5 \cdot 10^8/\bar{b}^6},$$

$$32.6\bar{b}^6 - 1920\bar{b}^3 + 4400 = 0,$$

$$\bar{b}^6 - 59\bar{b}^3 + 135 = 0,$$

$$\bar{b}^3 = \frac{59 \pm \sqrt{3481 - 546}}{2} = 56.6 \text{ in.}^3$$

$$\bar{b} = 3.84 \pm 0.06 \text{ in.}$$

(Compare this result with those in Examples 1 and 2, in Section 11.1.) Maximum deflection:

$$M_{max} = \frac{ql^2}{24},$$

$$\delta_{max} = \frac{q(l/2)^4}{8EI} = \frac{(ql^2/24)(l/2)^2}{2EI} \quad [36]$$

$$= \frac{ql^4}{EI}\left(\frac{1}{384}\right).$$

Substituting $I = b^4/12$,

$$\delta_{max} = \frac{ql^4}{32Eb^4}$$

$$(\bar{\delta}, s_\delta) = \frac{(\bar{q}, s_q)(\bar{l}, s_l)^4}{32(\bar{E}, s_E)(\bar{b}, s_b)^4}$$

Length RV estimate (Eq. 2.61)

$$(\bar{l}, s_l)^4 \approx (\bar{l}^4, 4\bar{l}^3 s_l) \text{ in.}^4,$$

$b - RV$ estimate

$$(\bar{b}, s_b)^4 \approx (\bar{b}^4, 4\bar{b}^3 s_b) \text{ in.}^4,$$

Mean deflection

$$\bar{\delta} = \frac{12.5(160)^4}{32 \cdot 1.5 \cdot 10^6 (3.84)^4},$$

$$\bar{\delta} = 0.783.$$

Standard deviation estimate—numerator s_N

$$s_N = \sqrt{(12.3)^2(16(160)^6(6)^2) + (160)^8(2.92)^2 + [16(160)^6(6)^2(2.92)^2]}$$

$$s_N = 2.76 \cdot 10^9$$

$$\bar{E}\bar{b}^4 = 1.5 \cdot 10^6 (3.84)^4 = 327 \cdot 10^6$$

$$s_{Eb^4} = \sqrt{(1.5 \cdot 10^6)^2 \left(\frac{\bar{b}^4}{45}\right)^2 + (\bar{b}^4)^2(0.03 \cdot 10^6)^2 + \left(\frac{\bar{b}^4}{45}\right)^2 (3 \cdot 10^4)^2}$$

$$= 9.78 \cdot 10^6$$

Deflection standard deviation

$$s_\delta = \frac{1}{32 \cdot 327 \cdot 10^6} \left[\frac{(8.19 \cdot 10^9)^2 (9.78 \cdot 10^6)^2 + (327 \cdot 10^6)^2 (2.76 \cdot 10^9)^2}{(327 \cdot 10^6)^2 + (9.18 \cdot 10^6)^2} \right]^{1/2}$$

$$s_\delta = 0.22 \text{ in.}$$

$$(\bar{\delta}, s_\delta) = (0.783, 0.22) \text{ in.}$$

CHAPTER 11

Cantilever Beams

The manner of treating cantilever beams is somewhat idealized, despite the representation of design parameters as random variables. For instance, fixity is assumed to be 100 percent.

In the example below, the cantilever is of rectangular cross section (Fig. 11.1) $R = 0.999$ ($z \approx 3$) specified.

The cross-sectional area RV estimate is

$$(\bar{A}, s_A) = (3\bar{b}, 3s_b)(\bar{b}, s_b) \text{ in.}^2$$

The moment of inertia RV estimate is

$$(\bar{I}_z, s_I) = \frac{(\bar{b}, s_b)(3\bar{b}, 3s_b)^3}{12} \text{ in.}^4 = \frac{27}{12}(\bar{b}, s_b)^4 \text{ in.}^4$$

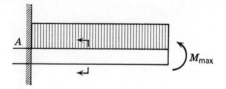

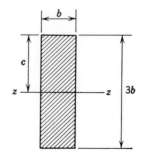

Figure 11.1 Beam cross section.

244

Maximum bending stress (at the point of support) is

$$\frac{M}{f} = \frac{I_z}{c}.$$

M is the moment random variable (see Fig. 11.2),

$$(\bar{M}, s_M) = (\bar{1}, s_1)(\bar{P}, s_p) = (20, 0.5)(40{,}000; 2000),$$

$c = 3b/2$, and the random variable is

$$(\bar{c}, s_c) = \tfrac{3}{2}(\bar{b}, s_b) \text{ in.}$$

The maximum stress random variable (estimate) is

$$(\bar{f}, s_f) = \frac{(\bar{M}, s_M)(\bar{b}, s_b)}{\tfrac{27}{12}(\bar{b}, s_b)^4} = \frac{2}{3}\frac{(\bar{M}, s_M)}{(\bar{b}, s_b)^3} \text{ psi.}$$

$$(\bar{f}, s_f) = \frac{(5.328 \cdot 10^5, 1.359 \cdot 10^4)}{(\bar{b}, s_b)^3} = \frac{(5.328 \cdot 10^5, 1.359 \cdot 10^4)}{(\bar{b}^3, 3\bar{b}^2 s_b)} \text{ psi.}$$

Having the applied stress random variable written as a function of b and yield is the failure criterion and the material of construction is aluminum alloy 2024-T6, then

$$(\bar{F}_{ty}, s_{F_{ty}}) = (64{,}100; 4300) \text{ psi.}$$

With the coupling formula (Eq. 4.6):

$$z = 3.0 = \frac{|\bar{f}_{ty} - \bar{F}_{ty}|}{\sqrt{s_{f_{ty}}^2 + s_{F_{ty}}^2}}.$$

This equation is solved as a quadratic in a power of b.

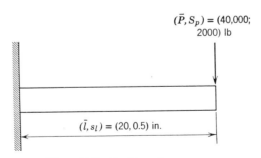

Figure 11.2 Cantilever beam.

246   Application

11.1 UNIFORMLY LOADED CANTILEVER

Example 1. From the data given, compute the cross section and tolerances of the cantilever beam in the sketch.

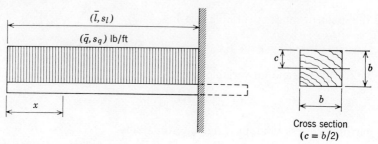

Cross section
($c = b/2$)

Given $(\bar{q}, s_q) = (150, 35)$ lb/ft,

$$R \text{ (specified)} - 0.9999 \ (z \approx 3.7)$$
$$(\bar{l}, s_l) = 160, 6) \text{ in.}$$

Allowable flexural strength

$$(\bar{M}_s, s_M) = (6000, 500) \text{ psi.}$$

Assume that the tolerance Δ on b is $\Delta = \bar{b}/60 \ (s_b = \bar{b}/60)$.

Solution. Moment:

$$M = \frac{ql^2}{2} \quad \text{(maximum stress at } x = 1\text{).}$$

Applied stress f_t:

$$f_t = \frac{M(b/2)}{b \cdot b^3/12} = \frac{6M}{b^3} = \frac{6[q(l/2/2)]}{b^3} = \frac{3ql^2}{b_3}.$$

Applied stress random variable:

$$(\bar{f}_t, s_f) = 3 \frac{(\bar{q}, s_q)(\bar{l}, s_l)^2}{(\bar{b}, s_b)^3}$$

$(\bar{l}, s_l)^2 = [\bar{l}^2 + s_l^2), \ (4\bar{l}^2 s_l^2 + s_l^4)]$ (Eqs. 3.17 and 3.21)
$(\bar{b}, s_b)^3 = (\bar{b}^3, 3\bar{b}^2 s_b)$ (Eq. 2.61)

$$3\bar{b}^2 s_b = \frac{\bar{b}^3}{60}$$

$$(\bar{b}, s_b)^3 = \left(\bar{b}^3, \frac{\bar{b}^3}{60}\right)$$

$$3\bar{q}(\bar{l})^2 = 3\bar{q}(\bar{l}^2 + s_l^2)$$

$$s_{3\bar{q}l^2} = 3\sqrt{\bar{q}^2 s_l^2(4\bar{l}^2 + 2s_l^2) + (\bar{l}^2 + s_l^2) s_q^2 + s_l^2(4\bar{l}^2 + 2s_l^2) \cdot s_q^2}.$$

Mean applied stress:

$$\bar{f}_t = \frac{3\bar{q}\bar{l}^2}{\bar{b}^3} = \frac{96 \cdot 10^4}{\bar{b}^3}.$$

Applied stress variance estimate:

$$s_{ft} = \frac{23 \cdot 61 \cdot 10^4}{b^3}$$

$$s_{ft}^2 = \frac{558 \cdot 10^8}{b^6}.$$

By utilizing Eq. 4.6 and squaring,

$$z^2 = 13.7 = \frac{[(6000b^3 - 96 \cdot 10^4)/b^3)]^2}{(25 \cdot 10^4 b^6 + 5.58 \cdot 10^{10})/b^6}$$

$$36 \cdot 10^6 b^6 - 1152 \cdot 10^7 b^3 + 9216 \cdot 10^8 = 3.42 \cdot 10^6 b^6 + 76.3 \cdot 10^{10}$$

$$b^6 - 354 b^3 + 5120 = 0$$

$$b^3 = \frac{354 \pm \sqrt{125,300 - 20560}}{2}$$

$$b^3 = 339 \text{ in.}^3$$

$$b = 6.47 \text{ in.}$$

Dimensions of cross section with tolerances are

$$b = 6.47 \pm 0.11 \text{ in.}$$

(See the results of Example 2 and Example 6, Section 10.7.)

Example 2. Consider the cantilever beam of Example 1. All parameters are as in Example 1 except loading (w), which increases uniformly from the unsupported end as shown in the sketch.

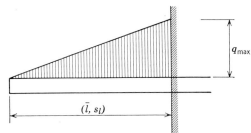

Given total load random variables,

$$(\bar{w}, s_w) = (\bar{q}, s_q)(\bar{l}, s_l)$$
$$= (150, 35)(\tfrac{160}{12}, \tfrac{6}{12}) \text{ lb}$$

(distributed as shown in the sketch).

Solution.

$$\bar{q} = \frac{q_{max}}{3}$$

Maximum moment:

$$M = \frac{ql^2}{3}.$$

Maximum stress:

$$f_t = \frac{M(b/2)}{b^4/12} = \frac{6M}{b^3} = \frac{6(ql^2/3)}{b^3} = \frac{2ql^2}{b^3}.$$

248 Application

Note. The stresses developed in this example identify with those of Example 1 except for difference by a constant $k = \frac{2}{3}$.

Thus (on substituting values into Eq. 4.6),

$$z^2 = 13.7 = \frac{[(6000b^3 - 64 \cdot 10^4)/b^3]^2}{(25 \cdot 10^4 b^6 + 2.48 \cdot 10^{10})/b^6}$$

By expanding and simplifying,

$$32.6 b^6 - 7680 b^3 + 7000 = 0$$

$$b^6 - 236 b^3 + 214 = 0$$

$$b^3 = \frac{236 \pm \sqrt{(235)^2 - 856}}{2}$$

$$b^3 = 235 \text{ in.}^3$$

$$b = 6.17 \text{ in.}$$

Dimensions of cross section with tolerances:

$$b = 6.17 \pm 0.10 \text{ in.}$$

11.2 PROPPED CANTILEVER

Next we examine the propped cantilever beam. The example problem is solved by the moment distribution method; first by the conventional approach, then by a probabilistic approach. The propped cantilever is statically indeterminate with one redundant constraint.

Example 1. Conventional Design (Fig. 11.3). Given specified tubular section,

safety factor = 1.5,

limit load = $P_L = 6670$ lb,

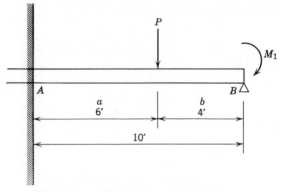

Figure 11.3 Example 1: cantilever.

ultimate load (P_U):
$$P_U = P_L \times SF = 6670 \times 1.5 = 10,000 \text{ lb,}$$
applied moment (end):
$$M = 1000 \text{ ft-lb.}$$

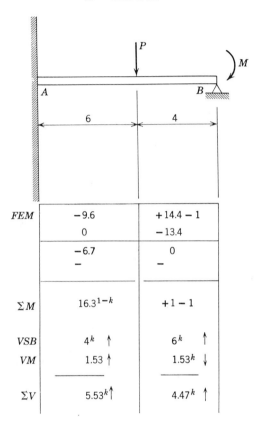

Computation of fixed end moments:
$$FEM_{AB} = \frac{Pab^2}{L^2} = \frac{10(6)(4)^2}{100}$$
$$= 9600 \text{ ft lb.}$$
$$FEM_{BA} = \frac{Pa^2b}{L^2} = \frac{10(6)^2(4)}{10}$$
$$= 14,400 \text{ ft-lb.}$$

(+) $M_{\max}$:
$$(4.47(4) - 1)1000 = 16,880 \text{ ft-lb.}$$

250 Application

Assuming $D/t = 100$ (where the beam cross section is a tube),

$$\text{section modulus} = S = \pi r^3/50, \text{ where } S = \frac{I}{c} = \frac{r^3 t}{r}; t = \frac{r}{50}.$$

$$S = \frac{M}{F_b} = \frac{16.88 \times 12}{165} = 1230 \text{ in.-lb}.$$

$$S = \pi r^3/50 = 1.23, r^3 = \frac{50 \times 1.23}{\pi} = 19.5 \text{ in.}^3$$

$$r = (19.5)^{1/3} = 2.69 \text{ in.}$$

$$\therefore D = 5.38 \text{ in.}$$

$$t = 0.054 \text{ in.}$$

Probabilistic Design.†

Given load RV,

$$(\bar{P}, s_p) = (6070, 200) \text{ lb}.$$

Applied moment:

$$(\bar{M}_1, s_{M_1}) = (607, 20) \text{ in-lb} \qquad \text{(standard deviation taken as 2.2\%)}.$$

Beam cross section—circular hollow tube.
Reliability (specified):

$$R = 0.9999, \text{ equivalent to } z \approx 4.$$

Solution. Fixed end moments:

$$M_{AB} = \frac{Pab^2}{l^2} = P \cdot \left(\frac{6 \cdot 4^2}{10^2}\right) = 0.96 \text{ lb/in}.$$

$$M_{BA} = \frac{Pba^2}{l^2} = P \cdot \left(\frac{4 \cdot (6)^2}{10^2}\right) = 1.44P \text{ lb/in}.$$

$M_1 = \dfrac{P_M \cdot 1 \cdot (4)^2}{5^2}$	Shear (B)	Shear (A)
$P_M = \dfrac{M_1 \cdot (5)^2}{(4)^2}$	$V_{S_B} = \dfrac{P \cdot 6}{10}$	$V_{S_A} = \dfrac{P \cdot 4}{10}$

Total shear (V) random variable:

$$(\bar{V}, s_V) = \frac{P \cdot 4}{10} + \frac{M_1 \cdot 5^2}{4^2}$$

$$= (\bar{P}, s_p) \cdot \left(\frac{b, s_b}{l, s_l}\right) + \frac{(\bar{M}_1, s_{M_1})[(\bar{b}, s_b) + (\bar{c}, s_c)]^2}{(\bar{b}, s_b)^2}.$$

† For moment distribution methods see [33].

Cantilever Beams 251

Maximum moment random variable:

$$(\bar{M}, s_M) = (\bar{V}, s_V) \cdot (\bar{b}, s_b) - (\bar{M}_1, s_{M_1}),$$

$$(\bar{M}, s_M) = \left\{ (\bar{P}, s_p) \cdot \frac{(\bar{b}, s_b)^2}{(\bar{l}, s_l)} + (\bar{M}_1, s_{M_1}) \frac{[(\bar{b}, s_b) + (\bar{c}, s_c)]}{(\bar{b}, s_b)} \right\} \text{ft-lb},$$

$$(\bar{l}, s_l) = (10, 0.003467) \text{ ft},$$

$$(\bar{b}, s_b) = (4, 0.0049) \text{ ft},$$

$$(\bar{c}, s_c) = (1, 0.003467) \text{ ft},$$

$$(\bar{P}, s_p) \frac{(\bar{b}, s_b)^2}{(\bar{l}, s_l)} \cdot 12 = (6070, 200) \cdot \frac{(4, 0.0049)^2}{(10, 0.003467)} \cdot 12 \text{ in-lb}.$$

Applying Eq. 2.61,

$$s_b{}^2 = 2\bar{b}s_b = 8(0.0049) = 0.0392,$$

$$(\bar{b}^2, s_b{}^2) = (16, 0.0392) \text{ ft},$$

$$\frac{(\bar{b}^2, s_b{}^2)}{(\bar{l}, s_l)} = \frac{(16, 0.0392)}{(10, 0003467)}$$

Applying Eq. 3.14,

$$s_{b^2/l} = \frac{1}{10} \cdot \left[\frac{10^2(0.0392)^2 + 16^2 \cdot (0.003467)^2}{10^2} \right]^{1/2}$$

$$= 0.01 [15.368 \cdot 10^{-2} + 0.306 \cdot 10^{-2}]^{1/2} = 0.0388 \text{ ft},$$

$$\left(\frac{\bar{b}^2}{\bar{l}}, s_{b^2/l} \right) = (1.6, 0.0388) \text{ ft},$$

$$(\bar{P}, s_p) \left(\frac{\bar{b}^2}{\bar{l}}, s_{b^2/l} \right) = (6070; 200)(1.6, 0.0388),$$

$$\bar{P} \cdot \frac{\bar{b}^2}{\bar{l}} = 9710 \text{ ft-lb}.$$

Applying Eq. 3.10,

$$s_{pb^2/l} = [(6070)^2 \cdot (0.0388)^2 + (200)^2 \cdot (1.6)^2]^{1/2} = 397 \text{ ft-lb}$$

$$(\bar{P}, s_p) \left(\frac{\bar{b}^2}{\bar{l}}, s_{b^2/l} \right) = (9710; 397) \text{ ft-lb}$$

$$(\bar{M}_1, s_{M_1}) = \frac{[(\bar{b}, s_b) + (\bar{c}, s_c)]}{(\bar{b}, s_b)}.$$

Applying Eq. 3.2,

$$s_{b+c} = \sqrt{(0.0049)^2 + (0.003467)^2} = 0.006,$$

$$(\bar{b} + \bar{c}, s_{b+c}) = (5, 0.006)$$

$$s_{b+c/b} = \frac{1}{4} \left[\frac{16 \cdot 36 \cdot 10^{-6} + 25 \cdot (3.467)^2 \cdot 10^{-6}}{4^2} \right]^{1/2}$$

$$= \tfrac{1}{16}[2.75 \cdot 10^{-4}] = 0.001036,$$

$$\frac{(5, 0.006)}{(4, 0.003467)} = (1.25, 0.001036),$$

$$(\bar{M}_1, s_{M_1}) \cdot (1.25, 0.001036) = (607, 20)(1.25, 0.001036).$$

252 Application

Applying Eq. 3.10 (neglecting the last term),

$$s_{(M)\,(1.25)} = \sqrt{(6.07)^2 \cdot 10^4 \cdot 10^{-6} + 400 \cdot (1.5625)}$$
$$= \sqrt{36 \cdot 8449 \cdot 10^{-2} + 625} = 25,$$
$$(M_1, s_{M_1}) = (759, 25) \text{ ft-lb},$$
$$(\bar{M}, s_M) = (9710, 397) + (759, 25)$$
$$= (10, 470, 398) \text{ ft-lb},$$
$$(\bar{M}, s_M) = (10,470; 398) \cdot 12 = (125, 640; 4776) \text{ in-lb},$$
$$\frac{I}{c} = \frac{M}{f}.$$

Applied stress

$$f = \frac{M}{(I/c)}$$

where $I = \pi r^3 t$ and $D/t = 2r/t = 100$; $t = r/50$:

$$A = 2\pi r t = \frac{\pi r^2}{25},$$

$$\frac{I}{c} = \frac{\pi r^3}{50},$$

$$f = \frac{(M) \text{ in-lb}}{\pi \bar{r}^3/50 \text{ in.}^3} = \frac{15 \cdot 9155 \cdot \bar{M}}{\bar{r}^3} \text{ psi}.$$

s_r is estimated as

$$s_r = 0.015\, r$$

Applying Eq. 2.61.

$$s_{r^3} = 3\bar{r}^2(0.015\bar{r}) = 0.045\bar{r}^3.$$

Applied stress random variable:

$$(\bar{f}, s_f) = 15.9155 \frac{(125, 640, 4776)}{(\bar{r}^3, s_{r}^{\,3})},$$

$$(\bar{f}, s_f) = \left(\frac{1.9996 \cdot 10^6}{\bar{r}^3}, \frac{4.816 \cdot 10^4}{\bar{r}^3}\right) \text{ psi}.$$

Allowable stress random variable:

$$(4130\ stl - 160^k\ HT.\ TRT.),$$

$$(\bar{F}, s_F) = 171,100;\ 4760)\ \text{psi}.$$

Substituting into Eq. 4.6,

$$4.0 = \frac{(1.9996 \cdot 10^6)/\bar{r}^3 - 1.71 \cdot 10^5}{\sqrt{[(4.816 \cdot 10^4)/\bar{r}^3]^2 + (4.76 \cdot 10^3)^2}}.$$

Expanding and simplifying,

$$0 = 137.17 - 23.681\bar{r}^3 + \bar{r}^6.$$

Solving the quadratic in $\bar{r}^3$,

$$\bar{r}^3 = \frac{23.681 \pm 3.441}{2} = 13.561,$$

$$\bar{r} = \sqrt[3]{13.561} = 2.385 \text{ in.},$$

$$\bar{D} = 4.77 \text{ in.},$$

$$t = 0.0477 \text{ in.},$$

$$A_{\text{st}} = \pi D t = 0.715 \text{ in.}^2,$$

$$A_{\text{conv}} = \pi 5.38 \cdot (0.054) = 0.9127 \text{ in.}^2,$$

$$\frac{A_{(\text{st})}}{A_{(\text{conv})}} = \frac{0.715}{0.9127} = 0.783.$$

CHAPTER 12

Column Design [41]

Columns are sometimes categorized, with overlap, into three classes:

1. *The short column.* The short column is relatively stiff with deflection under the influence of loads so small that it can be neglected in discussing the stresses. Failure is caused by *crippling*.
2. *The intermediate column.* With the intermediate column (due to flexibility), moments due to lateral deflection have a considerable effect on bending. Failure occurs when stresses approach (or exceed) the elastic limit.
3. *The slender column.* Failure occurs due to elastic instability of the column (stress within the elastic limit). Failure is by excessive lateral deflection.

Assumptions. With columns, the problem of instability appears. In order to examine the probabilistic behavior of columns in an orderly way, it is helpful to introduce idealizations which allow a simplification of the problem. By successively relaxing the assumptions in turn and interpreting the theoretical results, some insight into probabilistic column behavior is obtained.

First considered is the idealized column of nominally constant cross section axially compressed by random variable loads. Conventionally, the column is assumed to be constructed of homogeneous material. Actually, homogeneity is the limit at which material variability vanishes.

In addition to previous assumptions, the following are introduced.

1. Materials—ideally elastic.
2. No deviation from axial straightness.
3. Loads—applied along the centroidal axis.

Initial Imperfections. Conventionally, the stability of a system is tested by subjecting it to an infinitesimal disturbance. If the system returns to its original position upon removal of the disturbance, it is considered stable. Actually, we must work with collections of random variables, among which are concentricity and straightness. There is always the finite probability that a system will be unstable under specified loading. The most that can be expected is that the probability of instability is kept acceptably small.

Axial loading of a column is an idealization never wholly realized either as to direction or point of application. Therefore, in addition to axial compression, bending moments exist.

Symmetrical distribution of material (see Example 4) over a section is an idealization. Some degree of eccentricity is invariably present, and the amount is statistically distributed. With certain manufactured sections, such as tubes, the maximum amount of eccentricity (the 3σ interval) is specified.

A column with a straight neutral axis is an idealization. Manufacturers of structural sections commonly guarantee straightness within stated limits. The effect of initial curvature is discussed in detail.

12.1 THE COMPRESSION BLOCK

Possibly the simplest example of a compression element is that (Fig. 12.1) commonly called a "compression block." In the compression block stresses are compressive; bending and tension are essentially absent. It is assumed that loading is uniformly applied over the cross section. The failure mode is compressive yield or shear, depending on the material.

Under stress near or above the yield point of the material the compression block yields with accompanying increase in cross section. The stress (s) is given by

$$s = \frac{P}{A} \text{ psi,}$$

where P is the load and A the cross-sectional area. Since P is the random variable ($\bar{P}, s_p$) and A the random variable ($\bar{A}, s_A$), applied stress is a random variable.

$$(\bar{s}, \sigma_s) = \frac{(\bar{P}, s_p)}{(\bar{A}, s_A)} \text{ psi.}$$

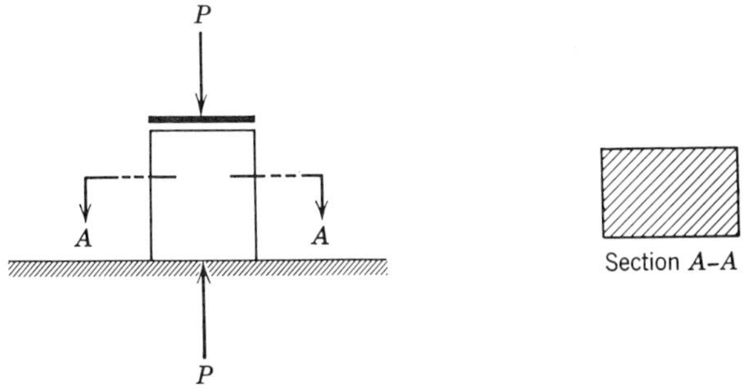

Figure 12.1 Compression block.

Mean stress ($\bar{s}$) is (Eq. 3.13)

$$\bar{s} = \frac{\bar{P}}{\bar{A}} \text{ psi}$$

and (Eq. 3.14) the standard deviation estimate

$$s_s = \frac{1}{\bar{A}} \left(\frac{\bar{A}^2 s_p^2 + \bar{P}^2 s_A^2}{\bar{A}^2 + s_A^2} \right)^{1/2} \text{ psi}.$$

Thus

$$(\bar{s}, s_s) = \frac{\bar{P}}{\bar{A}}, \frac{1}{\bar{A}} \left(\frac{\bar{A}^2 s_p^2 + \bar{P}^2 s_A^2}{\bar{A}^2 + s_A^2} \right)^{1/2} \text{ psi}.$$

Example 1. Stress Computation. Compute the applied stress random variable, given load RV — (10,000; 400) lb, cross section RV — (1.0, 0.05) in.² Mean stress estimate (Eq. 3.13):

$$\bar{s} = \frac{\bar{P}}{\bar{A}} = \frac{10,000 \text{ lb}}{1.0 \text{ in}^2.} = 10,000 \text{ psi}.$$

Stress standard deviation estimate (Eq. 3.14):

$$s_s = \frac{1}{1.0} \left[\frac{(1.0)^2 (500)^2 + (10^4)^2 (0.05)^2}{(1.0^2 + (0.05)^2} \right]^{1/2},$$

$$s_s = 706 \text{ psi}.$$

Applied stress random variable

$$(\bar{s}, s_s) = (10,000; 706) \text{ psi}.$$

12.2 THE SHORT COLUMN

In the studies of beams the forces acted transverse to the axis of symmetry; it is now assumed that a nominally prismatical bar is loaded in one of its planes of symmetry (see Fig. 12.2).

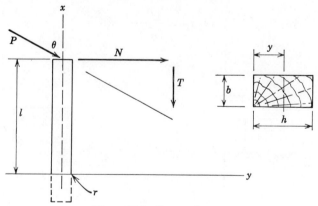

Figure 12.2 Short column.

The random variable $(\bar{P}, s_p)$ resolves into a vertical force variate $(\bar{T}, s_T)$ and a lateral force variate $(\bar{N}, s_N)$. $(\bar{T}, s_T)$ and $(\bar{N}, s_N)$ are determined by the magnitude of $(\bar{P}, s_p)$ and orientation $(\bar{\theta}, s_\theta)$. $N = P \sin \theta$ and $T = P \cos \theta$. From Section 2.9.2 let U and V be random variables such that $U = \sin \theta$ and $V = \cos \theta$. By applying Eq. 2.61, the stand deviations of U and V are $s_u = s_{(\sin \theta)} = s_\theta \cos \theta$, and $s_v = s_{(\cos \theta)} = s_\theta \sin \theta$. The mean value estimates of N and T are (Eq. 3.9)

$$\bar{N} = \bar{P}(\sin \bar{\theta}),$$

$$\bar{T} = \bar{P}(\cos \bar{\theta}).$$

The standard deviation estimates of T and N are (Eq. 3.10)

$$s_T = \sqrt{\bar{P}^2 s_\theta^2 \sin^2 \theta + \cos^2 \bar{\theta} s_p^2 + s_\theta^2 \cos^2 \bar{\theta} s_p^2},$$

$$s_N = \sqrt{\bar{P}^2 s_\theta^2 \cos^2 \theta + \sin^2 \bar{\theta} s_p^2 + s_\theta^2 \cos^2 s_p^2}$$

(θ is expressed in radians).

The stress due to $(\bar{T}, s_T)$ is the same at all cross sections in the column and is T/A. The stress random variable is

$$\frac{(\bar{T}, s_T)}{(\bar{A}, s_A)}.$$

The stress due to bending is the result of moments. Moments vary from 0 at the top of the column to Nl or (by Eqs. 3.9 and 3.10)

$$(\bar{M}, s_M) = (\bar{N}\bar{l}, \sqrt{\bar{N}^2 s_l^2 + \bar{l}^2 s_N^2 + s_N^2 s_l^2})\text{ in-lbs}$$

at the base. Maximum moment occurs at the base of a build in column. The stress (s) at a distance y above the base is

$$(\bar{s}, s_s) = -\frac{(\bar{T}, s_T)}{(\bar{A}, s_A)} - \frac{(\bar{N}, s_N)(\bar{l}, s_l)(\bar{y}, s_y)}{(\bar{I}, s_I)}.$$

For example, if the column is of rectangular cross section $(b \times h)$, with h parallel to the plane of bending (see Fig. 12.2), then mean area and mean moment of inertia are

$$\bar{A} = \bar{b}\bar{h}\text{ in.}^2,$$

$$\bar{I} = \frac{\bar{b}\bar{h}^3}{12}\text{ in.}^4,$$

and

$$y = \frac{h}{2}\text{ in.}$$

By applying Eqs 3.9 and 3.10,

$$(\bar{A}, s_A) = (\bar{b}\bar{h}, \sqrt{\bar{b}^2 s_h^2 + \bar{h}^2 s_b^2 + s_b^2 s_h^2}) \text{ in.}^2$$

Moment of inertia (I) is

$$(\bar{I}, s_I) = \frac{(\bar{b}, s_b)(\bar{h}, s_h)^3}{12} \text{ in.}^4$$

The random variable h^3 estimate is (Eq. 2.61) $(\bar{h}^3, 3\bar{h}^2 s_h)$. Applying Eqs. 3.9 and 3.10,

$$(\bar{I}, s_I) = \left(\frac{\bar{b}\bar{h}^3}{12}, \frac{1}{12}\sqrt{9\bar{b}^2\bar{h}^4 s_h^2 + \bar{h}^6 s_b^2 + 9\bar{h}^4 s_b^2 s_h^2}\right) \text{ in.}^4$$

The random variable y is

$$(\bar{y}, s_y) = \left(\frac{\bar{h}}{2}, \frac{s_N}{2}\right) \text{ in.}$$

Applied stress s is

$$s = \frac{T}{bh} + 6\frac{Nl}{bh^2}$$

at point r (Fig. 12.2).

$$(\bar{s}, s_s) = -\left[\frac{(\bar{T}, s_T)}{(\bar{b}, s_b)(\bar{h}, s_h)} + 6\frac{(\bar{N}, s_N)(\bar{l}, sl)}{(\bar{b}, s_b)(\bar{h}, s_h)^2}\right] \text{ psi.}$$

By applying Eq. 2.61, first taking partial derivatives with respect to s,

$$\frac{\partial s}{\partial T} = \frac{1}{bh}; \frac{\partial s}{\partial l} = \frac{6N}{bh^2}; \frac{\partial s}{\partial h} = \frac{T}{bh^2} + \frac{12Nl}{bh^3},$$

$$\frac{\partial s}{\partial N} = 6\frac{l}{bh^2}; \frac{\partial s}{\partial b} = \frac{T}{bh^2} + 6\frac{Nl}{b^2h^2},$$

$$s_s = \left[\frac{s_T^2}{\bar{b}^2\bar{h}^2} + 36\frac{\bar{l}^2}{\bar{b}^2\bar{h}^4}s_N^2 + 36\frac{\bar{N}^2}{\bar{b}^2\bar{h}^4}sl^2 + \left(\frac{\bar{T}}{\bar{b}\bar{h}^2} + \frac{6\bar{N}\bar{l}}{\bar{b}^2\bar{h}^2}\right)^2 s_b^2 + \cdots \right.$$

$$\left. + \left(\frac{\bar{T}}{\bar{b}\bar{h}^2} + 12\frac{\bar{N}\bar{l}^2}{\bar{b}\bar{h}^3}\right)^2 s_h^2\right]^{1/2}. \quad (a)$$

In Fig. 12.2 opposite r (where bending produces tensile stress) the sign of the second term in Eq. a is reversed.

Example 2. Short Column Analysis. For comparison, this example is first computed by conventional methods, then by probabilistic methods (see Fig. 12.3).

Column Design 259

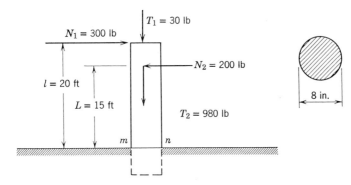

Figure 12.3 Stress in short column.

Conventional Approach. Compute the stress at m, given the data in Fig. 12.3,

$$M_{\max} = (20 \times 12 \times 300 - 15 \times 12 \times 200)\text{ in-lb}$$
$$= 36{,}00\text{ in-lb},$$
$$\text{thrust} = T = (980 + 30)\text{ lb} = 1010\text{ lb},$$

$$s = \frac{T}{A} + \frac{Nly}{I_z},$$

$$y = \frac{d}{2} \quad \text{and} \quad I_z = \frac{\pi r^4}{4},$$

$$s = \left[\frac{1010}{\pi d^2/4} + \frac{36{,}000 r}{\pi (d/2)^3 (r/4)}\right]\text{ psi},$$

$$s = \left[\frac{4 \cdot 1010}{\pi d^2} + \frac{32 \cdot 36{,}000}{\pi d^3}\right]\text{ psi},$$

$$s = 736\text{ psi}.$$

Probabilistic Approach. To compute the stress random variable at m, the following random variables are assumed.

$$(\bar{T}_1, s_{T1}) = (27, 1)\text{ lb},$$
$$(\bar{T}_2, s_{T2}) = (882, 33)\text{ lb},$$
$$(\bar{N}_1, s_{N1}) = (270, 10)\text{ lb},$$
$$(\bar{N}_2, s_{N2}) = (180, 7)\text{ lb},$$
$$(\bar{l}, s_l) = (20, 0.5)\text{ ft} = (240, 6)\text{ in.},$$
$$(\bar{L}, s_L) = (15, 0.5)\text{ ft} = (180, 6)\text{ in.},$$
$$(\bar{d}, s_d) = (8, 0.25)\text{ in.}$$

260 Application

Bending moment random variable estimate:

$$(\bar{M}, s_M) = [(\bar{N}_1, s_{N1})(\bar{l}, s_l) - (\bar{N}_2, s_{N2})(\bar{L}, s_L)] \text{ in-lb},$$
$$= [(270, 10)(240, 6) - (180, 7)(180, 6)] \text{ in-lb},$$
$$(\bar{M}, s_M) = (32,400; 340) \text{ in-lb},$$
$$(\bar{T}, s_T) = [(\bar{T}_1, s_{T1}) + (\bar{T}_2, s_{T2})] \text{ lb}$$
$$= [(27, 1) + (882, 33)] \text{ lb},$$
$$(\bar{T}, s_T) = (909, 33) \text{ lb}.$$

Mean stress ($\bar{s}$) estimate:

$$\bar{s} = \frac{4}{\pi}\left(\frac{\bar{T}}{\bar{d}^2}\right) + \frac{32}{\pi}\left(\frac{\bar{M}}{\bar{d}^3}\right),$$

$$\bar{s} = \frac{4}{\pi}\left(\frac{909}{64}\right) + \frac{32}{\pi}\left(\frac{32,400}{8^3}\right) \text{ psi},$$

$$\bar{s} = 663 \text{ psi}.$$

Standard deviation s_s estimate:

$$\frac{\partial T}{\partial s} = \frac{4}{\pi d^2},$$

$$\frac{\partial M}{\partial s} = \frac{32}{\pi d^3},$$

$$\frac{\partial d}{\partial s} = \frac{8T}{\pi d^3} + \frac{96M}{\pi d^4}.$$

By Eq. 2.61,

$$s_s = \left[\sum \left(\frac{\partial x_i}{\partial s}\right)^2 s_{x_i}^2\right]^{1/2},$$

where

$$x_1 = T, x_2 = M, x_3 = d,$$

$$s_s = \left[\frac{4^2(33)^2}{\pi^2(64)^2} + \frac{8^2 \cdot 4^2(1660)^2}{\pi^2 8^2} + \left(\frac{8 \cdot 909}{\pi \cdot 512} + \frac{96.23,400.}{\pi \cdot 4,096.}\right)\right]^{1/2},$$

$$s_s = 65.5 \text{ psi}.$$

The stress random variable is

$$(\bar{s}, s_s) = (663, 65.5) \text{ psi}.$$

12.3 SHORT COLUMN—ECCENTRIC LOADING

Eccentric loading produces bending combined with symmetrical axial loading. If column length is not great compared with the lateral dimensions, deflection is negligibly small compared with initial eccentricity. Thus, superposition of variates will be a usable method.

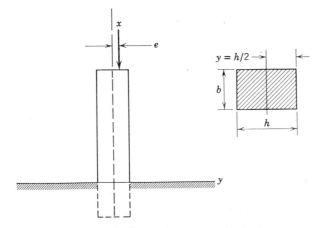

Figure 12.4 Short column: eccentric load.

For example, consider the case of compression due to a force $(\bar{P}, s_p)$ applied parallel to the column axis and with initial eccentricity e (Fig. 12.4). The force $(\bar{P}, s_p)$ produces the compressive stress random variable

$$-\frac{(\bar{P}, s_p)}{(\bar{A}, s_A)} \dagger$$

Compressive bending is produced by the couple

$$(\bar{P}, s_p)(\bar{e}, s_e).$$

The bending stress random variable is

$$\frac{(\bar{P}, s_p)(\bar{e}, s_e)(\bar{y}, s_y)}{(\bar{I}, s_I)}.$$

Total stress random variable is

$$(\bar{s}, s_s) = \left[-\frac{(\bar{P}, s_p)}{(\bar{A}, s_A)} - \frac{(\bar{P}, s_p)(\bar{e}, s_e)(\bar{y}, s_y)}{(\bar{I}, s_I)} \right] \text{psi}.$$

If a rectangular section (Fig. 12.2) is assumed, the expression for $(\bar{s}, s_s)$ is

$$(\bar{s}, s_s) = \frac{(\bar{P}, s_p)}{(\bar{b}, s_b)(\bar{h}, s_h)} - 12\frac{(\bar{P}, s_p)(\bar{e}, s_e)(\bar{y}, s_y)}{(\bar{b}, s_b)(\bar{h}, s_h)^3} \qquad (a)$$

$$= \frac{(\bar{P}, s_p)}{(\bar{b}, s_b)(\bar{h}, s_h)} - 6\frac{(\bar{P}, s_p)(\bar{e}, s_e)}{(\bar{b}, s_b)(\bar{h}, s_h)^2}.$$

† Tensile stress is considered positive.

The line of expected zero stress is located by setting the first term in Eq. *a* at zero, and solving for $(\bar{y}, s_y)$. The "line" of zero stress is a misnomer, for $(\bar{y}, s_y)$ is the random variable

$$(\bar{y}, s_y) = -\frac{1}{12}\frac{(\bar{h}, s_h)^2}{(\bar{e}, s_e)}.$$

Both mean and standard deviation of *y* decrease with larger eccentricity.

12.4 INTERMEDIATE COLUMNS—ECCENTRIC COMPRESSION

The bending of an intermediate column due to eccentric loading is discussed next. Deflection (δ) cannot be considered small compared to eccentricity (*e*). Assume eccentricity in the $x - y$ plane and that any deflection occurs in this plane. The bending moment, at any section $q - r$, will be (Fig. 12.5).

$$(\bar{M}, s_M) = -(\bar{P}, s_p)[(\delta, s_\delta) + (e, s_e) - (y, s_y)].$$

and

$$(\bar{M}, s_M) = -(\bar{P}, s_p)[(\delta + e - y), \sqrt{s_\delta^2 + s_e^2 + s_y^2}].$$

The differential equation of the deflection curve [36] is,

$$EI\frac{d^2y}{dx^2} = -M.$$

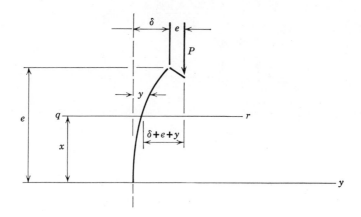

Figure 12.5 Intermediate column.

The second derivative is a random variable ([36] and Section 3)

$$\frac{1}{r} = \frac{M}{EI} \quad \text{or} \quad \frac{1}{(\bar{r}, s_r)} = \frac{(\bar{M}, s_M)}{(\bar{E}, s_E)(\bar{I}, s_I)},$$

$$\frac{1}{r} = -\frac{d^2y}{dx^2},$$

and

$$\overline{\left(\frac{d^2y}{dx^2}, s_{d^2y/dx^2}\right)} = \frac{(\bar{M}, s_M)}{(\bar{E}, s_E)(\bar{I}, s_I)},$$

and

$$\overline{\left(\frac{d^2y}{dx^2}, s_{d^2y/dx}\right)} = \frac{(\bar{P}, s_p)}{(\bar{E}, s_E)(\bar{I}, s_I)} [(\delta + \bar{e} - y), \sqrt{s_q^2 + s_e^2 + s_y^2}],$$

$$\rho^2 = \frac{P}{EI}, \quad [36]$$

$$(\bar{\rho}, s_p)^2 = \frac{(\bar{P}, s_p)}{(\bar{E}, s_E)(\bar{I}, s_I)},$$

and by substitution

$$\overline{\left(\frac{d^2y}{dx^2}, s_{d^2y/dx^2}\right)} + (\bar{P}, s_p)(y, s_y) = (\bar{P}, s_p)(\delta + \bar{e}, \sqrt{s_\delta^2 + s_e^2}).$$

Solution of the second-degree differential equation above for mean behavior conditions is

$$y = C_1 \sin \rho x + C_2 \cos \rho x + \delta + e,$$

and with the aid of the boundary conditions

$$(y) = 0 \quad \text{and} \quad \frac{d}{dx} = 0 \quad \text{at} \quad x = 0,$$

$$y = (\bar{\delta} + \bar{e})(1 - \cos \bar{\rho} x).$$

From the equation above deflection (δ) is

$$\delta = e \frac{(1 - \cos \rho l)}{\cos \rho l}.$$

The effect of deflection on the magnitude of bending moment random variable cannot be ignored in intermediate columns. Mean moment increases more rapidly than P, and the standard deviation also increases. Thus deflection is not proportional to P.

264 Application

The maximum moment $(\bar{M}, s_M)$ occurs at the base of the built-in column.

$$(\bar{M}, s_M) = (\bar{P}, s_p)(\bar{e}, s_e)\overline{(\sec \rho l}, s_{\sec \rho l})$$
$$= (P, s_p)\overline{(e + \delta}, \sqrt{s_e^2 + s_\delta^2}).$$

12.5 CRITICAL LOAD OR EULER LOAD

The deflection of an eccentrically loaded column increases rapidly as ρl approaches the value $\pi/2$ (the mean value). When the mean value of ρl becomes equal to $\pi/2$, mean moment becomes

$$M = Pe \sec (\rho l) = \infty$$

The mean value of critical load is,

$$\bar{P}_{cr} = \frac{\pi^2 \bar{E} \bar{I}}{4(\bar{l})^2} \dagger$$

$$(\bar{P}_{cr}, s_{P_{cr}}) = \frac{\pi(\bar{E}, s_E)(\bar{I}, s_I)}{4(\bar{l}, s_l)^2}.$$

for a column built in at the base (and assumed to possess absolute fixity).

Irrespective of how small the eccentricity e may be, large deflections are the result of loading to near the critical value. When deflections become large, moments at the base of a built-in column become large. Stresses also become large. Experimentally, elimination of all (geometric) eccentricity is virtually impossible. Consideration of the variability of point of load application leads to the same conclusions.

Thus column loading normally results in both compression and bending. As loads approach P_{cr} large deformations are always produced. These deformations usually exceed the elastic limit of the material. The statistical approach permits defining the probability of exceeding yield.

Example 3. Idealized Euler Column Design. In this example it is assumed that eccentricities are not present (see Fig. 12.6).

Design computations are by conventional and probabilistic methods.

Conventional Method. Compute the cross section of an Euler column from the data given. The column is a circular tube (see Fig. 12.6).

Solution. Ultimate load P_u is

$$P_u = P_L(SF) = 10{,}000 \text{ lb.}$$

Assume critical stress $s_{cr} = 40{,}000$ psi.

$$A = \frac{P_u}{s_{cr}} = \frac{10{,}000}{40{,}000} = 0.25 \text{ in.}^2$$

† The critical stress P_{cr}/A in column design is the maximum allowable stress. Thus in probabilistic design $(\bar{s}_{cr}, s_{s_{cr}})$ is the allowable stress and $(\bar{P}/\bar{A}, s_{P/A})$ is the applied stress.

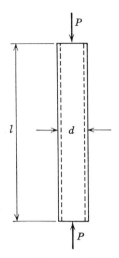

Figure 12.6 Euler column: idealized.

Assume that $D/t = 50$ and $A = \pi dt = \pi d^2/50$. Then

$$d^2 = \frac{50A}{\pi} = \frac{50(0.25)}{\pi} = 4 \text{ (approx)},$$

$$d = 2 \text{ in.} \quad \text{and} \quad t = 0.04.$$

Radius of gyration:

$$\rho = \left(\frac{I}{A}\right)^{1/2} = \left(\frac{0.126}{0.25}\right)^{1/2} = 0.707.$$

Slenderness ratio:

$$\frac{l}{\rho} = \frac{120 \text{ in.}}{0.707} = 170;$$

l/ρ is considered too large for acceptable design. As a second approximation, assume that

$$s_{cr} = 20{,}000 \text{ psi},$$

$$A = \frac{10{,}000}{20{,}000} = 0.5 \text{ in.}^2,$$

and

$$d^2 = 8 \text{ in.}^2, \quad d = 2.83 \text{ in.},$$

$$t = 0.057 \text{ in.}$$

Then

$$I_z = \pi(1.42)^3(0.057) - 0.507,$$

$$\rho = \left(\frac{0.507}{0.5}\right)^{1/2} = 1.0,$$

$$l/\rho = 120/1.0 - 120, \quad F_{cr} = 20{,}000 \text{ lb/in.},$$

266 Application

Thus
$$s_{cr} = F_{cr}.$$
The section to be used is
$$d = 2.83 \text{ in. (outside diameter)},$$
$$t = 0.057 \text{ in. (wall)}.$$

Probabilistic Method. The critical stress in a pin-ended Euler column is

$$s_{cr} = \frac{P_{cr}}{A} = \frac{\pi^2 E}{\left[\frac{l^2}{\rho}\right]^2}.$$

In this example, limit load is assumed to be a 3σ value and $s_P = 200$ lb.
$$(\bar{P}, s_P) = (6070, 200) \text{ lb}.$$

Column length is 120 in. with tolerance $\pm 0.375'' - (\pm 3\sigma_l)$. Thus
$$(\bar{l}, s_l) = (120, 0.125) \text{ in}.$$

Required probability of column adequacy is
$$R = 0.99997, \qquad (z \approx 4).$$

In probabilistic design of the Euler column applied stress and allowable stress are expressed in terms of the column radius r. Since $d/t = 50$, $r/t = 25$,

$$A = 2\pi r t = \frac{2\pi}{25}(r^2).$$

The approximate expression for moment of inertia I_z of a tubular section is (Section 7.4):
$$I_z = \pi r^3 t.$$
Radius of gyration ρ is (see Section 7.5)
$$\rho = \left(\frac{I}{A}\right)^{1/2} = \left(\frac{\pi r^3 t}{2\pi r t}\right)^{1/2} = \frac{r}{\sqrt{2}}.$$

Allowable stress $(\bar{S}, s_S)$ and critical stress $(\bar{s}_{cr}, s_{scr})$ are identical.
$$(\bar{S}, s_S) = (\bar{s}_{cr}, s_{scr}) = \left(\frac{\pi^2 \bar{r}^2 \bar{E}}{2\bar{l}^2}, s_s\right)$$

$s_S = 0.015\bar{r}$ (equivalent to tolerance 0.045 in. per in.—from Handbook information).

Applying Eq. 2.61,
$$s_{r^2} = 2\bar{r} s_r = 0.03\bar{r}^2.$$
The mean value of $\bar{S}$ is
$$\bar{s} = \frac{\pi^2 \cdot 30 \cdot 10^6}{2(120)^2}(0.03)\bar{r}^2 = 1.02812 \cdot 10^4 \bar{r}^2,$$
where
$$\frac{\pi^2 \bar{E}}{2} = 148 \cdot 10^6.$$

The standard deviation estimate (Eq. 3.14) is

$$s_S = \frac{148.044 \cdot 10^6}{14{,}400}\left[\frac{(1.44)^2 10^8 \cdot 9 \cdot 10^{-4} \bar{r}^4 + (9.6)^2 \bar{r}^4}{(14{,}400)^2}\right]^{1/2}$$

$$= 308.39 \bar{r}^2.$$

$$(\bar{S}, s_S) = (1.02812 \cdot 10^4 \bar{r}^2,\ 308.39 \bar{r}^2)$$

Applied stress $(\bar{s}, s_s)$ estimate:

$$(\bar{s}, s_s) = \frac{(\bar{P}, s_P)}{(\bar{A}, s_A)} = \frac{(6070;\ 200)}{\dfrac{2\pi}{25}\bar{r}^2,\ \dfrac{2\pi}{25}(0.03)\bar{r}^2}$$

$$(\bar{r}^2, s_r{}^2) = (\bar{r}^2, 0.03\bar{r}^2)\ \text{in.}^2$$

$$(\bar{l}^2, s_l{}^2) = (14{,}400, 9.6)\ \text{in.}^2$$

$$\bar{s} = \frac{6070}{0.2513\bar{r}^2} = \frac{24{,}154}{\bar{r}^2}\ \text{psi}.$$

By applying Eq. 3.14,

$$s_s = \frac{1}{0.2513\bar{r}^2} \cdot \left[\frac{(0.2513\bar{r}^2)^2(200)^2 + (6070.)^2(0.00754\bar{r}^2)^2}{(0.2513\bar{r}^2)^2}\right]^{1/2},$$

$$s_s = \frac{841.5}{\bar{r}^2}\ \text{psi},$$

$$(\bar{s}, s_s) = \left(\frac{24{,}154}{\bar{r}^2},\ \frac{841.5}{\bar{r}^2}\right)\ \text{psi}.$$

By substituting for allowable and applied stress in Eq. 4.6,

$$4.0 = \frac{24{,}154/\bar{r}^2 - 1.028 \cdot 10^4 \bar{r}^2}{[841.5^2/\bar{r}^2 + (308.39\bar{r}^2)^2]^{1/2}}.$$

Squaring both sides of the expression, then simplifying,

$$0 = \bar{r}^8 - 4.712\bar{r}^4 + 5.4125.$$

Solving the quadratic in $\bar{r}^4$,

$$\bar{r}^4 = 2.730\ \text{in.}^4,$$

$$\bar{r} = 1.286\ \text{in.},$$

$$\bar{d} = 2.572\ \text{in.},\ t = 0.051\ \text{in.}$$

Example 4. Effect of Eccentricity—Euler Buckling (see Fig. 12.7). The column in Example 3 is analyzed with all conditions identical except the requirement of cross-sectional concentricity. Eccentricity is introduced as a random variable. The 3σ value for eccentricity is set at 10 percent of column wall thickness†; $d/t = 50$ is specified.

$$(\bar{t}, s_t) = \frac{(\bar{r}_1, s_{r_1})}{25},$$

$$(\bar{r}_2, s_{r_2}) = (\bar{r}_1, s_{r_1}) - (\bar{t} - s_t) = \tfrac{24}{25}(\bar{r}_1, s_{r_1}).$$

† Alcoa structural handbook value.

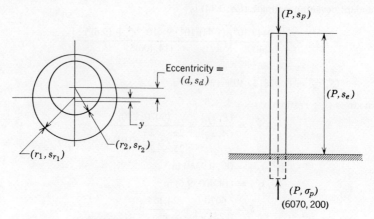

Figure 12.7 Euler column with eccentricity.

From Section 7.4 and [36], the moment of inertia of a circular section is

$$I = \frac{\pi r^4}{4},$$
$$I = \bar{I} + A d^2,$$
$$I = I_1 - (\bar{I} + A d^2),$$
$$I = \left[\frac{\pi r_1^4}{4} - \frac{\pi r_2^4}{4} - \pi r_2^2 d^2\right] \text{ in.}^4,$$
$$I = \frac{\pi}{4}(r_1^4 - (\tfrac{24}{25})^4 r_1^2) - \pi(\tfrac{24}{25})^2 r_1^2 d^2,$$
$$(\bar{I}, s_I) = \frac{\pi}{4}\left(\frac{58{,}849}{388{,}625}\right)(\bar{r}_1, s_{r_1})^4 - \frac{576\pi}{625}(\bar{r}_1, s_{r_1})^2(\bar{d}, s_d)^2.$$
$$= \pi(0.03766)(\bar{r}_1, s_{r_1})^4 - \pi(0.9216)(\bar{r}_1, s_{r_1})^2(\bar{d}, s_d)^2.$$

Eccentricity, $(\bar{d}, s_d) = 0.10(\bar{t}, s_t) = \dfrac{(\bar{r}_1, s_{r_1})}{250}$

$$= \pi(\bar{r}_1, s_{r_1})^4(0.03658 - 0.0000246),$$
$$(\bar{I}, s_I) = 0.03765\pi(\bar{r}_1, s_{r_1})^4.$$

Allowable stress, Euler column,

$$S = \frac{\pi^2 E I}{l^2 A} = \frac{(3.1416)^2 \cdot 30 \cdot 10^6 I}{l^2 A} \text{ psi},$$
$$A = \pi r_1^2 - \pi r_2^2$$
$$= \pi r_1^2 - \pi\left(\frac{24 r_1}{25}\right)^2,$$
$$A = \frac{49\pi}{625}(r_1^2) \text{ in.}^2$$

The column length random variable is

$$(\bar{l}, s_l) = (120, 0.125) \text{ in.}$$

To compute applied stress it is assumed that loading is applied uniformly around the periphery of the section. Applied stress is required at the point of potential maximum stress.

Load distribution around the section periphery is

$$\frac{\bar{P}, s_P}{(2\pi r_1, s_{2\pi r_1})}$$

Section area in square inches per inch of periphery in the neighborhood of minimum section thickness is

$$A = \frac{\pi r_1^2 - \pi(r_2 + r_1/250)^2}{2\pi r_1} \text{ in.}^2 \text{ per inch of periphery,}$$

$$r_2 = \tfrac{24}{25} r_1 \quad \text{(see Fig. 12.7),}$$

$$A = \frac{r_1^2 - [(241/250)r_1]^2}{2\pi r_1} = 0.03535 r_1 \text{ in.}^2 \text{ (per inch).}$$

Mean applied stress (estimate) is

$$\bar{s} = \frac{\dfrac{\bar{P}}{2\pi \bar{r}_1} \text{ lb/in.}}{0.03535 \bar{r}_1 \text{ in.}^2/\text{in.}} = \frac{4.5226 \bar{P}}{\bar{r}_1^2} \text{ psi}$$

Load random variable

$$(\bar{P}, s_P) = (6070, 200) \text{ lb} \qquad \text{(Example 3).}$$

Applied stress random variable in lb/in.² (per in. of periphery),

$$(\bar{s}, s_s) = \frac{4.5226(6070, 200)}{(\bar{r}_1, s_{r_1})^2},$$

$$\bar{s} = \frac{27{,}450}{\bar{r}_1^2}$$

$$(\bar{r}_1, s_{r_1}) = (\bar{r}_1^2, 0.03\bar{r}_1^2),$$

$$(\bar{P}, s_P) = (6070, 200).$$

Applying Eq. 3.14,

$$s_s = \frac{4.5226}{\bar{r}_1} \left[\frac{\bar{r}_1^2 (200)^2 + (6070)^2 (0.032 \bar{r}_1)^2}{(\bar{r}_1^2)^2} \right]^{1/2},$$

$$= \frac{1223}{\bar{r}_1^2}.$$

Thus the applied stress random variable estimate is

$$(\bar{s}, s_s) = \left(\frac{27{,}450}{\bar{r}_1^2}, \frac{1223}{\bar{r}_1^2} \right).$$

Allowable stress random variable (S).

$$(\bar{s}, s_S) = \frac{\pi^2 \bar{E}(\bar{I}, s_I)}{(\bar{l}, s_l)^2(\bar{A}, s_A)} = \frac{2.96089 \cdot 10^8 (0.03765\pi)(\bar{r}_1, s_{r_1})^4}{(120, 0.125)^2 \left(\frac{49\pi}{50}\right)(\bar{r}_1, s_{r_1})^2}$$

$$= 1.4219 \cdot 10^8 \frac{(\bar{r}_1^2, s_{r_1}^2)}{(14{,}400; 30)} = 1.4219 \cdot 10^8 \frac{(\bar{r}_1^2, 0.03\bar{r}_1^2)}{(14400, 30)},$$

$$\bar{s} = 9924.5 \bar{r}_1^2,$$

$$s_S = 9924.5 \left[\frac{(1.44)^2 \cdot 10^8 \cdot (103 \bar{r}_1^2)^2 + 30^2 \bar{r}_1^4}{(1.44)^2 \cdot 10^2}\right]^{1/2} = 295.8 \bar{r}$$

$$(\bar{S}, s_S) = (8625 \bar{r}_1^2, 260 \cdot \bar{r}_1^2) \text{ psi},$$

$$(\bar{s}, s_s) = \left(\frac{27{,}450}{\bar{r}_1^2}, \frac{1223}{\bar{r}_1^2}\right) \text{ psi}.$$

Employing Eq. 4.6 (with $z = 3.0$),

$$3.0 = \frac{27{,}450/\bar{r}_1^2 - 9924.5 \bar{r}_1^2}{\sqrt{(1223/\bar{r}_1^2)^2 + (295.8 \bar{r}_1^2)^2}}$$

Solving the quadratic in $\bar{r}_1^4$,

$$\bar{r}_1^4 = \frac{5.582 \pm \sqrt{31.16 - 30.32}}{2}$$

$$\bar{r}_1^4 = 3.249 \text{ in.}^4,$$
$$\bar{r}_1 = 1.342, \text{ tolerance} = \pm 0.08 \text{ in. (max.)},$$
$$\bar{d} = 2.684 \text{ in.}, \; t = 0.053 \text{ in.}$$

Percentage increase in diameter (mean) due to eccentricity,

$$(\text{Example 4/Example 3}) = \frac{0.112}{2.684} \cdot 100 = 4.1\% \quad (\text{increase}).$$

12.6 THE BEAM COLUMN

With the beam column, in addition to axial loading, it is necessary to consider the effect of simultaneous lateral loading. The expression for maximum allowable compressive stress (S) (Fig. 12.8) is

$$S = \frac{P}{A}\left(1 + \frac{e}{c} \sec\left[\frac{1}{2r}\left(\frac{P}{AE}\right)^{1/2}\right]\right), \dagger$$

ρ = radius of gyration,

e = eccentricity.

† [42].

Column Design 271

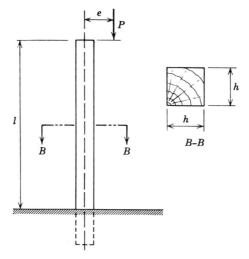

Figure 12.8

To compute s in terms of the beam column geometry, the secant function is represented as a series. Thus, if

$$u = \sin x = x - \frac{x^3}{3!} + \frac{x^5}{5!} + \frac{x^7}{7!} + \cdots,$$

$$\frac{\partial u}{\partial x} = \cos x = 1 - \frac{3x^2}{3!} + \frac{5x^4}{5!} - \frac{7x^6}{7!} + \cdots,$$

$$\cos x = 1 - \frac{x^2}{2} + \frac{x^4}{4!} - \frac{x^6}{6!} + \cdots.$$

Employing Eq. 2.61 yields

$$s_u{}^2 = \left(\frac{\partial u}{\partial x}\right)^2 s_x{}^2 = \left(1 - \frac{x^2}{2} + \frac{x^4}{4!} - \frac{x^6}{6!} + \cdots\right) s_x{}^2.$$

Consider a specific beam column of square cross section:

$$\text{cross-section area} = A = h^2,$$

$$\dagger\text{core radius} = c = \frac{h \cdot h^2/6}{h \cdot h} = \frac{h}{6},$$

$$\text{radius of gyration} = \rho = \left(\frac{I}{A}\right)^{\frac{1}{2}} = \frac{h}{2\sqrt{3}},$$

$$\text{moment of inertia} = I = \frac{h^4}{12}.$$

† See discussion, "Radius of the Core" [38].

272 Application

By substituting the expression for S is

$$S = \frac{P}{h^2}\left\{1 + \frac{6e}{h}\sec\left[\frac{\sqrt{3}}{h}\frac{1}{h}\left(\frac{P}{h^2E}\right)^{1/2}\right]\right\},$$

$$S = \frac{P}{h^2}\left\{1 + \frac{6e}{h}\sec\left[\frac{1}{h^2}\left(\frac{3P}{E}\right)^{1/2}\right]\right\}.$$

Let

$$u = \frac{1}{h^2}\left(\frac{3P}{E}\right)^{1/2}.$$

Then

$$\cos u > 1 - \frac{u^2}{2}, \tag{a}$$

$$\sec u < \left(1 - \frac{u^2}{2}\right)^{-1} = w_1. \tag{b}$$

Equation b yields a mean value estimate slightly larger than $\sec u$,

$$\cos u < 1 - \frac{u^2}{2} + \frac{u^4}{24}, \tag{c}$$

$$\sec u > \left(1 - \frac{u^2}{2} + \frac{u^4}{24}\right)^{-1} = w_2. \tag{d}$$

Equation d yields a mean value estimate slightly smaller than $\sec u$.

Thus the random variable $(\bar{w}_1, s_{w_1})$, as a factor in determining the maximum compressive stress random variable, introduces a small element of conservatism.

$$\bar{S} = \frac{P}{h^2}\left[1 + \frac{6e}{h}\left(1 - \frac{u^2}{2} + \frac{u^4}{24}\right)^{-1}\right],$$

$$\bar{S} = \frac{P}{h^2}\left[1 + \frac{6e}{h}\left(1 - \frac{Pl^2}{8EI} + \frac{P^2l^4}{384E^2I^2}\right)^{-1}\right],$$

$$\bar{S} = \frac{P}{h^2}\left\{1 + \frac{6e}{h}\left[1 - \frac{Pl^2}{8E(h^4/12)} + \frac{P^2l^4}{384E^2(h^8/144)}\right]^{-1}\right\},$$

$$= \frac{P}{h^2}\left\{1 + \frac{6e}{h}\left[1 - \left(\frac{3Pl^2}{2Eh^4}\right) + \left(\frac{3P^2l^4}{8E^2h^8}\right)\right]^{-1}\right\},$$

$$= \frac{P}{h^2}\left[1 + \frac{6e}{h}\left(\frac{8E^2h^8}{8E^2h^8 - 12Pl^2Eh^4 + 3P^2l^4}\right)\right],$$

$$\bar{S} = \frac{P}{h^2} + \left(\frac{48E^2eh^5P}{8E^2h^8 - 12Pl^2Eh^4 + 3Pl^4}\right).$$

As $e \to 0$, $s \to P/A$,

$$\bar{s} > \left(\frac{P}{h^2} + \frac{6Pe}{h^3}\right) = \frac{P}{h^2}\left(1 + \frac{6e}{h}\right),$$

$$\bar{s} < \frac{P}{h^2} + \frac{12eEhP}{2Eh^4 - 3Pl^2}.$$

Stress standard deviation estimate:

$$s = \frac{P}{A}\left[1 + \frac{6e}{h}\left(\frac{2}{2-u^2}\right)\right],$$

$$s = \frac{P}{h^2}\left[1 + \frac{6e}{h}\left(\frac{2}{2-Pl^2/4EI}\right)\right],$$

$$s = \frac{P}{h^2}\left[1 + \frac{6e}{h}\left(\frac{8EI}{8EI - Pl^2}\right)\right].$$

Partial derivatives (Eq. 2.61):

$$s = \left(\frac{P}{h^2} + \frac{12eEhP}{2Eh^4 - 3Pl^2}\right),$$

$$\frac{\partial s}{\partial e} = \frac{12EhP}{2Eh^4 - 3Pl^2},$$

$$\frac{\partial s}{\partial P} = \frac{1}{h^2} + \frac{(2Eh^4 - 3Pl^2)(12eEh) + (12eEhP)(3l^2)}{(eEh^4 - 3Pl^2)^2},$$

$$\frac{\partial s}{\partial h} = \frac{2}{h^3} + \frac{(2Eh^4 - 3Pl^2)(12eEP) - (12eEhP)(8Eh^3)}{(2Eh^4 - 3Pl^2)^2},$$

$$\frac{\partial s}{\partial l} = \frac{12eEhP(6Pl)}{(2Eh^4 - 3Pl^2)^2}.$$

Stress standard deviation:

$$s_s = \left[\left(\frac{\partial u}{\partial e}\right)^2 s_e^2 + \left(\frac{\partial u}{\partial P}\right)^2 s_P^2 + \left(\frac{\partial u}{\partial h}\right)^2 s_h^2 + \left(\frac{\partial u}{\partial l}\right)^2 s_l^2\right]^{1/2}.$$

CHAPTER 13

Torsion and Combined Torsion and Bending[†]

13.1 TORSION

The problem of design for torsional stressing is introduced by considering a circular shaft, built in at one end and twisted by a couple applied at the free end. For small angles of twist, distortion is essentially absent.

$$\tau = \tfrac{1}{2} G \theta d. \quad [36] \qquad (a)$$

τ is the shear stress.
G is the modulus of elasticity for shear.
d is the outside shaft diameter.
θ is the angle of twist per unit length.

The relationship between the applied twisting couple (M_t) and the stresses produced is

$$M_t = G\theta I_p.$$

I_p is the polar moment of inertia of the circular section,

$$I_p = \int_A r^2 \, dA.$$

Then

$$\theta = \frac{M}{GI_p}. \qquad (b)$$

Given a hollow, concentric, circular section, assume that

$$\frac{r_o}{r_i} = c \text{ (no variation in } c\text{)},$$

[†] Courtesy of Dr. Charles O. Smith, University of Detroit.

where r_o is the outside radius and r_i is the inside radius.

$$I_p = I_{p(\text{outer})} - I_{p(\text{inner})} = \frac{\pi}{4}(r_o^4 - r_i^4).$$

With $r_i = r_o/c$,

$$I_p = \frac{\pi}{4}\left(r_o^4 - \frac{r_o^4}{c^4}\right) = r_o^4 \frac{\pi}{4}\left(\frac{c^4 - 1}{c^4}\right).$$

Let

$$d = \frac{\pi}{4}\left(\frac{c^4 - 1}{c^4}\right).$$

Then,

$$I_p = dr_o^4.$$

By combining Eqs. a and b and substituting for I_p,

$$\tau = \frac{M_t r_o}{I_p} = \frac{M_t}{dr_o^3}.$$

If shearing stress governs, the applied stress random variable is

$$(\bar{\tau}, s_\tau) = \frac{(\bar{M}_t, s_{M_t})}{d(\bar{r}_o^3, s_{r_o^3})}.$$

Let $s_{r_o} = kr_o$.
The mean applied shear stress $(\bar{\tau})$ (Eq. 3.13) is

$$\bar{\tau} = \frac{\bar{M}_t}{d\bar{r}_o^3}.$$

The standard deviation estimate (s_τ) is, (Eq. 3.14)

$$s_\tau = \frac{1}{3\,dk\bar{r}_o^3}\left[\frac{\bar{M}_t^2(3\,dk\bar{r}_o^3)^2 + (d\bar{r}_o^3)^2 s_{M_t}^2}{(d\bar{r}_o^3) + (3\,dk\bar{r}_o^3)^2}\right]^{1/2}$$

$$s_\tau = \frac{1}{3\,dkr^3}\left[\frac{\bar{M}_t^2(9k^2) + s_{M_t}^2}{1 + 9k^2}\right]^{1/2}.$$

If the allowable shear stress random variable is $(\bar{\tau}_0, s_{\tau_0})$, (Eq. 4.6), then

$$z = \frac{|\bar{\tau}_0 - \bar{\tau}|}{\sqrt{s_{\tau_0}^2 + s_\tau^2}}.$$

Substituting for τ_0 and τ,

$$z^2 = \frac{\bar{\tau}_0 - (\bar{M}_t/d\bar{r}_o^3)}{s_{\tau_0}^2 + s_\tau^2}$$

$$z = \frac{(d\bar{r}_o^3 \bar{\tau}_0 - \bar{M}_t)^2/(d\bar{r}_o^3)^2}{s_{\tau_0}^2 + \frac{1}{3\,dk\bar{r}_o^3}[\bar{M}_t^2(9k^2) + s_{M_t}^2/1 + 9k^2]}.$$

Expanding and then collecting terms,

$$\bar{r}_o^6 + \bar{r}_{oo}^3 \left[\frac{2\bar{\tau}_0 \bar{M}_t}{d(z^2 s_{r_o}^2 - \bar{\tau}_0^2)} \right] + \frac{[9k^2(z^2-1)-1]\cdot \bar{M}_t^2 + z^2 s_{M_t}^2}{d^2(1+9k^2)(z^2 s_{r_o}^2 - \bar{\tau}_0^2)} = 0. \qquad (c)$$

The quadratic is solved for $\bar{r}_o^3$. For θ, rewriting Eq. b,

$$(\bar{\theta}, s_\theta) = \frac{(\bar{M}_t, s_{M_t})}{(\bar{G}, s_G)(\bar{I}_p, s_{I_p})}$$

$$(\bar{\theta}, s_\theta) = \frac{(\bar{M}_t, s_{M_t})}{(\bar{G}, s_G)(d\bar{r}_o^4, 4\, d\bar{r}_o^3 s_{r_o})}. \qquad (b')$$

By Eq. 2.61, $s_{r_o^4}$ estimate is

$$s_{r_o^4} = 4\, d\bar{r}_o^3 s_{r_o}.$$

To estimate $(\bar{\theta}, s_\theta)$ the denominator of (b') is computed as a product (Eqs. 3.9 and 3.10). Let $x = GI_p$; then

$$\bar{x} = G\, d\bar{r}_o^4$$

and

$$s_x = dr_o^4 \sqrt{16k^2 \bar{G}^2 + (1+16k^2)s_G^2},$$

$\theta = M_t/x$, and, by Eqs. 3.13 and 3.14,

$$\bar{\theta} = \frac{\bar{M}_t}{\bar{x}}.$$

$$s_\theta = \frac{\left\{ \dfrac{M_t^2[16k^2\bar{G}^2 + (1+16k^2)s_G^2 + \bar{G}^2 s_{M_t}^2]}{\bar{G}^2 + 16k^2\bar{G}^2 + (1+16k^2)s_G^2} \right\}}{d\bar{r}_o^4 \sqrt{16k^2\bar{G}^2 + (1+16k^2)s_G^2}}.$$

Total twist (ϕ) is a function of θ and length of shaft (l)

$$(\bar{\phi}, s_\phi) = (\bar{\theta}, s_\theta)(\bar{l}, s_1);$$

$\bar{\phi}$ is, by Eq. 3.9,

$$\bar{\phi} = \bar{\theta} \bar{l}.$$

The standard deviation (s_ϕ) is, by Eq. 3.10,

$$s_\phi = \sqrt{\bar{\theta}^2 s_1^2 + \bar{l}^2 s_\theta^2 + s_\theta^2 s_1^2}.$$

If shaft design is determined by θ, let θ_L denote allowable total twist and let θ denote applied twist, the result of loading. Applying Eq. 4.6 and expanding,

$$0 = \bar{r}_o^8 - \bar{r}_o^4 \left[\frac{2\bar{\theta}_L \bar{M}_t}{\bar{G}\, d(\bar{\theta}_L^2 - z^2 s_{\theta_L}^2)} \right] + \cdots$$
$$+ \left(\frac{\bar{M}_t^2 \{\bar{G}^2[16k^2(z^2-1)-1] + s_G^2(z^2-1)(1+16k^2)\} + \bar{G}z^2 s_{M_t}^2}{\bar{G}^2\, d^2(\bar{G}^2 + s_G^2)(1+16k^2)(-\bar{\theta}_L^2 + z^2 s_{\theta_L}^2)} \right).$$

The quadratic in $\bar{r}_o^4$ is solved.

Example 1. Design a hollow shaft of circular cross section, given the following data:

$$\frac{r_o}{r_i} c = 1.50,$$

applied torque = $(\bar{M}_t, s_{M_t}) = (120{,}000;\ 6000)$ in-lb,
allowable shear = $(\bar{\tau}_0, s_{\tau_0}) = (45{,}000;\ 2200)$ psi
required outside radius $(\bar{r}_o, s_{r_o})$ in.

Manufacturers tolerance on shaft r_o.

$$\pm 0.045 \text{ in./in.} \ (\pm 3\sigma)$$

The standard deviation estimate (s_{r_o}) is

$$s_{r_o} = 0.01 \bar{r}_o. \qquad (f)$$
$$k = 0.015.$$

Compute r_o by the conventional approach (safety factor = 1.0),

$$\tau = \frac{Mt}{dr_o^3}; \qquad d = \frac{\pi}{2}\left(\frac{c^4-1}{c^4}\right); \qquad c = 1.50,$$

$$d = \frac{\pi}{2} \cdot \frac{4.06}{5.06} = 1.26. \qquad (g)$$

$$r_o^3 = \frac{Mt}{d\bar{\tau}} = \frac{120{,}000}{1.26(45{,}000)} = 2.12; \qquad r_o = 1.28 \text{ in.}$$

Similarly, varying the safety factor,

Safety Factor	r_o^3	r_o
1.0	2.12	1.28 in.
1.5	3.17	1.47 in.
2.0	4.24	1.62 in.
2.5	5.24	1.74 in.
3.0	6.35	1.85 in.
4.0	8.45	2.04 in.
5.0	10.58	2.20 in.
6.0	12.70	2.33 in.

278 Application

Next, design a hollow cylindrical section to resist torosional loading by probabilistic methods. Shaft sizes are computed to satisfy the following specified probabilities.

	(Item)	(Item)^2	Probability
z	1	1	$R_1 = 0.8413$
z	2	4	$R_2 = 0.9772$
z	3	9	$R_3 = 0.9987$
z	4	16	$R_4 = 0.99997$
d	1.26	1.59	(see Eq. g)
k	0.015	$2.25 \cdot 10^{-4}$	(see Eq. f)
$\bar{\tau}_0$	45,000	$2025 \cdot 10^6 = 2.025 \cdot 10^9$	
s_{τ_0}	2,200	$4.84 \cdot 10^6$	
$\bar{M}_t$	120,000	$1.44 \cdot 10^{10}$	
s_{M_t}	6,000	$36 \cdot 10^6$	

The example calculations below are for $\bar{r}_o$ which satisfies $R_3 = 0.9987$ probability of adequacy.

The following computed values are needed:

$$2\bar{\tau}_0 \bar{M}_t = 2(45 \cdot 10^3)(120 \cdot 10^3) = 1.08 \cdot 10^{10},$$

$$z^2 s_{\tau_0}^2 = 9(4.84 \cdot 10^6) = 43.56 \cdot 10^6,$$

$$(z^2 s_{\tau_0}^2 - \tau_0^2) = -(2025 - 44)10^6 = -1981 \cdot 10^6,$$

$$9k^2 = 20.25 \cdot 10^{-4},$$

$$1 + 9k^2 = 1.002,$$

$$z^2 - (1 + 9k^2) = 7.998 \approx 8.0,$$

$$z^2 s_{M_t}^2 = 9(36 \cdot 10^6) = 216 \cdot 10^6$$

$$9k^2[z^2 - (1 + 9k^2)]\bar{M}_t = (2.025 \cdot 10^{-3})8(14.4 \cdot 10^9) = 234 \cdot 10^6,$$

The coefficients for Eq. c are

$$\frac{2\bar{\tau}_0 \bar{M}_t}{d(z^2 s_{\tau_0}^2 - \bar{\tau}_0^2)} = \frac{1.08 \cdot 10^{10}}{1.26(1.981 \cdot 10^9)} = -4.33,$$

$$\frac{9K^2[z^2 - (1 + 9k^2)]\bar{M}_t^2 + z^2 s_{M_t}^2}{9 d^2 k^2 (1 + 9k^2)(z^2 s_{\tau_0}^2 - \tau_0^2)} = -70.5,$$

$$\bar{r}_o^6 - 4.33 \bar{r}_o^3 - 70.5 = 0,$$

$$\bar{r}_o^3 = 10.83 \text{ in.}^3,$$

$$\bar{r}_o = 2.21 \text{ in.}$$

Summary

Reliability	(z)	r_o	s_{r_o}
0.8413	1	1.74 in.	0.025
0.9772	2	2.05 in.	0.031
0.9987	3	2.21 in.	0.033
0.99997	4	2.47 in.	0.037

Torsion and Combined Torsion and Bending   279

13.2 COMBINED STRESS (BENDING AND TORSION IN A CIRCULAR MEMBER)

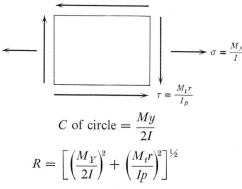

$$C \text{ of circle} = \frac{My}{2I}$$

$$R = \left[\left(\frac{M_Y}{2I}\right)^2 + \left(\frac{M_t r}{Ip}\right)^2\right]^{1/2}$$

R = maximum shear stress,

$$R^2 = \left(\frac{Mr_o}{2I}\right)^2 + \left(\frac{M_t r_o}{Ip}\right)^2, \quad 2I = I_p,$$

$$= \frac{r^2}{I_p^2}(M^2 + M_t^2).$$

As before, let

$$\frac{r_o}{r_i} = c \text{ (assume no variation in } c\text{)}.$$

Then

$$I_p = \frac{\pi}{2}(r_o^4 + r_i^4) = \frac{\pi}{2}\left(r_o^4 - \frac{r_o^4}{c^4}\right),$$

$$I_p = \frac{\pi}{2}r_o^4\left(\frac{c^4-1}{c^4}\right) = dr_o^4,$$

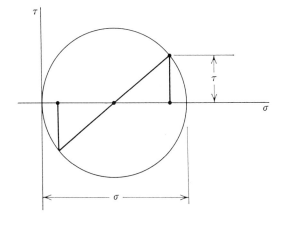

where

$$d = \frac{\pi}{2}\left(\frac{c^4 - 1}{c^4}\right),$$

$$\tau = \frac{r_o}{d\bar{r}_o^4}\sqrt{M^2 + M_t^2} = \frac{\sqrt{M^2 + M_t^2}}{d\bar{r}_o^3},$$

$$(\bar{\tau}, s_\tau) = \frac{(\overline{\sqrt{M^2 + M_t^2}}, s\sqrt{M^2+M_t^2})}{(d\bar{r}_o^3, 3\, d\bar{r}_o^2 s_{r_o})} \quad \text{(see Eq. 2.61).}$$

Let $s_r = k\bar{r}_o$

$$= \frac{(\overline{\sqrt{M^2 + M_t^2}}, s\sqrt{M^2+M_t^2})}{(d\bar{r}_o^3, 3\, dk\bar{r}_o^3)},$$

$$\overline{M^2} = \bar{M}^2 + s_M^2 \quad \text{(Eq. 3.17),}$$

$$s_{M^2} = \sqrt{4\bar{M}^2 s_M^2 + 2s_M^4} \quad \text{(Eq. 3.21),}$$

$$\overline{M_t^2} = \bar{M}_t^2 + s_M^2,$$

$$s_{M_t^2} = \sqrt{4\bar{M}_t^2 s_{M_t}^2 + 2s_{M_t}^4},$$

$$\overline{M^2 + M_t^2} = \bar{M}_t^2 + s_{M_t}^2 + \bar{M}^2 + s_M^2,$$

$$s_{(M^2+M_t^2)} = \sqrt{4M^2 s_M^2 + s_M^4 + 2s_M^4 + 4M_t^2 + 2s_M^2}.$$

Let

$$M = \sqrt{M^2 + M_t^2},$$

$$\overline{M^2} = \bar{M}^2 + s_M^2 + \bar{M}_t^2 + s_{M_t}^2,$$

$$s_{M^2}^2 = 4\bar{M}^2 \cdot s_M^2 + 2s_M^4 + 4\bar{M}_t^2 s_{M_t}^2 + 2s_{M_t}^4.$$

Equation 3.22, the expression for square root of a normal random variable, is

$$\mu_{\sqrt{\bar{M}}} = \{\tfrac{1}{2}[4(\bar{M}^2 + s_M^2 + \bar{M}_t^2 + s_{M_t}^2) - 2(4\bar{M}^2 s_M^4 + 2s_M^4 + \\ + 4\bar{M}_t^2 s_M^2 + 2s_{M_t}^4)]^{1/2}\}^{1/2}$$

$$= \{[(\bar{M}^2 + M_t^2)^2 + (\bar{M}^2 s_{M_t}^2 + \bar{M}_t^2 s_{M_t}^2 + s_M^2 s_{M_t}^2)]^{1/2}\}^{1/2},$$

$$\sigma_{\sqrt{\bar{M}}} = \{(\bar{M}^2 + s_M^2 + \bar{M}_r^2 + s_{M_t}^2) - [(\bar{M}^2 + \bar{M}_t^2)^2 \\ + 2(\bar{M}^2 s_{M_t}^2 + \bar{M}_t^2 s_M^2 + s_M^2 s_{M_t}^2)]^{1/2}\}^{1/2},$$

$$(\bar{\tau}, s_\tau) = \frac{(\bar{M}, s_M)}{(d\bar{r}_o^3, 3\ dk\bar{r}_o^2)},$$

$$\bar{\tau} = \frac{\bar{M}}{d\bar{r}_o^3},$$

$$s_\tau = \frac{1}{dr_o^3}\left(\frac{\bar{M}^2 9\ d^2 k^2 \bar{r}_o^6 + d^2 \bar{r}_o^6 s_M^2}{d^2\bar{r}_o^6 + 9\ d^2 k^2 \bar{r}_o^4}\right)^{1/2},$$

$$s_\tau = \frac{1}{dr_o^3}\left(\frac{9k^2\bar{M}^2 + s_M^2}{1 + 9k^2}\right)^{1/2}.$$

$(\bar{\tau}_A, {}^s\tau_A).$

By applying Eq. 4.6,

$$z = \frac{\bar{\tau}_A - \tau}{\sqrt{s_{\tau_A}^2 + 2_\tau^2}}.$$

Substituting values and squaring,

$$z^2 = \frac{(\bar{\tau}_A\, d\bar{r}_o^3 - \bar{M})/d\bar{r}_o^3}{\left[\dfrac{s_\tau^2\, d^2\bar{r}_o^6 + (9k^2\bar{M}^2 + s_M^2)/(1 + 9k^2)}{d^2\bar{r}_o^6}\right]}$$

$$-\frac{2\bar{\tau}_A\bar{M}}{d(\bar{\tau}_A^2 - z^2 s_{\tau_A}^2)}\, r_o^3 + \frac{\bar{M}^2[1 - 9k^2(z^2 - 1)] - z^2 s_M^2}{d^2(1 + 9k^2)(\tau_A^2 - z^2 s_{\tau_A}^2)} = 0.$$

Example 1. Compute the tube dimensions given the following data:

allowable stress $(\bar{\tau}_A, s_{\tau_A}) = (65{,}000;\ 3200)$ psi,
bending moment $(\bar{M}, s_M) = (60{,}000;\ 300\ )$ in-lb,
twisting moment $(\bar{M}_t, s_{M_t}) = (40{,}000;\ 200)$ in-lb.

Assume that

$$c = \frac{\bar{r}_o}{\bar{r}_i} = 1.5;\quad d = \frac{2}{\pi}\left(\frac{c^4 - 1}{c^4}\right) = \frac{\pi}{2}\frac{4.06}{5.06} = 1.26$$

Radius $(\bar{r}_o, s_{r_o})$:

Specified reliability: $\quad R = 0.999 \quad z \approx 3. \quad k = 0.015.$

Solution.

$$\bar{M}^2 = 36 \cdot 10^8,\ s_M^2 = 9 \cdot 10^6$$
$$\bar{M}_t^2 = 16 \cdot 10^8,\ s_{M_t}^2 = 4 \cdot 10^6.$$

Applying Eq. a',

$$\bar{M} = [(52 \cdot 10^8)^2 + 2(144 \cdot 10^{14} + 144 \cdot 10^{14} + 36 \cdot 10^{12})]^{1/4}$$
$$= 7.216 \cdot 10^4\ \text{in-lb}$$
$$s_M = [36 \cdot 10^8 + 9 \cdot 10^6 + 16 \cdot 10^8 + 4 \cdot 10^6 + 52 \cdot 10^8]^{1/2} = 2600\ \text{in-lb}.$$

Applying Eq. c',

$$\bar{r}_o^6 - \frac{2.65 \cdot 10 \cdot 7.216 \cdot 10^4}{1.26[41.33 \cdot 10^8]} r_o^3 + \frac{52.06 \cdot 10^8[1 - 0.002(8)] - 0.56 \cdot 10^6}{1.59(1.002)[41.33 \cdot 10^8]} = 0,$$

$$\bar{r}_o^6 - 1.8\bar{r}_0^3 + 0.772 = 0,$$

$$\bar{r}_o = 0.97 \text{ in.}$$

Summary

Computed $\bar{r}_o$ utilizing average values, 0.96 in.
$\bar{r}_o$ utilizing extreme values, 1.09 in.
$\bar{r}_o$ with reliability $R = 0.999$, 0.97 in.

CHAPTER 14

Statistical Study of Distortion

The versatility of probabilistic methods is now demonstrated by computing the probability of a specified amount of distortion in a flat plate of hypothetical elemental makeup. The loading is thermal.

The model as idealized is composed of a large number of cubic elements arranged in layers, rows, and columns. Each cubic element is given the property of anisotropic thermal expansion along one of its principal axes. The direction of anisotropic expansion is in a plane parallel to a surface of the plate. The anisotropic random orientation is either row-wise or column-wise (see Fig. 14.1).

Under specified thermal loading (ΔT), the probability of exceeding a specified amount of distortion (0.4μ) is small. It is shown, by probabilistic considerations, that cubic elements represent the model most conducive to critical distortion.

14.1 MODEL AND INITIAL ASSUMPTIONS

Basic Geometry. The model plate is square, and 100 centimeters along each side with uniform five centimeter thickness.

Each identical cubic element measures 10 μ along all edges (see Fig. 14.2).

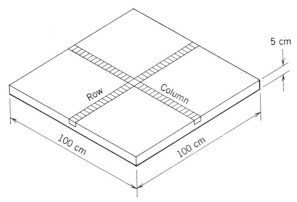

Figure 14.1   Model plate.

283

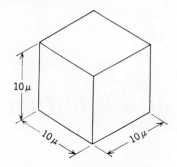 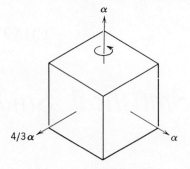

Figure 14.2 Typical cubic element. **Figure 14.3** Anisotropic thermal property.

Statistical Relationships. Statistical relationships among elements, rows of cubes, layers of rows, plates made up of layers and, finally, sets of plates are involved. In each row are 10^5 elements, in each layer are 10^5 rows and in each plate are $5 \cdot 10^3$ layers. With each plate is associated a neutral plane, above which are 2500 layers and below which are 2500 layers.

Each element is visualized as free to rotate about a principal axis perpendicular to the neutral plane of the plate. Any cube may assume anisotropic row-wise ($\rightarrow$) or column-wise ($\uparrow$) orientation of the coefficient of thermal expansion ($\frac{4}{3}\alpha$ versus 1α),† (see Fig. 14.3).

The orientation of each element is random and independent of the orientation of any other element. Consequently, rows, layers, and plates are statistically independent.

Additional assumptions: (a) initially flat surface planes (at temperature T_0), prior to ΔT thermal loading; (b) no residual stresses; and (c) distortion limited to simple bending.

14.2 STATISTICAL NATURE OF THE PROBLEM

The Anisotropic Property. Recall that the direction in which the $\frac{4}{3}\alpha$ property acts may be parallel to the axis of a column ($\uparrow$) or to a row of cubic elements ($\rightarrow$). Since orientation is random with two equally likely and mutually exclusive directions, $p(\frac{4}{3}\alpha \rightarrow) = p(\alpha \rightarrow) = \frac{1}{2}$. The phenomenon is repetitive, from cube to cube (in statistical language, a Bernoulli trail).

Distribution of the Anisotropic Random Variable (Fig. 14.4). Appearance of the $\frac{4}{3}\alpha$ property in row-wise orientation ($\rightarrow$), from the conditions described, is a random variable that follows a binomial distribution (see Section 2.2.3).

† α denotes the coefficient of thermal expansion.

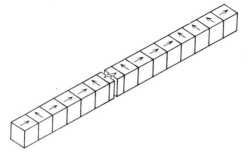

Figure 14.4 Typical row of cubic elements.

The binomial density is (Eq. 2.17)

$$f(x) = \binom{n}{x}(p)^{n}(1-p)^{(n-x)}.$$

For a typical row the parameters are $n = 10^5$ and $p = \frac{1}{2}$.

Because of the large n (10^5 cubes per row) and because $p = \frac{1}{2}$, this binomial distribution closely approximates a normal distribution (see Section 2.3.8) with density

$$f(x) = \frac{1}{s_x\sqrt{2\pi}} \exp\left[-\frac{1}{2}\left(\frac{\bar{x}-x}{s_x}\right)^2\right] \quad \text{(Eq. 2.28).}$$

With sample moments (Fig. 14.5)

$$\bar{x} = np = 10^5 \cdot (\tfrac{1}{2}) = 5 \cdot 10^4$$
$$s_x = \sqrt{n \cdot p \cdot (1-p)} = [10^5 \cdot (\tfrac{1}{2})^2]^{1/2} = 158$$
$$s_x \approx 160$$

For the $\pm s_x$ interval (Fig. 14.5) the range is 49,984 to 50,160 and for the $\pm 2s_x$ interval the range is 49,680 to 50,320. The distribution is characterized by a large mean value ($\bar{x} = 5 \cdot 10^4$) and relatively small standard deviation ($s_x = 160$).

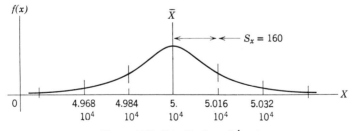

Figure 14.5 Distribution of $\tfrac{4}{3}\alpha$.

14.3 MECHANICS OF THE DISTORTION

The Unbalancing Forces. It is postulated that bending, when found, results from an unbalance of forces about the neutral plane. In this model force differences are traceable to a predominance of $\frac{4}{3}\alpha \rightarrow$ cubes and thermal loading. The amount of bending produced by a given numerical predominance of $\frac{4}{3}\alpha \rightarrow$ cubic elements is a function of distances from the neutral axis. Maximum effect results from forces in the outermost layers.

Critical Distortion D_{cr} (Fig. 14.6). The critical value of distortion (induced by ΔT) is specified as $D_{cr} = 0.4 \, \mu = 0.4(10^{-4} \text{ cm}) = 4 \cdot 10^{-5} \text{ cm}$

The coefficients of thermal expansion are

$$\alpha = 7.2 \cdot 10^{-6} \, \mu \text{ per } \mu \text{ per deg F.}$$
$$\tfrac{4}{3}\alpha = 9.6 \cdot 10^{-6} \, \mu \text{ per } \mu \text{ per deg F.}$$

Assumed thermal loading:
$$\Delta T = 10°\text{F}.$$

Edge length L of a typical cube:
$$L = 10 \, \mu = 10(10^{-4} \text{ cm}) = 10^{-3} \text{ cm}$$
$$(\mu = 10^{-6} \text{ m} = 10^{-4} \text{ cm}).$$

Linear Expansion Differential. Linear expansion ΔL, for a cube with coefficient $\alpha \rightarrow = 7.2 \cdot 10^{-6} \, \mu$ per μ per deg F is

$$\Delta L_1 = 7.2 \cdot 10^{-6} \cdot (10)(10) = 7.2 \cdot 10^{-4}.$$

For a cube with linear coefficient $\frac{4}{3}\alpha \rightarrow = 9.6 \cdot 10^{-6}$, it is

$$\Delta L_2 = 9.6 \cdot 10^{-6} \cdot (10)(10) = 9.6 \cdot 10^{-4} \, \mu.$$

Length differential ΔL is

$$\Delta L = \Delta L_2 - \Delta L_1 = (9.6 - 7.2)10^{-4} = 2.4 \cdot 10^{-4} \, \mu.$$

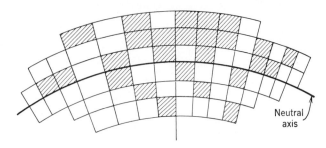

Figure 14.6 Convex distortion.

Unit of Force Magnitude. The force magnitude which leads to bending (deflection) specified as critical ($0.4\,\mu$) is now computed. Utilizing Hooke's law, the force (K) developed by a single element, due to differential expansion $\Delta L = \delta$, is (see Section 8.3):

$$\delta = \frac{KL}{AE}.$$

$\delta = \Delta L = 2.4 \cdot 10^{-8}$ cm
$A = (10 \cdot 10^{-4} \text{ cm})^2 = 10^{-6} \text{ cm}^2$
$E = 30 \cdot 10^6$ psi
$ = 19.61 \cdot 10^6 (1.02 \cdot 10^{-3}) \text{ g/cm}^2$
$ = 20 \cdot 10 \text{ g/cm}^2$
$L = 10\,\mu = 10^{-3}$ cm

$$K = \frac{\delta AE}{L} = \frac{2.4 \cdot 10^{-8} \cdot 10^{-6} \cdot (2 \cdot 10^9)}{10^{-3} \text{ cm}} = 4.8 \cdot 10^{-2} \text{ g}.$$

Critical Distortion Moment. Bending produced by uniformly distributed loading provides a reasonable analogy to the bending produced by internal forces and moments.

Critical deflection $= y = 0.4\,\mu = 4 \cdot 10^{-5}$ cm (Fig. 14.7 and 14.8)

$$y = \frac{5ql^4}{384EI_z} \quad \text{(Section 10.6, Eq. i),}$$

$$I_z = \frac{bh^3}{12} \text{ cm}^4,$$

$$I_z = \frac{100(5)^3}{12} = 1.04 \cdot 10^3 \text{ cm}^4;$$

q is the distributed load intensity in grams per centimeter.

$$q = \left(\frac{384}{5}\right)\frac{EI_z y}{l^4} \text{ g/cm},$$

$$q = \frac{384(20 \cdot 10^8)(1.04 \cdot 10^3) 4 \cdot 10^{-5}}{5(10^2)^4} = 63.9 \text{ g/cm}.$$

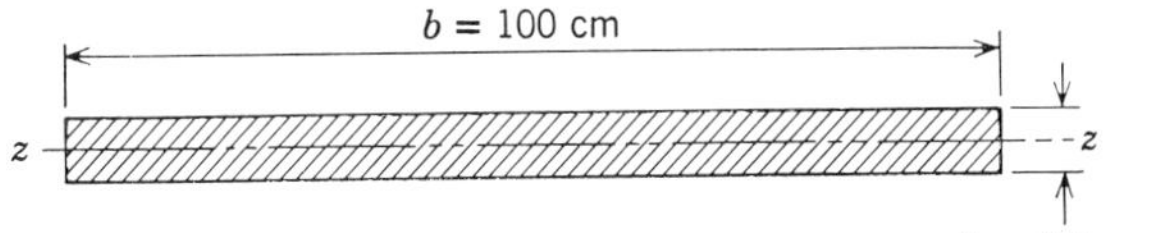

Figure 14.7 Beam model.

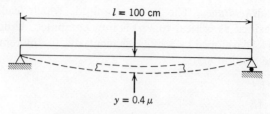

Figure 14.8 Beam cross section.

The required moment unbalance ΔM for critical deflection is (Section 10.3)

$$\Delta M = \frac{ql^2}{4} = \left(\frac{6.39 \cdot 10}{4}\right) \cdot (100)^2 = (1.60 \cdot 10^5) \text{ g/cm}.$$

The expression above describes the minimum condition for $0.4 \, \mu$ deflection.

14.4 LIKELIHOOD OF CRITICAL DISTORTION

Force Distribution Random Variables. Examining the likelihood of critical distortion requires the concept of a set (S) of plates. The forces tending to be unbalancing are neutralized in a theoretically average plate. There exists however, a finite probability that, in the set (S), are plates in which the forces are not in balance.

The internal force, developed above the neutral axis, is the product of the number of rows containing excess numbers of force producing cubes, y, multiplied by the number of cubes in the row contributing to the force, x, multiplied by the increment of force per cube, K. Then $Z = XY$ and $F = KZ$.

Y is a binomial random variable (over a set of plates). Y describes the number of rows in which more cubes are $\frac{4}{3}\alpha \rightarrow$ than are $\alpha \rightarrow$ (a positive or negative number). The set of members of Y (rows) is confined to the rows above the neutral axis of a plate or

$$n = (2500 \text{ layers})(10^5 \text{ rows/layer}) = 2.5 \cdot 10^8 \text{ rows}.$$

The probability that the number of $\frac{4}{3}\alpha \rightarrow$ cubes exceeds the numbers of $\alpha \rightarrow$ cubes is equal to the probability that the number of $\alpha \rightarrow$ cubes exceeds the number of $\frac{4}{3}\alpha \rightarrow$ cubes in any row. Thus the average value, $\bar{y}$,

$$p(4/3\alpha \rightarrow) = \tfrac{1}{2},$$
$$p(\alpha \rightarrow) = \tfrac{1}{2},$$
$$y = \frac{2.5 \cdot 10^8 p(4/3\alpha \rightarrow) - 2.5 \cdot 10^8 p(\alpha \rightarrow)}{2.5 \cdot 10^8} = 0.$$

The standard deviation estimate of Y is

$$s_y = \sqrt{np(1-p)} = \sqrt{2.5 \cdot 10^8 (\tfrac{1}{2})(\tfrac{1}{2})} = 7910.$$

$$(\bar{y}, s_y) = (0, 7910).$$

The normal random variable $(\bar{y}, s_y)$, provides a description of Y over the face of a cross section above the neutral plane, of each of the set (S) of plates.

To define the variate X, requires the following. Let ϕ = the number of $\alpha \rightarrow$ cubes in a row, and θ = the number of $\tfrac{4}{3}\alpha \rightarrow$ cubes in a row. Then

$$\phi + \theta = 10^5 \text{ cubes.}$$

$X = \theta - \phi$ is the random variable that describes the distribution of force producing unbalance due to $\tfrac{4}{3}\alpha \rightarrow$ cubes $> \alpha \rightarrow$ cubes in a row.
The mean value estimate is

$$\bar{x} = \frac{10^5 p(4/3\alpha \rightarrow) - 10^5 p(\alpha \rightarrow)}{10^5} = 0,$$

and the standard deviation estimate is

$$s_x = \sqrt{np(1-p)} = \sqrt{10^5 (\tfrac{1}{2})(\tfrac{1}{2})} = 160.$$

$$(\bar{x}, s_x) = (0, 160).$$

The product variate (Z) (Fig. 14.9) is computed (Eqs. 3.9 and 3.10) as

$$\bar{z} = \bar{x}\bar{y} = (0)(0) = 0$$

and

$$s_z = \sqrt{\bar{x}^2 s_y^2 + \bar{y}^2 s_x^2 + s_x^2 s_y^2},$$

$$s_z = \sqrt{0^2 s_y^2 + 0^2 s_x^2 + (7.9 \cdot 10^3)^2 (160)^2} = 12.6 \cdot 10^5.$$

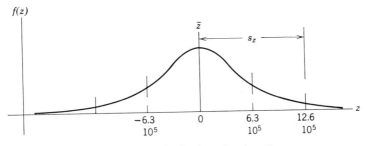

Figure 14.9 Distribution of variate Z.

Computation Probability of Critical Distortion. Figure 14.9 shows random variable (Z). The lever is determined by considerations of symmetry to be 1.25 cm.

$$\Delta M = (1.25)(4.8 \cdot 10^{-2})(12.6 \cdot 10^5)(N) = 7.56 \cdot 10^4 N, \qquad (a)$$

$$\Delta M = \frac{ql^2}{4} = 1.60 \cdot 10^5 \text{ g/cm}. \qquad (b)$$

By equating Eqs. *a* and *b* the probability of critical distortion $p(D = 0.4\,\mu)$ is computed:

$$7.36 \cdot N = 1.60 \cdot 10.$$

Condition for Maximum Distortion.

CASE 1. In equal distortion inducing forces, negative on one side of the neutral plane and positive on the opposite side, a couple is produced and

$$N = \frac{1.60 \cdot 10}{2(7.56)} = 1.06 \text{ (standard deviations)}.$$

From Table 2.4, with $N = z = 1.06$, it is seen to be equivalent to a probability of $(1 - 0.7108)/2$:

$$p_1 = p_2 = 0.1446,$$

and by Eq. 2.14

$$p(D \gtrless 0.4\,\mu) = p_1 p_2 = 0.0209.$$

CASE 2. Distortion-producing forces confined to one side of the neutral plane:

$$N = \frac{16.0}{7.56} = 2.12$$

and

$$p_1 = \tfrac{1}{2}.$$

From Table 2.4 $N = z = 2.12$ is equivalent to $(1 - 0.966)/2$:

$$p_2 = 0.017.$$

By Eq. 2.14

$$p(D \gtrless 0.4\,\mu) = p_1 p_2 = 0.0085.$$

CASE 3. Distortion-producing forces exist on both sides of the neutral plane. The dominating force must neutralize the lesser force before acting with the residual to induce distortion. Assume a lesser force equivalent to 1.0 standard deviation frequency, or

$$p_1 = \frac{1 - 0.6827}{2} = 0.1586.$$

The number of standard deviations associated with p_2 is $1.0 + 2.12 = 3.12$:

$$p_2 = \frac{1 - 0.9982}{2} = 0.0009.$$

By Eq. 2.14

$$p(D \gtrsim 0.4\,\mu) = p_1 p_2 = 0.00014.$$

CASE 4. Unequal distortion-producing forces (of opposite sign). A negative force equivalent to 1.0 standard deviation frequency is assumed on one side:

$$p_1 = 0.1586.$$

The number of standard deviations associated with p_2 is $2.12 - 1.0 = 1.12$:

$$p_2 = \frac{1 - 0.7372}{2} = 0.1314,$$

and by Eq. 2.14

$$p(D \geq 0.4\,\mu) = p_1 p_2 = 0.0208.$$

Examination of the results of Cases 1 through 4 reveals that the maximum probability of distortion is associated with equal couple-producing forces of opposite sign. For the specified amount of distortion specified as critical $(0.4\,\mu)$ the maximum probability of occurrence is indicated as

$$p(D \geq 0.4\,\mu) = 0.0209.$$

14.5 MOST SEVERE CONFIGURATION

The model (Fig. 14.10) is used in this study. The two possible orientations of the anisotropic property are

$$p(4/3\alpha \rightarrow) = \tfrac{1}{2}$$
$$p(\alpha \rightarrow) = \tfrac{1}{2} \quad (a)$$

Consider an element of hexagonal configuration (Fig. 14.11). The three possible orientations of the $4/3\alpha \rightarrow$ may be stated as follows.

1. $p(4/3\alpha + \alpha \cos 60° + \alpha \cos 60°) = \tfrac{1}{3}$.
2. $p(\alpha + 4/3\alpha \cos 60° + \alpha \cos 60°) = \tfrac{1}{3}$.
3. $p(\alpha + \alpha \cos 60° + 4/3\alpha \cos 60°) = \tfrac{1}{3}$.

$$p(4/3\alpha + 2\alpha \cos 60°) + 2p(\alpha + 7/3\alpha \cos 60°) = \tfrac{1}{3} + \tfrac{2}{3} = 1$$
$$p(\tfrac{14}{6}\alpha) + 2p(\tfrac{13}{6}\alpha) = \tfrac{1}{3} + \tfrac{2}{3} = 1$$
$$p(\tfrac{14}{6}\alpha) = \tfrac{1}{3}$$
$$p(\tfrac{13}{6}\alpha) = \tfrac{2}{3} \quad (b)$$

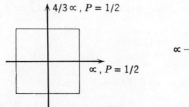 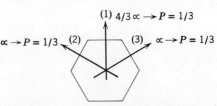

Figure 14.10 Model cubic element face.

Figure 14.11 Model: hexagonal element.

An examination of Eqs. *a* and *b* reveals that the square configuration is more conducive to distortion than the hexagonal configuration. The coefficient differential $(4/3\alpha - \alpha) = \frac{1}{3}\alpha$ for the square is larger than the coefficient differential

$$\left(\tfrac{14}{6}\alpha - \tfrac{13}{6}\alpha\right) = \tfrac{1}{6}\alpha$$

for the hexagon. Frequency (of differential increments) is larger in the case of the square configuration. Similarly, the octagonal model upon analysis, leads to the same conclusions. Conclusions may be generalized to polygons of *n* sides.

CHAPTER 15

Electromechanical Devices

15.1 STATISTICAL APPROACH TO ANALYSIS [44] [45]

Some of the important relationships in electrical engineering are given by relatively simple mathematical expressions, such as the following:

1. Ohm's law: $I = V/R$.
2. Series circuit: $R_t = R_1 + R_2 + \cdots + R_n$.
3. Parallel circuit:

$$\frac{1}{R_t} = \frac{1}{R_1} + \frac{1}{R_2} + \cdots + \frac{1}{R_n}:$$

For the special case of turn resistors in parallel:

$$R_T = \frac{R_1 R_2}{R_1 + R_2}.$$

4. Kirchhoff's law:

$$\sum_{}^{I} I = 0; \sum_{}^{V} V = 0.$$

5. Power: $W = VI = V^2/R = I^2 R$.
6. Power inductance:

$$V = L\frac{dI}{dt}; L_t = L_1 + L_2 + \cdots + L_n.$$

7. Capacitor:

$$I = C\frac{dV}{dT}; \frac{1}{C_T} = \frac{1}{C_1} + \frac{1}{C_2} + \cdots + \frac{1}{C_n}.$$

For the special case of two capacitors in series:

$$C_T = \frac{C_1 C_2}{C_1 + C_2}$$

8. Transformer:

$$\frac{V_1}{V_2} = \frac{I_2}{I_1}, \quad \frac{N_1}{N_2} = \left(\frac{Z_1}{Z_2}\right)^{1/2}.$$

9. $v = V \sin(\omega t + \theta)$.
10. $X_L = 2\pi f L$; $z = R + Jx$.
11. $X_o = \frac{1}{2}\pi f c$.
12. $f_r = \frac{1}{2}\pi\sqrt{LC}$.
13. In series RLC: (see sketch)

$$Z = R + J(X_L - X_c)$$

$$z = R + JWL - 1/wc$$

$$I = \frac{V}{R} + J(X_L - X_c)$$

14. $z_0 = 273 \ln\left(\frac{a}{b}\right)$.

Treating mathematical expressions such as those above as relationships among random variables is straightforward. For instance, the value of two resistors in parallel is computed in Example 1, Section 3.3.3. An interesting result in Example 1 of Section 3.3.3 is that mean resistance (R_T) reduces in value more slowly than does standard deviation (s_{R_T}). In other words, it could in some applications be profitable to achieve a desired resistance by coupling several resistors parallel to minimize variability. Clearly, similar observations apply to capacitors in series.

Example 1. Consider two capacitors C_1 and C_2, where (see sketch)

$$C_1 = 0.02 \pm 20\% \text{ F},$$
$$C_2 = 0.025 \pm 20\% \text{ F}.$$

Then

$$\Delta C_1 = \pm 0.004 \text{ F},$$
$$\Delta C_2 = \pm 0.005 \text{ F}.$$

C_1 and C_2 are coupled in series, as shown in the sketch. Compute the random variables C_T,

$$\frac{1}{C_T} = \frac{1}{C_1} + \frac{1}{C_2}; \quad C_T = \frac{C_1 C_2}{C_1 + C_2}. \tag{a}$$

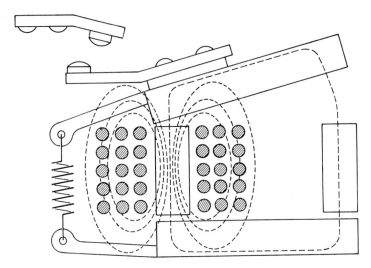

Figure 15.1 Typical electromechanical device.

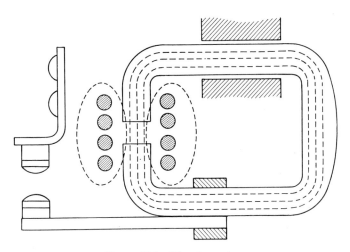

Figure 15.2 Hingeless relay.

Let

$$\Delta C_1 = 3 s_{C_1}; \quad s_{C_1} = \frac{\Delta C_1}{3} = 0.00133 \text{ F},$$

$$\Delta C_2 = 3 s_{C_2}; \quad s_{C_2} = \frac{\Delta C_2}{3} = 0.00166 \text{ F},$$

$$\bar{C}_T = \frac{\bar{C}_1 \bar{C}_2}{\bar{C}_1 + \bar{C}_2} = \frac{(0.02)(0.025)}{(0.02)+(0.025)} = \frac{5 \cdot 10^{-4}}{4.5 \cdot 10^{-2}} = 1.11 \cdot 10^{-2} \text{ F},$$

$$\frac{\partial C_T}{\partial C_1} = \frac{\bar{C}_2{}^2}{(\bar{C}_1 + \bar{C}_2)^2} = 0.31; \quad \frac{\partial C_T}{\partial C_2} = \frac{\bar{C}_1{}^2}{(\bar{C}_1 + \bar{C}_2)^2} = 0.197,$$

and by Eq. 2.61,

$$s_{C_T} = 5.26 \cdot 10^{-4},$$

$$(\bar{C}_T, s_{C_T}) = (1.11 \cdot 10^{-2}, 5.26 \cdot 10^{-4}) \text{ F}.$$

The remainder of this section is devoted to the analysis of a simple electromagnetic device (Figs. 15.1 and 15.2). Two design exercises are also included. In the first design problem, probability of adequate performance is specified and the number of coil windings optimized. In the second design exercise, probability is specified and the voltage requirements optimized.

15.2 ELECTROMECHANICAL SWITCHES

Problems in electromechanical design (specifically relays) are now briefly discussed.

Most switching systems are electrical; their elements are either electronic or electromechanical. Electromechanical switches provide adequate speed for most purposes. The operating element in switching relays is usually an electromagnet which is operative when the winding circuit is closed and inoperative when the winding circuit is open.

In relay design most technical problems relate to either mechanics or electricity and magnetism. The parameters of design are random variables.

Relays are of two major classifications: (1) the neutral relay in which armature motion is independent of the polarity of the signal, and (2) the polarized relay which is responsive to the polarity of the applied signal in the direction of motion in the force developed. The example in Fig. 15.3 is a dc neutral design.

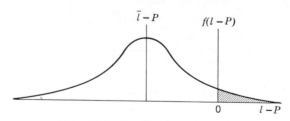

Figure 15.3 Coupling formula density.

The design process involves consideration of three sets of random variable parameters.

1. Those associated with the mechanical circuit (e.g., armature and spring).
2. Those associated with the magnetic circuit (e.g., coil, permeability characteristics, magnetic reluctance, and ampere turns).
3. Those associated with the electrical circuit (e.g., coil resistance, impressed voltage, and current).

Two different approaches to the subject of magnet design are discussed in the literature. The approach used here employs a magnetic circuit modified to provide for accurately determining effective values of lumped parameters. This provides a form of magnetic circuit treatment which is experimentally valid and gives a satisfactory basis for estimating the parameters from design dimensions and tolerances.

The first objective is to satisfy requirements for satisfactory contact performance. These requirements involve the resistance versus-force relation ships of contact alloys and their variations with surface conditions. The resultant force requirements depend upon the mechanics of the contact springs.

15.3 STATISTICAL CONSIDERATIONS

Mechanical Circuit. In the mechanical circuit are included (a) the contacts, (b) the armature operating in the magnetic field, and (c) the spring. Mechanical parameters that result from physical geometry, such as beam length, cross-section area, moment of inertia, and contact area, are clearly random variables. The strength properties are random variables (Fig. 15.4).

Magnetic Circuit. Magnetic field strength is a function of coil characteristics, magnetic characteristics of the core material, the return circuits and gap characteristics.

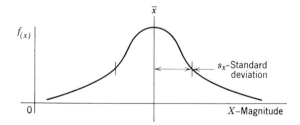

Figure 15.4 Strength parameter distribution.

Electrical Circuit. Impressed voltage and coil resistance (hence current) are random variables. Voltage is a random variable due to source variations, while resistance variations result from variability in conductor resistivity, variability in geometry (length and cross section), and variabilities in characteristics over time.

Since the mechanical, magnetic, and electrical parameters which determine the characteristics of each circuit are functionally related, the circuit characteristics determined by each class of random variables are random variables, and the performance of the device is a random variable. In the design and/or analysis process, relations of the following kind are needed.

$$\text{(allowable stress)} \; R \; \text{(applied stress)}$$
$$\text{(load function)} \; R \; \text{(pull function)}$$

where R is a probabilistic coupling (Eq. 4.6).

15.4 MECHANICAL CIRCUIT

A specific electrical relay design is analyzed. The design, however, may be more sophisticated than that presented here. It may take into account dynamic effects, thermal effects, satisfaction of time requirements, and so on.

The mechanical circuit accounts for the total load variate $(\bar{P}, s_p)$, which equals contact force variate $(\bar{P}_1, s_{P_1})$ plus deflection force variate $(\bar{P}_2, s_{P_2})$.

Contacts. Because a relay is essentially a switch, the first consideration is proper operation of its contacts. Satisfactory contact performance is attained by providing mechanical functionings that are acceptable with respect to (a) motion, (b) mounting, (c) contact force, and (d) actuation.

As electrical circuit components, contacts are two-valued resistances. When closed, the resistance must be small. The statistical description of closed contacts should combine small average resistance with small relative variance. Figure 15.5 shows the type of distribution needed. Conventionally, the resistance across closed contacts is held to a fraction of one ohm. With open contacts, the resistance must be of the order of several megohms. To satisfy the requirements of low resistance, the contacts are pressed together

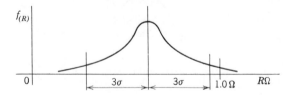

Figure 15.5 Across contact resistance distribution.

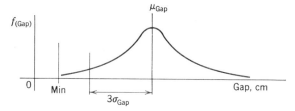

Figure 15.6 Open-gap distribution.

by a force variate which (on the low side of the distribution) exceeds some minimum value, with high probability.

Open-gap spacing must exceed some minimum value (see Fig. 15.6). Clearly, the open-gap state or the closed-gap state must be maintained. The transition from one state to another must be abrupt and free from oscillations. Departure from the condition described represents defective performance or the end of useful life.

The minimum (-3σ value) force between contacts† is established as 2 kdyn, with estimated standard deviation of 0.2 kdyn. As shown in Fig. 15.7, the force distribution is

$$(\bar{P}_1, s_{P_1}) = (2.6, 0.20) \text{kdyn}.$$

The hingeless, close—differential relay analyzed here is an instrumentation type (maximum switching capacity 0.3 A) designed with a performance objective of 10^6 cycles of operation. The contact design is shown in Fig. 15.8.

With the contact force component of loading estimated as $(\bar{P}_1, s_{P_1})$, the next estimation is the force required to depress the spring.

Spring Design. The spring is shown in Fig. 15.9. In addition to supporting the contacts and functioning as the armature, the spring provides the return path in the magnetic circuit. The spring-armature is supported from point A to point B. The main deflection results from bending, beginning at point A and continuing around the semicircle and along the leaf to E. Vertical deflection results from bending in the semicircle and in the cantilever.

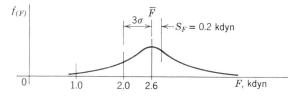

Figure 15.7 Contact force distribution.

† [45].

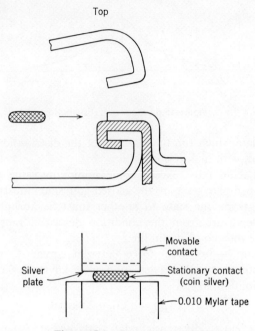

Figure 15.8 Contact geometry.

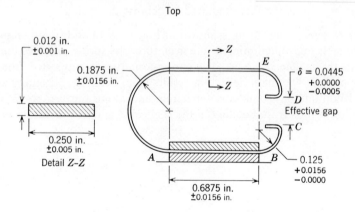

Figure 15.9 Cantilever leaf spring.

300

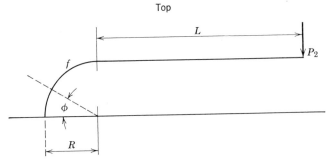

Figure 15.10 Bending of a bent bar.

Computation of random variable $(\bar{P}_2, \sigma_{P_2})$.

    Required. The load, P_2, to close the effective gap,
    Known.   a. Semicircle radius, R random variable.
                  b. Cross-sectional area, A random variable.
                  c. Modulus of elasticity, E random variable.
                  d. Length of cantilever random variable.
                  e. Effective gap, δ random variable.

Deflection, δ_1, in half of semicircle circle spring (see Fig. 15.10);

$$\delta_1 = \frac{1}{EI_z} \int_0^{\pi/2} M \frac{dM}{dP_2} R \, d\phi,$$

where

$$M = P_2(R \cos \phi + L) = P_2 R \cos \phi + P_2 L$$

and

$$\frac{dM}{dP_2} = R \cos \phi + L,$$

$$\delta_1 = \frac{1}{EI_z} \int_0^{\pi/2} (P_2 R \cos \phi + P_2 L)(R \cos \phi + L) R \, d\phi,$$

$$\delta_1 = \frac{P_2 R}{EI_z} \left(\frac{R^2 \pi}{4} + 2RL + \frac{L^2 \pi}{2} \right).$$

Deflection of the semicircle, δ_2,

$$\delta_2 = 2\delta_1.$$

Deflection of the cantilever δ_3:

$$\delta_3 = \frac{P_2 L^3}{3EI_z}.$$

Total deflection, δ:

$$\delta = 2\delta_1 + \delta_3 = \frac{2P_2 R}{EI_z} \left(\frac{\pi R^2}{4} + \frac{\pi L^2}{2} + 2RL \right) + \frac{P_2 L^3}{3EI_z}.$$

302 Application

Solving for $\bar{P}_2$,

$$\delta = \frac{2P_2 R}{EI_z}\left[\frac{\pi R^2}{4} + \frac{\pi L^2}{2} + 2RL\right] + \frac{2P_2 L^3}{6EI_z},$$

$$\delta = \frac{2P_2}{EI_z}\left[\frac{\pi R^3}{4} + 2R^2 L + \frac{\pi RL^2}{2} + \frac{L^3}{6}\right],$$

$$\bar{P}_2 = \frac{\delta EI_z}{2\left[\dfrac{\pi R^3}{4} + 2R^2 L + \dfrac{RL^2}{2} + \dfrac{L^3}{6}\right]}. \qquad (a)$$

Note.

1. The numerator and denominator in (a) are independent multivariates.
2. The denominator is a polynomial in random variables R and L each of which appears in more than one term. The terms cannot be treated as a statistically independent.

The standard deviation estimate, s_{P_2}, is computed as a quotient (Section 2.9.2 and Eq. 2.61).

$$P_2 = \frac{y}{x} = \frac{\delta EI_z}{2(\pi R^3/4 + 2R^2 L + RL^2/2 + L^3/6)}.$$

Consider the numerator, $y = \delta EI_z$. The gap random variable $(\bar{\delta}, s_\delta)$ is independent of the modulus of elasticity $(\bar{E}, s_E)$. Moment of inertia $(\bar{I}, s_I)$ is independent of both. In fact, the three random variables are mutually independent. Thus s_y is computed as a triple product of variates δ, E and I_z; $\bar{y} = (\bar{\delta}\bar{E}\bar{I}_z)$; s_y is computed by Eq. 3.10; s_x is computed by partial derivative methods.

$$x = 2\left(\frac{\pi R^3}{4} + R^2 L + \frac{RL^2}{2} + \frac{L^3}{6}\right).$$

By applying Eq. 2.61 let $L = x_1$ and $R = x_2$.

$$\frac{\partial x}{\partial L} = 2R^2 + RL + \frac{L^2}{2}; \frac{\partial x}{\partial R} = \frac{3\pi R^2}{4} + 4RL + \frac{L^2}{2},$$

$$s_x = 2\left[\left(2\bar{R}^2 + \bar{R}\bar{L} + \frac{\bar{L}^2}{2}\right)^2 s_L^2 + \left(\frac{3\pi \bar{R}^2}{2} + 4\bar{R}\bar{L} + \frac{\bar{L}^2}{2}\right) s_R^2\right]^{1/2}.$$

Total load, P, is the sum: $P = P_1 + P_2$. By Eqs. 3.1 and 3.2

$$\bar{P} = \bar{P}_1 + \bar{P}_2 \quad \text{and} \quad s_P = \sqrt{s_{P_1}^2 + s_{P_2}^2}.$$

Now the load random variable P_2 (Figs. 15.8 and 15.9) is computed. In Fig. 15.9, the gap is given as

$$\delta = 0.0445 \begin{array}{c} +0.0000 \\ -0.0005 \end{array} \text{in.} = 0.04475 \pm 0.00025 \text{ in.}$$

Changing units,

$$\delta = 0.1137 \pm 0.00064 \text{ cm.}$$

With tolerance a 2σ interval, the variate estimate is

a.
$$(\bar{\delta}, s_\delta) = (0.1137, 0.00032) \text{ cm.}$$

Moment of inertia, $I_z = bh^3/12$, for the spring cross section (Fig. 15.9)

$$b = 0.25 \pm 0.005 \text{ in.} = 0.635 \pm 0.013 \text{ cm}$$
$$h = 0.012 \pm 0.001 \text{ in.} = 0.0305 \pm 0.0025 \text{ cm}$$

$$\bar{I}_z = \frac{0.635(0.0305)^2}{12} \text{ cm}^4 = 1.501 \cdot 10^{-6} \text{ cm}^4.$$

By applying the formulas for powers and then products

$$s_b = \frac{0.013}{2} \text{ cm} = 0.006 \text{ cm}, \qquad s_h = \frac{0.0025}{2} \text{ cm} = 0.00125 \text{ cm}.$$

By Eq. 2.61

$$s_h{}^3 = 3\bar{h}^2 s_h = 3.488 \cdot 10^{-6} \text{ cm}^3$$
$$\bar{h}^3 = (0.0305)^3 = 28.36 \cdot 10^{-6} \text{ cm}^3.$$

Applying Eq. 3.10,

$$s_{I_z} = \tfrac{1}{12}\sqrt{(0.635)^2(3.49 \cdot 10^{-6})^2 + (28.36 \cdot 10^{-6})^2(6.5 \cdot 10^{-3})^2} \text{ cm}^4$$
$$= 1.845 \cdot 10^{-8} \text{ cm}^4.$$

b.
$$(\bar{I}_z, s_{I_z}) = (1.501 \cdot 10^{-6}, 1.845 \cdot 10^{-7}) \text{ cm}^4.$$

Modulus of elasticity, E, of the material of the spring (Armco M6, 0.012 in. thick oriented electrical steel) is (modulus of elasticity, psi, in the direction of rolling):

$$E = 19 \cdot 10^6 \pm 5\% \ (3\sigma \text{ level}) \text{ psi}$$
$$E = 19 \cdot 10^6 \pm 9.5 \cdot 10^5 \text{ psi.}$$

Mean value and standard deviation estimates of E are

$$\bar{E} = 19 \cdot 10^6 \text{ psi} \quad \text{and} \quad s_E = \frac{9.5 \cdot 10^5}{3} \text{ psi} = 3.166 \cdot 10^5 \text{ psi}$$

and

c.
$$(\bar{E}, s_E) = (19 \cdot 10^6, 3.166 \cdot 10^5) \text{ psi}.$$

Random variable Y is

$$(\bar{y}, s_y) = (\bar{\delta}, s_\delta)(\bar{I}_z, s_{I_z})(\bar{E}, s_E)$$
$$(\bar{y}, s_y) = (0.1137, 0.00032)(1.501 \cdot 10^{-6}, 1.845 \cdot 10^{-7}) \cdots$$

and
$$(13.36 \cdot 10^8, 2.216 \cdot 10^7) \text{ g cm}^3$$

$$\bar{y} = (0.1137)(1.501 \cdot 10^{-6})(13.36 \cdot 10^8) \text{ g cm} = 222.4 \text{ g cm}.$$

Computing s_y involves two steps. First, estimate s_{I_z} is computed:

$$s_{\delta I_z} = [(0.1137)^2(1.845 \cdot 10^{-7})^2 + (1.501 \cdot 10^{-6})^2(3.2 \cdot 10^{-4})^2]^{\frac{1}{2}}$$
$$= 2.094 \cdot 10^{-8},$$

$$\delta I_z = (0.1137)(1.5 \cdot 10^{-6}) = 1.705 \cdot 10^{-7},$$

and

$$s_y = [(13.36 \cdot 10^8)^2(2.094 \cdot 10^{-8})^2 + (1.705 \cdot 10^{-7})^2(2.216 \cdot 10^7)^2]^{\frac{1}{2}}$$
$$s_y = 26.94 \text{ g cm}^3,$$

$$(\bar{y}, s_y) = (222.4, 26.94) \text{ g cm}^3.$$

The denominator variate X is

$$\bar{x} = 2\left(\frac{\pi \bar{R}^3}{4} + \bar{R}^2 L + \frac{\bar{R} L^2}{2} + \frac{L^3}{6}\right),$$

$$s_x = 2\left[\left(2R^2 + RL + \frac{L^2}{2}\right)s_L^2 + \left(\frac{3\pi R^2}{4} + 4RL\frac{L^2}{2}\right)s_R^2\right]^{\frac{1}{2}}.$$

The radius variate $(\bar{R}, s_R)$ is (Fig. 15.9)

$$R = 0.1875 \pm 0.0156 \text{ in.} = 0.476 \pm 0.0396 \text{ cm}.$$

With tolerance equal to 3σ, the radius variate estimate is

$$(\bar{R}, s_R) = (0.476, 0.0132) \text{ cm}.$$

The cantilever length variate $(\bar{L}, s_L)$ is (Fig. 15.9)

$$L = 0.6875 \pm 0.0156 \text{ in.} = 1.746 \pm 0.0396 \text{ cm}$$

and
$$(\bar{L}, s_L) = (1.746, 0.0132) \text{ cm}.$$

Computing the value of partial derivatives $\partial x/\partial L$ and $\partial x/\partial R$:

$$\frac{\partial x}{\partial L} = \left(2\bar{R}^2 + \bar{R}L + \frac{L^2}{2}\right) \text{cm}^2 = 0.453 + 0.831 + 1.529 = 2.813 \text{ cm}^2$$

and

$$\left(\frac{\partial x}{\partial L}\right)^2 = (2.813)^2 = 7.913 \text{ cm}^4$$

$$\frac{\partial x}{\partial R} = \left[\frac{3\pi\bar{R}^2}{4} + 4\bar{R}L + \frac{L^2}{2}\right] = 0.534 + 3.324 + 1.529 = 5.387 \text{ cm}^2$$

and,

$$\left(\frac{\partial x}{\partial R}\right)^2 = (5.387)^2 = 29.020 \text{ cm}^4$$

$$s_L^2 = (0.0132)^2 = 1.742 \cdot 10^{-4} \text{ cm}^2$$
$$s_R^2 = (0.0132)^2 = 1.742 \cdot 10^{-4} \text{ cm}^2$$

Substituting into Eq. 2.61:

$$s_x = [2(7.913 + 29.020)(1.742 \cdot 10^{-4}) \text{ cm}^6]^{1/2} = 0.160 \text{ cm}^3$$

$$\bar{x} = 2\left[\frac{\pi\bar{R}^3}{4} + \bar{R}^2L + \frac{\bar{R}L^2}{2} + \frac{L^3}{6}\right] \text{cm}^3 = 4.185 \text{ cm}^3$$

$$(\bar{x}, s_x) = (4.185, 0.160) \text{ cm}^3$$

$(\bar{P}_2, s_{P_2})$, by Eqs. 3.13 and 3.14 is

$$(\bar{P}_2, s_{P_2}) = \frac{(\bar{y}, s_y)}{(\bar{x}, s_x)} = \frac{(222.4, 26.94) \text{ g cm}^3}{(4.185, 0.160) \text{ cm}^3} = (53.14, 6.75) \text{ g.}$$

From Section 14.4,

$$(\bar{P}_1, s_{P_1}) = (2.6, 0.20) \text{ g.}$$

Total load is the sum of contact pressure and the force required to depress the spring ($P = P_1 + P_2$):

$$(\bar{P}, s_P) = (55.74, 6.76) \text{ g.}$$

15.5 ELECTRICAL CIRCUIT

In this section, the random variables in the electrical circuit (specifically the relay coil) are examined.

Application

Voltage Variate. To estimate the current random variable, $(\bar{I}, s_I)$, the voltage, E, and resistance, R, must be described. The source of power is a battery which delivers 30 volts dc current. The voltage standard deviation, s_E, is estimated as 2 percent of the nominal voltage. Thus, the voltage random variable is $(\bar{E}, s_E) = (30, 0.6)$ V.

Coil Resistance Variate. Coil resistance is a random variable determined by the variable resistance of the wire and the length random variables. The relay coil wire is No. 40 AWG copper wire. In Table 12.5 of [26], the resistance (ohms per inch) characteristics for No. 40 AWG copper wire are given as

$$\rho = \begin{cases} \text{minimum} = 0.08328 \ \Omega/\text{in.,} \\ \text{average} = 0.09050 \ \Omega/\text{in.,} \\ \text{maximum} = 0.09739 \ \Omega/\text{in.} \end{cases}$$

The mean and standard deviations are

$$\bar{\rho} = 0.0905 \ \Omega/\text{in.}$$

and

$$s_\rho = \frac{0.00679}{3} = 0.00226 \ \Omega/\text{in.,}$$

$$(\bar{\rho}, s_\rho) = (0.0905, 0.00226) \ \Omega/\text{in.}$$

The wire length is a random variable. This variable results from (1) wire diameter and insulation thickness variability, and (2) the spool geometry. For No. 40 AWG enamel—coated copper wire, the diameter characteristics (Table 12.6 of [45]) are

$$\text{overall diameter } (OD) = \begin{cases} \text{minimum} = 0.0032 \text{ in.,} \\ \text{average} = 0.0035 \text{ in.,} \\ \text{maximum} = 0.0038 \text{ in.} \end{cases}$$

The overall diameter random variable is

$$(\overline{OD}, s_{OD}) = (0.0035, 0.0001) \text{ in.}$$

Coil spool geometry is shown in Fig. 15.11:

$$\bar{t}_L = \text{average number of turns/layer} = \frac{\text{average spool width}}{\text{average wire diameter}},$$

$$\bar{t}_L = \frac{0.297 \text{ in.}}{0.0035 \text{ in.}} = 84.86$$

$$s_{t_L} = \frac{1}{(0.0035)^2} [(0.297)^2 (0.0001)^2 + (0.0035)^2 (0.003)^2]^{1/2},$$

$$s_{t_L} = 2.51.$$

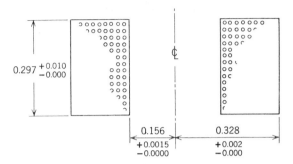

Figure 15.11 Relay coil geometry.

Thus
$$(\bar{i}_L, s_{t_L}) = (84.86, 2.51).$$

The specified total number of turns N is 3200. The number of layers (of turns), however, is a random variable, L_N.

$$(\bar{L}_N, s_{L_L}) = \frac{(3200, 0)}{(84.86, 2.51)} = (37.71, 1.12).$$

Length of wire in an average turn is (see Fig. 15.11):

$$(\bar{L}_t, s_L) = 2\pi \left[(0.156, 0.0005) + \frac{(37.71, 1.12)}{2} (0.0035, 0.0001) \right]$$

$$= (1.37, 0.018) \text{ in.}$$

Total wire length in a coil is

$$(\bar{L}, s_L) = 3200(1.37, 0.018) \text{ in.} = (4384; 57.6) \text{ in.}$$

The resistance variate RL is

$$(\bar{R}, s_R) = (0.0905, 0.00226)(4384; 57.6) = (396.7, 11.17) \, \Omega.$$

The voltage variate estimate is $(\bar{E}, s_E) = (30, 0.6)$ V.

Coil Current Variate Computation. By Ohm's law ($I = E/R$), the coil current variate is (Eqs. 3.13 and 3.14)

$$(\bar{I}, s_I) = \frac{(\bar{E}, s_E)}{(\bar{R}, s_R)} = \frac{(30, 0.6)}{(396.7, 11.17)} = (0.0756, 0.0026) \, a.$$

15.6 MAGNETIC CIRCUIT [45]

Magnetic field equations determine the field energy in terms of the parameters which characterize them. The physical dimensions and material

308 Application

properties of the parts comprising the system, and the electrical currents flowing within it. These relations (among random variables) are expressed in terms of B and H vectors which define the magnetic field:

$$B = \text{induction (flux density)}$$
$$H = \text{field intensity}$$

Reluctance Variate Estimation. Figure 15.12 (the magnetic circuit):

L_1 = center line length of the return path, centimeters,
L_2 = center line length of the coil core, centimeters,
a = cross-sectional area of the return path and coil core, square centimeters,
x = main gap separation (when armature is operated), centimeters,
A = effective pole face area multiplied by μ_0 (μ_0 = permeability of air), square centimeter.

These random variables must be estimated in order to compute pulling force F;

$$F = \frac{2\pi(NI)^2\dagger}{A\left(R_0 + \dfrac{x}{A}\right)^2}.$$

See Fig. 15.9 for dimensions of the magnetic flux return path.

The gap with the relay coil energized is 0.018 in. ± 0.0005 in. (Fig. 15.12) expressed as a random variable,

$$(\bar{x}_G, s_{x_G}) = (0.0457, 0.00042) \text{ cm}.$$

The variate a is estimated from the following information:

$$\text{thickness} = 0.012 \pm 0.0001 \text{ in.}$$
$$\text{width} = 0.250 \pm 0.0005 \text{ in.}$$

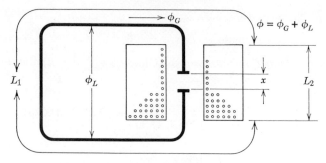

Figure 15.12 Electromagnetic flux paths.

† See [45], p. 60.

Then, by Eqs. 3.9 and 3.10:

$$(\bar{a}, s_a) = (0.019, 0.00053) \text{ cm}^2.$$

The variate L_2 is estimated (from dimensions in Fig. 15.11):

$$(\bar{L}_2, s_{L_2}) = (0.754, 0.008) \text{ cm}.$$

The variate L_1 is estimated (from dimension in Fig. 15.9) and $(\bar{L}_2, s_{L_2})$:

$$(\bar{L}_1, s_{L_1}) = (6.141, 0.122) \text{ cm} - (0.754, 0.008) \text{ cm} = (5.387, 0.1223) \text{ cm}.$$

Estimation of the variate A requires, in addition to consideration of the pole face area, consideration of the effects of probable misalignment of the pole faces (see Fig. 15.13). The shaded area in Fig. 15.13 indicates the effective pole face area, Ka. K is estimated as

$$(\bar{K}, s_K) = (0.85, 0.05).$$

Thus

$$\bar{A} = 0.85 \, \mu_0 a$$

and at the 3σ values

$$A_{\max} = \mu_0 a$$

and

$$A_{\min} = 0.70 \, \mu_0 a.$$

Hence, by applying Eqs. 3.9 and 3.10

$$(\bar{A}, s_A) = (0.85, 0.05)(\bar{a}, s_a)$$
$$= (0.85, 0.05)(0.019, 0.00053).$$

Thus

$$(A, s_A) = (0.016, 0.0001) \text{ cm}^2.$$

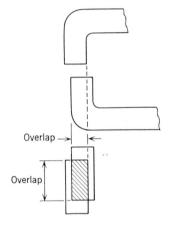

Figure 15.13 Effective pole face area.

The permeability random variable μ (Armco oriented electrical steel—M6) is estimated from the following data† :

$$x_{(\mu \max)} = 12{,}000; \; x_{(\mu \min)} = 5000,$$
$$\bar{x}_\mu = 8500 \quad \text{and} \quad 3s_\mu = 3500.$$

Thus

$$(\bar{x}_\mu, s_\mu) = (8500; 1170).$$

† See [45], p. 340.

The closed-gap reluctance $\mathcal{R}_0$ is

$$\mathcal{R}_0 = \frac{L_1}{\mu_a} + \frac{L_2}{\mu_a} + \frac{x_G}{A},$$

$$(\mathcal{R}_0, s_{\mathcal{R}_0}) = \frac{(\bar{L}, s_{L_1})}{(\bar{x}_\mu, s_\mu)(\bar{a}, s_a)} + \frac{(\bar{L}_2, s_{L_2})}{(\bar{x}_\mu, s_\mu)(\bar{a}, s_a)} + \frac{(\bar{x}_G, s_{x_G})}{(\bar{A}, s_A)}.$$

By applying Eqs. 3.1 and 3.2 to the first two terms on the right

$$\frac{(\bar{L}_1, s_{L_1}) + (\bar{L}_2, s_{L_2})}{(\bar{x}_\mu, s_\mu)(\bar{a}, s_a)} = \frac{(5.387, 0.1223) + (0.754, 0.008)}{(8500, 1170)(0.019, 0.00053)}.$$

Then Eqs. 3.13 and 3.14 yields

$$\frac{(6.141, 0.122)}{(161.5, 22.7)} = (0.038, 0.0054). \tag{a}$$

The third term on the right (by Eqs. 3.13 and 3.14) is

$$\frac{(\bar{x}_G, s_{x_G})}{(\bar{A}, s_A)} = \frac{(0.0457, 0.00042)}{(0.016, 0.0001)} = (2.856, 0.316). \tag{b}$$

By combining (a) and (b) (Eq. 3.1 and 3.2)

$$(\mathcal{R}_0, s_{\mathcal{R}_0}) = (2.894, 0.3162).$$

The unoperated gap variate X is specified as 0.0445 ± 0.0005 in. (Fig. 15.9):

$$(\bar{x}, s_x) = (0.113, 0.0013) \text{ cm}.$$

By applying Eqs. 3.13 and 3.14 to determine x/A (see pulling force F p. 308),

$$\frac{(\bar{x}, s_x)}{(\bar{A}, s_A)} = \frac{(0.113, 0.0013)}{(0.016, 0.0001)} = (7.06, 0.87).$$

Random variable (by Eqs. 3.1 and 3.2):

$$\left(\mathcal{R}_0 + \frac{x}{A}\right) = (2.894, 0.3162) + (7.06, 0.087) = (9.95, 0.328).$$

Pulling force variate estimate:

$$F = \frac{2(NI)^2}{A(\mathcal{R}_0 + x/A)^2}\left(\frac{1}{980}\right).$$

With $3.265 = 3200/980$ and $2\pi(3200) = 20{,}096\dagger$:

$$(\bar{F}, s_F) = 20{,}096 \cdot \frac{(0.0743, 0.00237)^2 \cdot 3.265}{(0.016, 0.0001)(9.95, 0.328)^2}$$

$$= (20{,}096) \cdot \frac{(0.00571, 0.00035)}{(1.584, 0.105)} \cdot 3.265$$

$$(\bar{F}, s_F) = (236.3, 21.38) \text{ g}.$$

15.7 RELIABILITY ESTIMATION

With the pulling force random variable

$$(\bar{F}, s_F) = (236.3, 21.38) \text{ g}$$

and the total load random variable (Section 13.4)

$$(\bar{P}, s_P) = (55.79, 6.76) \text{ g}$$

substituting into the coupling formula, Eq. 4.6,

$$z = \frac{|236.3 - 55.79|}{\sqrt{(21.4)^2 + (6.76)^2}} = \frac{180.5}{22.45} = 8.04.$$

From tables of normal probability functions (see Table 2.4 or [21])

$$R > 0.999999.$$

15.8 FATIGUE CONSIDERATIONS

An estimate of the level of stress of the beam with contacts closed and the cantilever stressed in bending reveals that the level of stress (tensile or compressive) in the outer fibers is approximately

$$1.2 \cdot 10^4 \text{ psi}.$$

At this relatively low level of stressing, the probability of a problem due to fatigue, at 10^6 stress reversals, is very small.

15.9 DESIGN TO SPECIFIED RELIABILITY

Example 1. If reliability is specified (as $R \geq 0.999$) and all other design requirements (except voltage E) remain the same as before, a simple probabilistic design problem exists. The load random variable is

$$(\bar{P}, s_P) = (55.74, 6.76) \text{ g}. \tag{a}$$

† $N = 3200$.

312 Application

Force required to depress the spring and exert the required contact pressure is unchanged. The pulling force variate $(\bar{F}, s_F)$ is now written as a function of E:

$$F = \frac{2\pi N^2 I^2}{A(\mathcal{R}_0 + x/A)^2} \cdot \frac{1}{980} = \frac{2\pi N^2}{A(\mathcal{R}_0 + x/A)^2} \cdot \left(\frac{E}{R}\right)^2 \cdot \frac{1}{980},$$

$$F = \frac{2\pi N^2}{980} \cdot \frac{E^2}{A(\mathcal{R}_0 + x/A)^2 \cdot (R)^2} = 65{,}650 \cdot \frac{E^2}{A(\mathcal{R}_0 + x/A)^2 R^2}.$$

From Section 13.5 the Coil Resistance random variable is

$$(\bar{R}, s_R) = (396.7, 11.17)\ \Omega$$

By employing Eq. 2.61

$$(\bar{R}, s_R)^2 + (R^2, 2\bar{R}s_R) = (1.574 \cdot 10^5;\ 8862).$$

By first employing Eq. 2.61, then Eqs. 3.9 and 3.10,

$$\overline{[A(\mathcal{R}_0 + x/A)^2}, s_{A(\mathcal{R}_0 + x/A)^2}] = (1.584, 0.105).$$

The product variate estimate of R^2 and $A(\mathcal{R}_0 + x/A)$ is (Eqs. 3.9 and 3.10),

$$(2.493 \cdot 10^5;\ 2.167 \cdot 10^4).$$

Thus the pulling force random variable as a function of the variate, E, is

$$(\bar{F}, s_F) = \frac{(65{,}650;\ 0)}{(2.493 \cdot 10^5;\ 2.167 \cdot 10^4)} \cdot (\bar{E}, s_E)^2,$$

$$(\bar{F}, s_F) = (0.2633, 0.0229)(\bar{E}, s_E)^2$$

$$= (0.2633, 0.0229)(\bar{E}^2, 0.2\bar{E}^2).$$

By Eq. 2.61

$$s_E^2 = 2\bar{E}s_E. \qquad s_E \approx 0.1\ \bar{E}\text{(specified)}.$$

Applying Eqs. 3.9 and 3.10,

$$(\bar{F}, s_F) = (0.2633\bar{E}^2;\ 0.0574\bar{E}^2)\ \text{g}$$

and

$$(\bar{P}, s_P) = (55.79;\ 6.76)\ \text{g}.$$

By substituting values into Eq. 4.6 with $z \approx 3.0$,

$$z = 3.0 = \frac{|\bar{P} - \bar{F}|}{\sqrt{s_P^2 + s_F^2}} = \frac{|55.79 - 0.2633\bar{E}^2|}{\sqrt{(6.76)^2 + (0.0574\bar{E}^2)^2}},$$

$$0 = 0.03967\bar{E}^4 - 29.379\bar{E}^2 + 2711.2,$$

$$\bar{E}^2 = 632.5;\ 108,$$

$$E = 25.1\ \text{V}.$$

Note. 30 V in the original analysis.

$$s_E = 2.5\ \text{V}.$$

The results are not unexpected, because conventional design tends to be conservative. If voltage were held at the original values, it is to be expected that other parameter values could be somewhat reduced (see the following example).

Example 2. With design conditions the same (except for the number of turns, N, in the relay coil) and reliability 0.999, the problem is that of optimizing the number of coil windings. This is computed as follows: wire length and resistance variates are functions of the number of turns, N. Thus the resistance random variable is

$$(\bar{R}, s_R) = N[(0.0887 + 0.000117N); \sqrt{(0.0036)^2 + (0.00000045N^2)}].$$

The current random variable estimate is

$$(\bar{I}, s_I) = \frac{(30, 0.6)}{(\bar{R}, s_R)}.$$

Thus

$$\bar{F} = \frac{3.713}{(0.0887 + 0.000117N)^2},$$

$$s_F = \left[\frac{1.36 \cdot 10^{-3} + 1.70 \cdot 10^{-7}N + 1.134 \cdot 10^{-9}N^2}{(0.0887 + 0.000117N)^2}\right]^{1/2},$$

$$(\bar{P}, s_P) = (55.74, 6.76) \text{ g}.$$

Substituting values into the coupling formula, Eq. 4.6, with $z \approx 3.0$,

$$3.0 = \frac{\left|\dfrac{3.713}{(0.0887 + 0.00017N)^2} - 55.74\right|}{\left[\dfrac{1.36 \cdot 10^{-3} + 1.70 \cdot 10^{-7}N + 1.134 \cdot 10^{-9}N^2}{(0.0887 + 0.00017N)^2} + (6.76)^2\right]^{1/2}}.$$

By expanding and then simplifying

$$0 = 2.09 \cdot 10^{13} - 1.50 \cdot 10^9 N - 9.96 \cdot 10^6 N^2 + 3.00 \cdot 10^2 N^3 + N^4.$$

The fourth-degree polynomial in N is transformed to a reduced fourth-degree polynomial in u by a change of variable,† $u = N + a_1/n$, with

$$a_1 = 300 \quad \text{and} \quad n = 4,$$

$$\frac{a_1}{n} = 75 \quad \text{and} \quad N = u - 75.$$

The reduced equation in u becomes

$$0 = u^4 - 9.99 \cdot 10^6 u^2 + 2.09 \cdot 10^{13}.$$

By solving the quadratic in u^2

$$u = 2652$$

and

$$N = 2577.$$

(N in the original analysis was 3200.)

† See *Theory of Equations*, C. C. MacDuffee, John Wiley and Sons, Inc., New York, 1954.

References

[1] *Safety, Safety Factors, and Reliability of Mechanical Systems*, A. M. Freudenthal, Columbia Univeristy.
[2] *Saturn Launch Vehicle Reliability Study*, Arinc Research Corporation, Publication No. 141-2-199, December 20, 1960.
[3] E. T. Haire, *Structures Reliability Report*, Martin Marietta Corporation, Aerospace Division, ER 11862, December 1962.
[4] D. Devine et al., *Reliability Analysis of Hazardous Fittings*, North American Aviation, LAD, July 1962.
[5] H. Hilton, and M. Feigen, *Minimum Weight Analysis Based on Structural Reliability*, J. Aerospace Sci., **27 (9)** (September 1960).
[6] Dr. H. L. Leve, *A Reliability Framework for Structural Design*, Douglas Aircraft Company, Inc., Structural Mechanics Division, 1962.
[7] Dr. J. H. DeHardt, *A Statistical Approach to Structural Reliability*, North American Aviation and California State College, Los Angeles, LCM 63-2, July 1963.
[8] Dr. J. R. Benjamin, *Statistics for Civil Engineers*, Stanford University, unpublished manuscript, 1964.
[9] I. Bouton, *Fundamental Aspects of Structural Reliability*, Aerospace Eng., **21 (6)** (June 1962).
[10] G. E. Ingram, *A Basic Approach for Structural Reliability*, 11th National Symposium on Reliability and Quality Control, Miami Beach, Fla., January 1965.
[11] Dimitri Kececioglu, and David Cormier, *Designing a Specified Reliability into a Component*, Third Annual Aerospace and Maintainability Conference, Washington D.C., June 1964.
[12] G. P. Wadsworth, and J. G. Bryan, *Probability and Random Variables*, McGraw-Hill, New York, 1960.
[13] E. Parzen, *Modern Probability Theory and its Applications*, Wiley, New York, 1960.
[14] W. Feller, *An Introduction to Probability Theory and its Application*, Wiley, New York, 1957.
[15] M. Loeve, *Probability Theory*, Van Nostrand, Princeton, N.J., 1963.
[16] Birkhoff and MacLane, *A Survey of Modern Algebra*, Macmillan, New York, 1959.
[17] J. M. H. Olmsted, *Real Variables*, Appleton-Century-Crofts, New York, 1959.
[18] A. M. Mood, *Introduction to the Theory of Statistics*, McGraw-Hill, New York, 1950.
[19] H. D. Brunk, *Mathematical Statistics*, Ginn, New York, 1960.
[20] L. G. Parratt, *Probability and Experimental Error in Science*, Wiley, New York, 1961.

References

[21] *Tables of Normal Probability Functions*, National Bureau of Standards, Applied Mathematics Series 23, U.S. Government Printing Office, Washington D.C., June 1955.

[22] E. B. Haugen, *The Algebra of Normal Functions*, SID 64-1598, NAA/S & ID, November 1964.

[23] E. B. Haugen, *A Statistical Algebra for Engineering Applications*, RSR-2, NAA/S & ID, January 1964.

[24] A. S. Merrill, Frequency Distribution of an Index When Both the Components Follow the Normal Law, *Biometrics*, **20Z**, 55–63 (1928).

[25] L. A. Aroian, *Tables and Percentage Points of the Distribution Function of a Product*, Systems Development Laboratory, Hughes Aircraft Company.

[26] J. B. Scarborough, *Numerical Mathematical Analysis*, John Hopkins Press, Baltimore, Md. 1962.

[27] W. E. Milne, *Numerical Calculus*, Princeton University Press, Princeton, N.J., 1949.

[28] D. Teichroew, *A History of Distribution Sampling Prior to the Era of the Computer and its Relevance to Simulation*, J. Amer. Statist. Assoc. (March 1965).

[29] C. R. Wylie, *Advanced Engineering Mathematics*, McGraw-Hill, New York, 1960.

[30] J. M. Hammersly, and D. C. Handscomb, *Monte-Carlo Methods*, Wiley, New York, 1964.

[31] R. L. Mador, *Synthesis and Analysis of Reliability Confidence Levels*, General Dynamics/Astronautics, December 1962.

[32] C. Dicks, and S. Wilson, *Structural Reliability the General Engineering Design Approach*, 11th National Symposium on Reliability and Quality Control, Miami Beach, Fla., January 1965.

[33] E. F. Bruhn, *Analysis and Design of Aircraft Structures*, Tri-State Offset Company, Cincinnati, Ohio, 1958.

[34] J. E. Hayes, *Structural Design Criteria for Boost Vehicles by Statistical Methods*, NAA/S & ID, March 4, 1965.

[35] E. B. Haugen, and H. Mikasa, *Statistical Methods Applied to Design of Tension Elements*, NAA/S & ID, June 1964.

[36] S. Timoshenko, *Strength of Materials*, Van Nostrand, Princeton, N.J., 1948.

[37] E. B. Haugen, and H. Mikasa, *Statistical Method Applied to Structural Design of Simple Beams Subjected to Concentrated Lateral Loading*, NAA/S & ID, June 1964.

[38] E. B. Haugen, *Statistical Method Applied to Structural Design of Simple Beams Subjected to Distributed Loads*, SID 65-121, NAA/S & ID, January 21, 1965.

[39] A. G. Sines, and J. L. Waisman, *Metal Fatigue*, McGraw-Hill, New York, 1959.

[40] E. B. Haugen, *Statistical Strength Properties of Common Metal Alloys*, NAA/S & ID, SID 65-1274, October 30, 1965.

[41] E. B. Haugen, *Probabilistic Design of Columns*, NAA/S & ID, SID 65-957, April 1965.

[42] Timoshenko and Gere, *Theory of Elastic Stability*, McGraw-Hill, New York, 1961.

[43] E. B. Haugen, *Analysis of Electromagnetic Devices by Probabilistic Methods*, NAA/S & ID, January 1966.

[44] E. B. Haugen, *Relay Design by Statistical Methods*, 14th Annual National Relay Conference, Oklahoma State University, Stillwater, Okla., April 1966.

[45] R. L. Peek, and H. N. Wagar, *Switching Relay Design*, Van Nostrand, Princeton, N.J., 1955.

Index

Abelian group, 140
Accuracy of Monte Carlo method, 178
Addition of normal random variables, commutative properties, 137
  of gamma random variables, 41
  of linear combinations, 109
  of normal random variables, 84, 106
  with one variate negative, 106
Additive inverse, in normal function algebra, 114
Adequacy of component, probabilistic, 145
Alexander, Madeline, 118
Algebra of normal functions, 105
Allowable and applied stress, generally distributed, 148
Allowable stress, random variable, 208
  shear, 275
Analogue of real numbers, the set S, 139
Angle of twist variate, 274
Anistropic coefficient, random variable, 284
  thermal expansion, 284
Applied stress random variable, 208
Assembly stress factor, 9
Associative law, in normal function algebra, 137
Average of x, continuous distributions, 64
  discrete distributions, 64
Averages, arithmetic, 63
Axioms of probability theory, 13

Basic distribution transformation, 79
Basic probability concepts, 13
Beam, built-in ends, 241
  columns, 254
  conventional design, 226
  distributed loads, 226
  probabilistic theory, 218, 228, 245
  simple, concentrated loads, 217
Bending and torsion, probabilistic, 278

Bending moment, curve, 232
  signs, 229
Benjamin, Jack R., 47
Beta, density function, 43
  distribution, 42
Binary operations, correlated, difference functions, 131
  product functions, 132
  quotient functions, 132
  sum functions, 131
  summary, independent and correlated, 123
  special combinations, 131
Binomial, mean of, 67
  variance of, 67
Binomial density, anisotropic, 285
Binomial distribution, definition, 25
Bivariate, distribution, 56
  normal density function, 72
  normal distribution, 73
  normal moment generating function, 74, 99

Camp-Meidall inequality, 68
Cantilever beam, 244
  design of (probabilistic), 250
  stress variate, 245
  uniform load variate, 246
Capacitor, random variable, 293
  in series (multivariate), 293
Central limit theorem, 53
Central tendency in values, 65
Centroids, distribution of, 200
Chi-square distribution, 40
Classical theory, of probability, 15
Closed gap reluctance variate, 310
Closure, in statistical algebra, 138
  absence of, in rigorous sense, 138
  difference variates, 154
  linear function variates, 77
Coefficient of thermal expansion, 284

317

318 Index

Coefficient of variation, 135
Coil, resistance variate, 306
  spool geometry variates, 306
Coil current random variable computation, 307
  products of normal variates, 135
  quotients of normal variates, 135
Column design (probabilistic), 254
  initial assumptions, 254
  initial curvature, statistical, 255
  initial imperfection, statistical, 254
Combinatorial operation, 89
Combined stress, in probabilistic design, 278
Commutative law in statistical algebra, 137
Compensation in normal approximation of binomial distribution, 54
Complex numbers, statistical analogy, 139
Component, reliability of, 146
Compression block (probabilistic), 255
Computation, of probability, 35
  of probability of critical distortion, 290
Condition for maximum distortion (statistical), 290
Constraints, on products of normal variates, 134, 135
  on quotients of normal variates, 135
  on squares of normal variates, 134
  see Table 3.1, 134
Contact relay variate, 298
Convolution, 81
  of difference of random variables, 154
  of sum of random variables, 81
  theorem, 81
Correlated combinations of random variables, 131
Correlated differences of random variables, 114
Correlated functions, general, 134
Correlated products of random variables, 118
Correlated sums of random variables, 111
Correlation, in beam design (probabilistic), 221
  among variates, 111
Couple notation, normal variates, 3
Coupling equation (probabilistic), 150
Crippling, failure mode, 254
Critical distortion (probabilistic), 286

  moment variate, 287
Critical load variate, 264
Critical stress, definition, 264
Cross section, area variate, 210
  generalization, 214
Crude Monte Carlo simulation, 194
Cumulative distribution function, definition, 24

Deflection, computation (statistical), 239
  effect on moment (probabilistic), 262
  maximum (random variable), 238
  uniform loaded beam, 238–241
  variate, 238–241
Degenerate random variables, 136
De Moivre, 44
Density, concept of, 36
  as defining function, 38
  function, normal distribution, 47
  function, gamma distribution, 39
  of powers of random variables, 81
  properties of, 37
  of squares of random variables, 79
Derivation, normal density, 44
  normal distribution, 44
  Poisson distribution, 29
Design, coil turns variate, 311
  optimum voltage variate, 311
  uniform loaded beam, 236
  variables, 2
Design of beams (probabilistic), with concentrated load variates, 222
  with single load variate, 220
Determination, reliability of normal random variables, 148
Deviation, 91
Difference, independent normal random variables, 113
Difference formula, by convolutions, 151
Difference function, derivation for normal variates, 148
Direct simulation, Monte Carlo, 183
Discrete random variable, definition, 23
Distribution, definition, 24
  of functions of variates, 75
  of square of normal random variable, 81
  statistical study, 284
Distributive law in statistical algebra, 139
Division, with lognormal random variables, 52
  with normal random variables, 120

Index 319

Double interpolation, 170
Double numerical integration of normal random variables, 176

E (expectation), 14
Eccentricity, statistical consideration of, 255
Elastic limit, probabilistic, 264
Electrical circuit random variables, 298
Electromechanical devices, 293
  switches, 295
Element length, consideration, 213
Elongation, δ, random variable, 213
Engineering random variables, 23
Equality, set S of normal random variables, 136
Equidistant ordinates, 162
Equivalence relations in statistical algebra, 136
Error expression, derivation, 170
Errors, in Simpson's rule, 167
Estimating variance, 205
Estimation of area, by relative frequency, 16
Estimator, biased, 59
  consistent, 59
  definition, 58
Euler buckling, effect of eccentricity variate, 267
Euler column, allowable stress variate, 268
  conventional design, 264
  design for eccentricity variate, 267
  minimum section thickness variate, 269
  probabilistic design, 266
Euler load, statistical description, 264
Evaluation, Poisson from normal distribution, 40
Events, complementary, 14
  definition, 14
  elementary, 14
  intersection of, 17
  mutually exclusive, 17
  union of, 17
$E(x)$, expected value of x, 61
$E(x^2)$, expected value of x^2, 126
$E(x^3)$, expected value of x^3, 126
$E(x^4)$, expected value of x^4, 126
$E(x^r)$, expected value of x^r, 70
Expected value, of continuous random variable, 61
  definition of, 61
  of discrete random variable, 61
Experiment, definition of (in statistics), 14
Exponential distribution, 39
  density function, 39
Extreme value probability, 68

Failure governing stress, 9
  mode of component, 9
  probability of component, 5
Field, number, definition of, 139
First moment (statistical), 69
Force distribution random variables, 288
Force system, elements of, 199
Force variate expressions, 200
Forward interpolation, numerical, 159
Freudenthal, A., 1
Function, simple (single values), monotonic, 75
  monotonically decreasing, 76
  monotonically increasing, 75
Functionally related random variables, 4

Gamma approximation, of binomial distribution, 42
  of normal distribution, 55
Gamma density function, 39
Gamma distribution, 40
  integrability of, 40
  properties of, 40
Gamma function (Γ), 41
Gap (relay), random variable, 308

Hammersly, J. M., 178
Handscomb, D. C., 178
Heat treatment (HT) stress factors, 9
Hey, G. B., 179
Hit or miss Monte Carlo, 186
Homogeneous material, 254
Hooke's law, 287

Idealized loading (discrete valued), 226
Imbedded numbers, in the set S, 139
Independence, definition, 22
  n events, 22
Independent binary operations, 106
Independent events, 19
Integral domain, definition, 105
Integration of normal density function, 47
Interference theory, see Fig. 1.3, 5

320 Index

Intermediate column, probabilistic design, 254
  deflection variate, 262
  eccentric compression variate, 262
Internal force random variable, 219

Joint cumulative distribution, 57
Joint distribution, 57
  from marginal distribution, 57
Joint moment generating function, definition, 72
Joint probability, definition, 22

Kirchoff's laws, 293

Lateral deflection variate, 254
Law, of compound probability, 22
  of large numbers, 68
  of total probability, 21
Laws of combination, in algebra of normal functions, 136
Lehmer, 181
Likelihood of critical distortion, 288
Limiting forms, 54
Limit of convergence, 140
Linear combination, of normal random variables, 109
Linear expansion, differential, 286
Line of zero stress, 262
Load, factors, 9
  part of span (variate), 234
Lognormal, density function, 49
  distribution, 47
Log of random variables, 92

Magnetic circuit, statistical description, 297
Magnitude versus frequency relationships, 2
Manufacturing stress factors, 9
Margin of safety (MS), 1
Mathematical considerations in probabilistic design, 13
Maximum fiber stress variate, 221
Maximum likelihood (estimator), 89
  of function combinations, 89
  method of, 59
Mean, of binomial distribution, 67
  of gamma distribution, 64
  of lognormal distribution, 52
  of normal distribution, 67
  of Poisson distribution, 66
Mean square deviation, 65
Mean value, 3
  estimator, 90
Measurable parameter, in design, 3
Measure of dispersion, of distribution, 65
Mechanical circuit (statistical description), 297
Mechanical elements variates, 199
Mechanism of distortion, 286
Median of lognormal distribution, 52
Membrane stress random variable, 215
Method of computing moments, 89
Midsquare method for random numbers, 180
Mode, of failure (component), 9
  of lognormal distribution, 52
  of normal distribution, 46
Moment, about arbitrary point, 69
  definition, 61, 69
  of function of random variable, 88
  maximum variate, 224
  mean, 224
Moment generating function, 70, 98
  of bivariate normal, 99
  in normal function algebra, 123
  rth derivative of, 70
  trivariate normal distribution, 99
  univariate normal distribution, 98
Moment generating function methods, 96
  for mean value of distribution, 93
  for n variable functions, 94
  for single variable functions, 94
  for standard deviation of distribution, 96
  for variance of distribution, function of n random variables, 94
  function of one random variable, 97
Moment generating functions, for independent normal random variables, difference, 124
  product, 125
  root, 127
  square, 126
  sum, 124
Moment of inertia, polar, 274
  random variable, 201
  variate, 221
Moments, multivariate distribution, 72
  of one random variable, 69
  of Poisson distribution, 70
Monotonic increasing, definition, 75

Index 321

Monte Carlo, general principles, 194
  methods for problem solution, 178
  simulation, of a product of random variables, 186
    of a quotient of random variables, 186
    of a square of a random variable, 186
Multiforce loading variate, 225
Multiplication of lognormal random variables, 52
  of normal random variables, 116
Multiplicative law for independent events, 23
Multivariates, distribution of, 56
Murthy, V. M., 119
Mutually exclusive events, definition, 21

n-dimensional density functions, 57
Negative correlation of random variables, 111
Newton's formula for forward interpolation, 159
No-failure probability of component, 146
Nonmonotonic functions, 81
Nonrectangular distribution sampling, 182
Normal approximation, of binomial, distribution, 54
  distortion model, 286
Normal density function, 47
Normal distribution, 44
  properties, 46
Normalizing transformation for distributions, 55
Normal sums, correlated random variables, 111
  independent random variables, 113
Number field, 105
Numerical integration, applied to variates, 161
  of exponential distribution function, 165
  of gamma distribution function, 166
  of normal distribution function, 164
Numerical methods in probabilistic design, 158

Ohm's law, statistical expression, 293
Open interval, as range of distribution, 35
Order relations, in algebra, of normal random variables, 136
Overdesign by conventional methods, 1

Parallel axis theorem, for moment of inertia, 203
Parallel circuit variate, 293
Partial derivative methods, determination of variance, for log functions, 90
  for moments, 90
  for products, 129
  for quotients, 129
  for sums and differences, 129
  for trigonometric function, 90
$p(E)$ for event E, 14
$p(E_1 | E_2)$ for event E_1 given E_2, 19
Permeability random variable, 309
Pillai, C. S., 119
Point estimation, of parameter, 58
  definition, 58
Poisson approximation of binomial distribution, 29
  conditions for, 29
  derivation of, 29
  distribution, 29
Poisson probability function, 30
Polynomial representation of function, 158
Population of values of variate, 3
Positive correlation of random variables, 111
Predictable distribution of values, 3
Probabilistic approach, reasons for, 7
Probability, basic theorems, 21
  conditional, 18
  definition of, 14
  density function definition, 24
  joint, 19
  of no failure, 146
  properties, 19
  total, 19
  transformation, 79
Products, independent random variables, 114
  limit conditions, 117
  of lognormal random variables, 52
  normal random variables, 116
  of variate and inverse, 133
Properties of E, 95
Properties of mean, 64
Propped cantilever, beam, design of, 248
  maximum moment, 251
  probabilistic design, 250
Pseudo-random numbers, 180
  by congruential methods, 181
Pulling force of solenoid, 308

322 Index

Quotients, approximation, closure, 121
  correlated random variables, 123
  distribution of normal random variables, 86
  independent normal random variables, 120
  limit conditions, 120
  of lognormal random variables, 52
  of moments of random variables, 126

Radius, of core, 271
  of gyration variate, 204, 265
Random number, definition, 180
Random number generation, by arithmetic processes, 180
  by physical processes, 180
Random variable, continuous, 24, 32
  definition, 3, 23
  differentiable, 36
  discrete, 23
Range, continuous distribution, 32
  discrete distribution, 33
Range of definition, 38
$R(E)$, relative frequency expectation, 15
$R(E_1, E_2)$, 18
Reaction, computation of variates, 228
  with uniform load variates, 226
  variable, 218
Real numbers, 2
  algebra of, 2
Rectangular distribution, 38
  as special case of beta, 43
Reflexive property in normal function algebra, 136
Relative efficiency, Monte Carlo, 194
Relative frequency, conditional, 18
  properties of, 17
  theory, 15
Reliability, of correlated normal random variables, 156
  definition, 5
  of nonnormal distributions, 154
  predictions, 10
  by transform method, 154
  versus shaft diameter, 277
Reliability change, with $\overline{F}_{tu}$, 212
  with S_p, 211
Reluctance random variable estimation, 308
Remainder term, in double numerical interpolation, 171
Representation, of parameters, single valued, 2
Resisting couple variate, 225
rth moment of distribution, 69

s_x, σ_x; sample and universal standard deviation, 3
Safety factor (SF), 1
  definition, 4
  fallacy in use of, 5
Safety margin (SM), 8
Sample mean, value of, distribution, 53
Sample of finite size, 3
Scarborough, James B., 158
Secant formula, 271
Second moment, of distribution, about mean value, 70
Section modulus variate, 227
Series circuit random variable, 293
Set of couples, normal random variables, 89
Shaft, design (probabilistic) for torsion, 278
Shaft diameter versus safety factor, 277
Shapes of Poisson distribution, 32
Shear, uniform load, 229
  variate, 220
Shearing force, signs, 230
Shear modulus of elasticity, 274
Shear stress random variable, 275
Short column, 254
  analysis, conventional, 259
  probabilistic, 259
  eccentric load variate, 260
Simple integration (numerical), 162
Simpson's rule, 163
  applications (to random variables), 164
Skewed distribution, 39
Slender column, probabilistic design, 254
Slenderness ratio, 265
Smith, C. O., 274
Spring design, probabilistic, 299
Spring variate, 300
Standard deviation, 3
  of binomial distribution, 67
  definition, 65
  estimation from tolerances, 222
  of gamma distribution, 66
  of normal distribution, 67
Standard normal distribution, table of values, 48
Statically indeterminate element, 248

Statistic, definition, 3
Statistical inference, 145
Statistical methods, practical results, 9
Step function, 25
Stirling's approximation, 44
Stress and strength, distributions, 5
Stress concentration factor, 9
Stresses in beams, 219
Subtraction, normal random variables, 113
Summary, binary operations, 133
Sums, gamma random variables, 41
  independent normal random variables, 107
Surface stress factor, 9
Symmetric property of normal functions, 137

Taylor expansion, 91
Tchebysheff's inequality, 68
Temperature stress factors, 9
Tension element, analysis, 210
  design, probabilistic, 208
  elongation random variable, 213
Theorem of mean, 168
Tolerance, 220
  manufacturing, 222
Torsion, variate, 274
  and bending variate, 274
Total probability, of mutually exclusive events, 21
Transformer, analytic expression, 294
Transitive property of normal functions, 137

Trial, definition of (in probability theory), 14
Trivariate moment generating function, 96
Trivariate normal moment generating function, 100
Twisting couple, random variable, 274
Two-way differences in numerical integration, 171

Unbalancing force, probabilistic, 286
Unit, force magnitude, 287
Unity, in algebra of normal functions, 138
Univariate distribution, 56
Univariate normal moment generating function, 98

Vallee-Poussin equation, 168
Variance, of binomial distribution, 67
  definition, 65
  estimator, 91
  of gamma distribution, 66
  of normal distribution, 67
  of Poisson distribution, 66
Vector space, properties of set S, 140
Voltage variate, 306

Weddel's rule, 174
Wire length variate, 306

x, μ_x: sample and universal mean value, 3

Zero, in normal function algebra, 138
Zero probability, 36

MAR 14 1969

BROOKLYN PUBLIC LIBRARY
3 4444 04736 1217

DISCARDED

DISCARDED

3233680

Brooklyn Public Library,
Central Service
Grand Army Plaza
Brooklyn, New York 11238